Parasitic Insects

Parasitic Insects

R. R. ASKEW, B.SC., D.PHIL., F.R.E.S.
Lecturer in Zoology at the University of Manchester

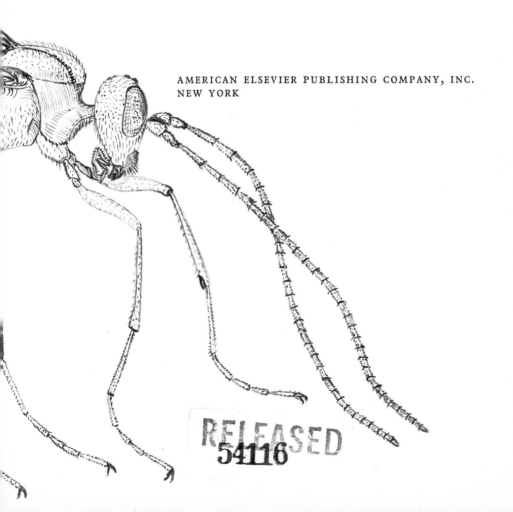

AMERICAN ELSEVIER PUBLISHING COMPANY, INC.
NEW YORK

First American Edition 1971
Library of Congress Catalog Card Number 73-137683
International Standard Book Number 0-444-19629-3

Published in the United States by
American Elsevier Publishing Company, Inc.
52 Vanderbilt Avenue, New York, New York 10017
Printed by C. F. R. Woodward (Bath) Ltd.

Preface

ONE TENTH of all animal species are parasitic insects. Many quite different insect groups have adopted a parasitic existence, each coming to terms with the demands made upon it by this mode of life in a different manner within the limits imposed by its phylogeny, morphology, life cycle, and behaviour.

So far as I am aware, no book has previously been written that is devoted exclusively to parasitic insects. Most texts on parasitology, if insects are mentioned at all, progress little beyond accounts of fleas, lice, and other medically important groups. Protelean parasitic insects (parasitoids) are included in C. P. Clausen's excellent work *Entomophagous Insects*, now almost thirty years old, but for an account of all parasitic insects, there is no single volume to which one can turn.

This book is thus an attempt to fill a gap in entomological literature. I have had in mind, whilst writing, mainly university undergraduate and postgraduate students specialising in entomology, economic entomology, ecology, parasitology, and associated subjects. However it is hoped that others who are interested by parasitic insects, and especially those such as agriculturalists and veterinarians who come into contact with them professionally, will find it of some value as a reference text. Parasitology is essentially a part of ecology, and it is the ecological approach that I have adopted.

Because the subject is such a very broad one, it is discussed principally within the framework provided by zoological classification, but reference is made, where appropriate, to principles of wider application and to analogous situations within other groups. In making general statements, there is a conflict between complete accuracy and presenting a readable account unencumbered by a profusion of qualifications. When discussing a group in which only a small fraction of species are at all well known, time may show that I have erred a little on the side of too sweeping generalisations.

The only research I have been personally engaged in has been centred upon the parasitic Hymenoptera, and my experience of other groups has been limited to sporadic collecting and observations when the opportunity has arisen. I have had to lean very heavily, therefore, upon specialist papers and books. In a book of this size, a comprehensive review of the subject's literature would have been impossible. However, from those works that are cited, and the lists of references that they contain, the student should be able to make a good start on a search of the literature in a chosen topic.

I am much indebted to the following persons for checking and criticising drafts of those chapters related to their particular interests: Mr J. G. Blower, Miss Theresa Clay, Dr L. M. Cook, Dr M. E. G. Evans, Dr G. R. Gradwell, Dr M. W. R. de V. Graham, Mr A. M. Hutson, Dr G. J. Kerrich, Dr Miriam Rothschild, Mr K. G. V. Smith, and Professor G. C. Varley. I alone, of course, am responsible for errors, omissions, and other shortcomings. My thanks also to those authors and publishers who have generously given me permission to reproduce their illustrations. Mr Allan Brindle, Keeper of Entomology in the Manchester Museum, very kindly provided me with several of the specimens from which illustrations were prepared. Dr D. W. Yalden advised me on certain problems relating to birds and mammals. Finally, I thank my wife who has been a constant source of help and encouragement, has provided time for me to work by shouldering extra responsibilities, and has assisted in the preparation of many of the figures.

Hale, Cheshire   R. R. A.
1971

Contents

To My Parents

Introduction

As USUALLY DEFINED, parasitism is a relationship between two species in which one, the parasite, obtains its nutritional requirements from the body material of the other, the host. The parasite lives on or in the body of the host, and the host receives no benefits from the association although it is not usually destroyed. Often included in a definition is the prolonged nature of the association, extending over a large part of the parasite's life cycle and frequently involving it in profound morphological adaptations to its way of life. Such a description accommodates typical parasites, for example flukes and lice, but in order to cover other associations commonly found among insects and regarded as parasitic, it has to be either enlarged and qualified, or other terms must be employed. This is very often the difficulty with biological definitions. They are attempts to categorise, for ease of reference, phenomena which, being the products of evolution, are rarely discrete enough to allow for ready categorisation. For the purposes of this book a very broad view of parasitism is taken, and a range of exceedingly diverse associations will therefore be described.

Among insects there are ectoparasites which live upon the outside of their hosts, and endoparasites which live inside. Consider some of the ectoparasites that feed upon the bodies of vertebrates. Lice (Mallophaga and Anoplura) have become entirely committed to parasitism, both immature and adult stages being completely dependent upon their bird or mammal host on which the entire life cycle is passed, and their morphology is much adapted to fit them for this environment. Fleas (Siphonaptera) are rather less committed in their adult stage, passing more freely from host to host and often spending long periods apart from a host; their larval life also is spent away from the host's body. At the other end of the range are mosquitoes and some other blood-sucking flies which feed from the host only for brief periods and often only in the female sex. These have been described as predators, impudent predators perhaps, but I prefer to regard them as parasites whose association with the host is ephemeral but nevertheless often obligatory since many are unable to produce eggs prior to a blood meal. In these instances, morphological modification for parasitism is mostly confined to the structure of the mouthparts. As we pass down this scale from prolonged to brief contact between parasite and host, so also, in general, do we tend to pass down a scale of host relationships from those insects confined to a single host species (host-specific), or at most to a group of allied host species (oligophagous), to those which have a wide range of hosts (polyphagous).

A few insects that are not normally or obligatorily parasitic occasionally behave as ectoparasites if the opportunity is offered. Such insects are facultative parasites. For instance, bugs of the family Anthocoridae, which usually feed as predators of small insects which they rapidly kill, sometimes take a vertebrate blood meal should the chance arise. Their piercing and sucking mouthparts make this possible. I have even seen an aphid with its proboscis inserted into the body of an aphidivorous hover-fly larva!

Among endopterygote insects, that is those whose life history includes a complete metamorphosis between the grub-like larva and the winged adult, there has developed in several groups parasitism by the larval stages. Parasitism by juvenile stages is termed protelean parasitism. With the exception of a few Diptera whose larvae parasitise vertebrate hosts, the hosts of protelean insect parasites are other invertebrates, usually insects. Protelean parasites are most frequently endoparasites, but ectoparasitism is not rare, especially when the host is in a protected situation such as a plant mine or gall. Often only one parasite larva is capable of development in or on each host, and usually it consumes nearly all except the integument of the host. Nearly always the host dies very soon after the parasite has completed feeding, most of its vital organs having been devoured in the final stages of parasitism. Sometimes, in fact, the host is killed by the ovipositing female 'parasite' so that the larvae feed only on dead tissue and are therefore saprovores.

In the eventual death of the host, parasitic insect larvae do not fit the usual definition of a parasite and a special term, parasitoid, has been devised for them (Reuter 1913). Parasitoid is an 'adjective . . . characterizing a range of feeding behavior intermediate between the parasite and predaceous ends of the behavioral continuum' (Knutson and Berg 1966). A parasitoid at first feeds like a parasite, being adapted to living in intimate physical association with its host, and only after it has extracted all the nourishment that it requires from the host's living body does it eventually destroy the host. The term parasitoid, however, will not be used in this book since it too embraces a wide range of phenomena and is little more precise than the term parasite. Braconidae (Hymenoptera) of the tribe Euphorini parasitise adult insects and, exceptionally, hosts may survive parasitism to later reproduce. There is little to distinguish the behaviour of these Hymenoptera from that of typical parasites. Some groups of parasitic Hymenoptera include species which are ectoparasites and others which are endoparasites. In most cases it seems likely that endoparasitism has developed from ectoparasitism. Ectoparasitic larvae very occasionally feed upon more than one host individual, for instance the chalcid *Eurytoma rosae*, which, in the course of its larval life in the many-celled rose gall of *Diplolepis rosae*, may eat several host larvae. The larvae of those Diptera and Hymenoptera that live in egg masses of locusts or spiders generally consume the major part of the host brood. These larvae are really behaving as predators, consuming a succession of prey organisms, but since they are often closely allied to parasitic (or parasitoid) species, they are not excluded from this book.

Also included are the special associations called inquilinism and social parasitism. The female inquiline lays her egg in the larval cell of a host species, and the resulting inquiline larva feeds upon the food supplies originally destined for the host larva. The host larva often either dies from starvation or is directly killed by the inquiline. Social parasitism is an extension of this habit with a social insect serving as the host species. The social parasite diverts to its own use the production of the workers of the host species.

In recent years there has been a tendency amongst some entomologists to describe as parasites phytophagous larvae living concealed in plant tissue, for example leaf-mining flies (Agromyzidae) and gall-wasps (Cynipidae), but this seems to be widening the concept of parasitism too much, and these insects are not included in this work.

Distribution of Parasitic Species in the Orders of Insects

The number of orders in which living insects are grouped has increased and become rather variable in recent years as some of the old orders, for instance the Orthoptera, have been split up into smaller orders, and other adjustments have been suggested which have not yet found acceptance with all entomologists. The classification adopted in this work is shown below. Orders known to contain parasitic species are asterisked. The approximate world number of described species in each order is given, together with the approximate or estimated number of parasitic species.

APTERYGOTA
No metamorphosis, primitively wingless.

Order	No. of world spp.	No. of parasitic spp.
	2800	0

EXOPTERYGOTA
Incomplete metamorphosis, usually without pupal stage. Nymphs (larvae) resemble adults (Figure 1a) and wings develop externally.

Order	No. of world spp.	No. of parasitic spp.
Ephemeroptera (mayflies)	1500	0
Odonata (dragonflies, damselflies)	5000	0
Plecoptera (stoneflies)	1600	0
Orthoptera *s.l.* (grasshoppers, crickets, cockroaches, mantids, stick insects, etc.)	22 000	0

Order	No. of world spp.	No. of parasitic spp.
Isoptera (termites)	1900	0
*Dermaptera (earwigs)	1100	11 (mammal ectoparasites)
Embioptera (webspinners)	150	0
Psocoptera (booklice, psocids)	1100	0
Zoraptera (zorapterans)	22	0
*Mallophaga (biting lice)	2800	2800 (mammal and bird ecto-parasites)
*Anoplura (sucking lice)	300	300 (mammal ectoparasites)
Thysanoptera (thrips)	3200	0
*Hemiptera (bugs, aphids, scales, etc.)	55 000	100 (mammal and bird ecto-parasites)

ENDOPTERYGOTA

Complete metamorphosis with a pupal stage. Larvae bear no resemblance to adults and wings develop internally (Figure 1).

Order	No. of world spp.	No. of parasitic spp.
*Neuroptera and Megaloptera (alderflies, lacewings, etc.)	5000	190 ('parasites' of sponges, spiders' egg cocoons)
*Coleoptera (beetles)	300 000	5400 (invertebrate protelean parasites, mammal ecto-parasites)
*Strepsiptera (stylops)	300	300 (insect protelean parasites)
Mecoptera (scorpionflies)	350	0
Trichoptera (caddisflies)	4500	0
*Lepidoptera (butterflies, moths)	120 000	60 (insect protelean parasites, mammal ectoparasites)
*Diptera (flies)	90 000	11 000 (invertebrate and verte-brate protelean parasites, vertebrate and insect ectoparasites)
*Siphonaptera (fleas)	1800	1800 (mammal and bird ecto-parasites)
*Hymenoptera (bees, wasps, ants, ichneumons, etc.)	200 000	100 000 (insect protelean parasites)

There are over 820 000 described species of insects (and undoubtedly many more await description) and of this number at least 15% are parasitic. This means that in the region of 10% of all animal species are parasitic insects. Arndt (1940) calculated that in Germany there are over 40 000 animal species and that parasitic Hymenoptera contribute no less than 15% to this total. From the point of view of number of species alone parasitic insects are obviously an ecological group of great significance.

Some proposed or fairly recently effected changes in nomenclature affect orders which contain parasitic species. The Anoplura are sometimes called Siphunculata, and Siphonaptera used to be Aphaniptera and before that, Suctoria.

Mallophaga and Anoplura have been united in the order Phthiraptera. Their general body form is similar in many ways, but their mouthparts, and therefore manner of feeding, are usually quite different, and most authorities keep them separate.

Considered here with the Dermaptera is *Hemimerus*, a genus which has been shown recently (Popham 1961b) to have few affinities with other earwigs and for which the order Hemimerina has been proposed.

Crowson (1955) recognises, with good reason, the Strepsiptera as a super-family Stylopoidea within the suborder Polyphaga of the Coleoptera. However, since they are so biologically distinct from other beetles, it is preferred here to retain the old classification and grant them ordinal status.

Classification of Diptera and Hymenoptera

The figures for the number of parasitic species in each of the insect orders clearly show that the majority are contained in the orders Diptera and Hymenoptera. Both of these are large orders containing some very specialised species. Since they figure prominently in this book, it is opportune to present here their major subdivisions.

> Order DIPTERA
> > Suborder Nematocera (midges, mosquitoes, etc.)
> > Suborder Brachycera
> > Suborder Cyclorrhapha ('higher' flies)
> > > Series Aschiza
> > > Series Schizophora
> > > > Section Acalypterae
> > > > Section Calypterae
> > > > Section Pupipara

A number of recent authors unite the suborders Brachycera and Cyclorrhapha under the name of the former.

> Order HYMENOPTERA
> > Suborder Symphyta (sawflies)

Suborder Apocrita
Division Aculeata (bees, wasps, ants)
Division Parasitica (ichneumons, chalcids, etc.)

Types of Parasitic Insect

In spite of the diversity of ways in which an insect may be parasitic, it is possible to recognise two quite clearly defined types:

(i) Those in which the adults, and sometimes their immature stages as well, are parasitic. The host, nearly always a bird or mammal, is not killed. e.g. fleas, lice, mosquitoes.

(ii) Those in which only the immature stages are parasitic, the adult being free-living (protelean parasites). The host is nearly always another insect which is almost invariably killed by the parasite (=parasitoid), e.g. ichneumons. Less commonly the host is a vertebrate which is not usually killed, e.g. bot flies. Social parasites and inquilines are included with the protelean parasites, and so too are the Strepsiptera even though in this order the adult female does not leave the body of its host.

The book is divided into two sections which correspond with these two types of parasitic insect. The first group includes Exopterygota (both nymphs and adults) together with adults of some Endopterygota; the second contains only larvae of Endopterygota. The great majority of parasitic insects are embraced by the diagram of generalised host relationships printed below.

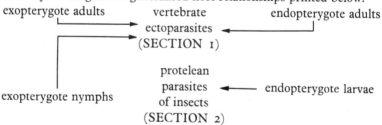

ADULT INSECTS REPRODUCE, usually sexually, and they disperse, and these functions require a degree of mobility in the outside world. For this reason adult insects cannot be endoparasites, confined within the body of their host; they are all ectoparasites. The nearest approach to endoparasitism is made perhaps by lice of the genus *Piagetiella* which inhabit the pouches of pelicans. The normal hosts are birds and mammals on which the ectoparasites find a warm and relatively constant environment amongst feathers and fur. Only one group, the dipterous family Braulidae, lives upon the bodies of invertebrates, but here also the hosts, bees, are furry and usually slightly warmer than the ambient temperature.

Insect ectoparasites are always much smaller than their hosts, on which they inflict little serious direct damage. Their food is usually the host's blood, but

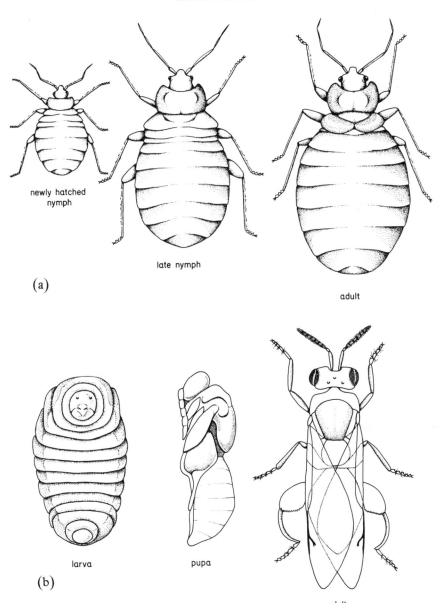

(a)

newly hatched
nymph

late nymph

adult

(b)

larva

pupa

adult

Figure 1. (a) Stages in the life cycle of an exopterygote parasite, the bed bug. All stages except the egg are ectoparasitic. (b) Stages in the life cycle of an endopterygote parasite or parasitoid, a chalcid. The adult is free-living and lays eggs inside the host by means of an ovipositor. The resulting larva feeds upon the tissues of the host and eventually kills it.

some eat hair, feathers, or skin debris. Characteristically they are flattened insects. This allows them to pass easily through the host's pelage or plumage, and it also makes their removal by the host, a hazard of great importance, more difficult. Their eyes are often absent or reduced, and they have sometimes lost their wings. A tough and thick cuticle resists the scratching or preening of the host. Strong claws on the legs make dislodgement unlikely, and often there are series of backwardly directed spines, termed combs, that serve a similar purpose. Combs have been developed independently in fleas, nycteribiids, streblids, polyctenid bugs, and a parasitic beetle; they are never found in non-parasitic insects.

Among endopterygote insects, with the exception of a few Coleoptera, it is only the adults which can be morphologically modified in these ways to equip them for an ectoparasitic existence, and the larval stages are free-living or passed within the body of the female parent. In exopterygotes, however, often the entire life cycle is spent on the host, since both nymphs and adults have basically the same body form, and both can become equally well adapted morphologically to an ectoparasitic life.

Also falling within this group of adult parasites is a series of Diptera, and also a few Hemiptera, which lead a predominantly independent existence and only sporadically come on to the body of a host for the purpose of taking a blood meal. Again, vertebrates are the usual hosts, although some Ceratopogonidae (Diptera) feed on other insects. In these species visible adaptations towards parasitism are confined more or less to the structure of the mouthparts.

The adults of several parasitic Hymenoptera feed upon fluids obtained from hosts and could be regarded as behaving parasitically, but this is secondary to their protelean parasitism and they are considered in the second section of this book.

Vertebrate blood is the diet of most adult parasites. The Diptera Pupipara, all blood-feeders, have been able to suppress a free-living larval stage altogether. The larva instead is nurtured within the mother's body and extruded only when ready to pupate. In these flies vertebrate blood is the only food taken by an individual in the course of its lifetime. Vertebrate blood would therefore seem to be nutritionally rich. It is, however, deficient in certain vitamins. How then do those insects which feed on nothing but blood, for example Anoplura, *Cimex*, some *Triatoma*, and *Glossina*, as well as the Pupipara, obtain their vitamins? The answer seems to be that in their bodies they have groups of cells called mycetomes which harbour yeast-like organisms, and these symbionts synthesise the necessary vitamins. The mycetomes are in the intestine of Hippoboscidae, the dorsal part of the abdomen unconnected with the gut in Nycteribiidae, in the wall of the mid-gut in *Glossina*, and amongst fatty tissue in Cimicidae. The micro-organisms are transmitted from one generation to the next, in *Glossina* and Pupipara by way of the mother's 'milk' glands to the larva, and in Anoplura through the ovary to the egg. Mycetomes are not found in mosquitoes, Tabanidae, *Stomoxys*, or fleas which feed on blood as adults but all of which have alternative food as larvae. If the mycetome is removed from the louse, *Pediculus*, its growth is severely inhibited.

Section 1

Insects with Parasitic Adults

1. *Mouthparts of Parasitic Adult Insects*

THE DIVERSITY OF DIETS exploited by insects is remarkable, and it has been made possible largely through the adaptability of their mouthparts. Yet, whilst varying in details of structure from group to group, insect mouthparts remain referable to a basic pattern. This fundamental, unspecialised type is adapted for biting and chewing. It is found, for example, in the locust (Figure 2) and cockroach, and consists of three types of segmental appendages (Figure 3). The segment lying immediately behind the embryological mouth,

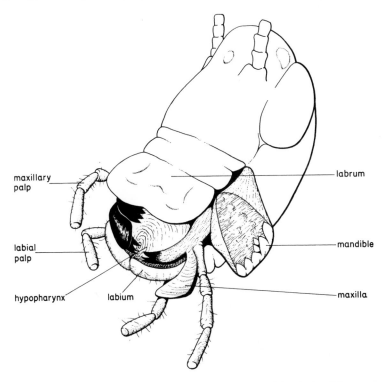

Figure 2. Head of a locust to show the spatial relationship of its mouthparts.

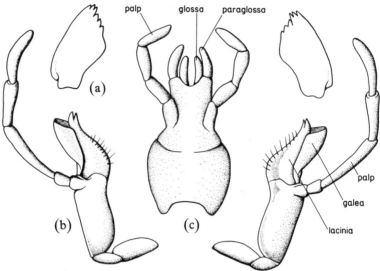

Figure 3. Biting and chewing mouthparts of an orthopteroid insect (a) Mandibles. (b) Maxillae. (c) Labium.

the fourth head segment, carries a pair of mandibles, the fifth segment carries a pair of maxillae, and the sixth segment bears the labium, which is compounded from a pair of appendages.

The mandibles of a cockroach are massive structures. Each has a distal series of medially-directed teeth and a proximal grinding molar area. The mandibles are operated by powerful muscles which swing them transversely. Behind and inside the mandibles is a pair of maxillae. Each maxilla carries a five-segmented sensory palp and terminates in two pieces, an inner lacinia armed with sharp teeth and an outer, protective, hood-like galea. The labium constitutes much of the posterior wall of the head. It is made up of a pair of fused appendages and bears a pair of short, three-segmented palps, and distally four small lobes, an inner pair of glossae and an outer pair of paraglossae. The mandibles, maxillae, and labium are members of a homologous series derived ancestrally from locomotory appendages. In addition, there are two other structures closely associated with these mouthparts. Anteriorly they are shielded by a roughly rectangular sclerite, the labrum, which is hinged to the lower edge of the clypeus, and between the maxillae is the median hypopharynx which is a tongue-like organ carrying the opening of the common salivary duct. All these structures enclose a preoral space at the back of which, just in front of the true mouth, is another cavity, the cibarium. In feeding, saliva is first poured on to the food which is then pierced by the teeth of the laciniae and lifted on to the anterior surface of the hypopharynx. The hypopharynx then pushes the food between the teeth and molar processes of the mandibles for mastication. Finally, the liquid mass is sucked into the cibarium and thence

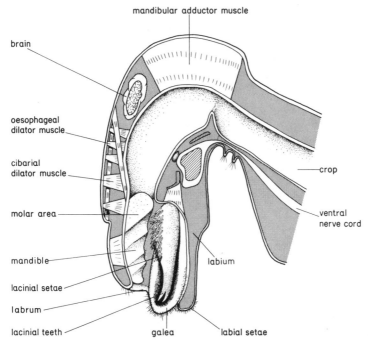

Figure 4. View of the right side of the head and neck of a cockroach as seen from the mid-line (from Popham 1961a).

through the mouth into the crop by cibarial and oesophageal dilator muscles (Figure 4) (Popham 1961a).

The closest approach to this basic mouthpart structure among adult ectoparasitic insects is found amongst those which do not suck blood but instead feed upon feathers, fur, or skin debris. Even so, this diet demands conspicuous modifications. In the biting lice (Mallophaga) the large, strong mandibles dominate the other parts and are used to cut off pieces of feather or hair. The maxillae and labium are reduced in size and they play probably only a minor role in the mechanics of feeding since the length of fibre cut off by the mandibles is pushed into the mouth by the labrum (Rothschild and Clay 1952).

Ectoparasites, which are blood-suckers, are unable to feed on solid particles, and their mouthparts are drawn out to form a proboscis. This pierces the host's skin and forms a vessel up which blood is sucked. Very often the proboscis is exceedingly fine and this reduces the chances of it touching a nerve-ending during penetration and alerting the host. The proboscis has been developed in a number of ways, the same functions being performed by different components in different insects, emphasising the versatility of these structures. The cibarium in sucking insects performs the important function of pumping up the liquid food. It is situated further inside the head capsule, being a more or less discrete chamber and not merely a space between the bases of the mouthparts as it is in the cockroach. When sucking up liquid, the volume of the cibarium is increased by contraction of its dilator muscles which originate on the clypeus.

In Anoplura (Figure 5) the hypopharynx, two maxillae, and the labium are in the form of long, thin stylets. Palps are absent, which renders interpre-

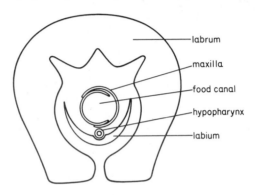

Figure 5. Transverse section through the mouthparts of a sucking louse (Anoplura).

tation of the parts difficult. At rest the stylets are retracted into a pocket inside the head with their apices protected dorsally and laterally by the snout-like labrum. The labrum is armed with recurved teeth, and when the louse feeds these teeth fasten into the host's skin, anchoring the insect whilst the labium, which terminates in three serrated lobes, is driven into a small blood vessel. Saliva is passed into the wound from the salivary duct in the hypopharynx, and blood is drawn up into the gut by way of a food canal formed by the two transversely-rolled maxillae which are fused ventrally. Mandibles have been lost by most adult lice and are represented only by rudiments in the embryo. They are still present, however, in adult *Haematopinus*.

As an example of blood-sucking Hemiptera, *Cimex*, the bed bug, may be taken. The proboscis, again, is made up of very long, thin stylets which to-gether constitute the fascicle, but its composition is quite different from that of the Anoplura. To begin with, the labrum is undeveloped, and its function as a stylet sheath is taken over by the ventrally-situated labium. The labium is relatively soft and does not enter the wound during feeding, serving instead merely as a support and guide for the stylets. It is jointed and folded back (Figure 6) as the stylets are driven home. The role of anchorage in *Cimex* is undertaken by the stylet mandibles which have sharp, backwardly-directed serrations at their apices (the mandibles in Anoplura are usually absent). The remaining stylets are the maxillae. These are locked together by grooves and

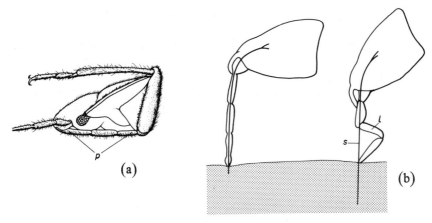

(a)

(b)

Figure 6. The bed bug (*Cimex*). (a) Head and prothorax with proboscis *p* in the resting position. (b) Diagrammatic representation of stylet penetration. The labium *l* is folded back whilst supporting the stylets *s* with its tip (after Kemper 1932).

ridges running their length and they enclose two canals formed by opposing grooves on their inner faces (Figure 7). The larger, dorsal canal is the food canal up which blood is drawn, and its diameter is just large enough to allow two erythrocytes to pass up together. The smaller, ventral canal is the salivary canal. The hypopharynx is not extended to form a stylet in the Hemiptera; instead it opens into the proximal end of the maxillary salivary canal. Saliva is forced down the hypopharynx and salivary canal and into the wound by the action of a powerful salivary pump.

The four stylets (the fascicle) are so exceedingly fine that they must act in concert in order to penetrate vertebrate skin to the depth of the capillary bed. Their action has been studied by Dickerson and Lavoipierre (1959). The

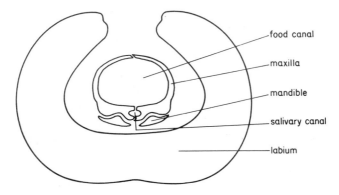

food canal

maxilla

mandible

salivary canal

labium

Figure 7. Transverse section through the mouthparts of *Cimex* (Hemiptera).

lip-like tips of the labium grasp and steady the fascicle. The mandibles move in rapid, alternating movements, projecting slightly in front of the maxillae and, with their teeth, cutting a path for the maxillae. The mandibles slide over the maxillae, held against them by the system of grooves and ridges, and by the surface tension of the saliva. The fascicle continually probes about in all directions until a suitable blood vessel is encountered. Haemorrhages are caused by the probing movements of the fascicle, but the bug seldom feeds at these pools. When a small blood vessel is found, it is probably only the maxillae which enter its lumen. In triatomine bugs the mandibles penetrate the host's skin only very superficially, where they serve as anchors for the very flexible maxillae which are thrust deep.

The bending and twisting probing movements of the fascicle of a blood-sucking bug have all the appearance of muscular movements. This is not possible, however, since the only muscles associated with the stylets are inserted right at their bases in the head capsule. When one stylet is slid forward in front of the others, its tip is easily deflected to one side by resistance encountered in the host tissue. It then serves as a guide to the other stylets when they are thrust to the same depth.

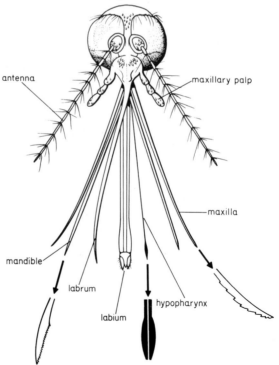

Figure 8. Frontal view of the head of a female mosquito with mouthparts displayed.

The Diptera have mouthparts in the form of a proboscis which is more varied in structure than in any other insect order. It is used for sucking or sponging up fluids, sometimes for piercing as well, and exceptionally is adapted for chewing. Of the three suborders of Diptera the Nematocera are the most generalised and the proboscis consists maximally of seven components. This number is reduced in the other two suborders.

In a blood-sucking nematoceran, for example a female mosquito, the labium functions as a sheath, as in Hemiptera, and it supports the elongated stylets which consist of a pair of mandibles, a pair of maxillae, the hypopharynx, and the labrum (or labrum-epipharynx) (Figure 8). This latter is grooved ventrally to form a tube up which blood may be drawn (Figure 9), a

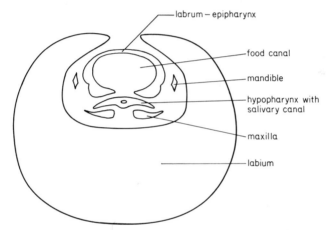

Figure 9. Transverse section through the mouthparts of a female mosquito (Diptera Nematocera).

function which in bugs is served by the maxillae. The maxillae and mandibles of the female mosquito are very fine and apically they are serrated. They are the cutting organs. The hypopharynx is also much elongated, in contrast to its condition in bugs, and it carries the saliva, which contains an anticoagulant, into the wound. Only female mosquitoes are able to take blood meals, the males lacking mandibles. In other groups of Nematocera the proboscis is much shorter than in mosquitoes and its components are more robust. In Simuliidae the mandibles have a scissor-like movement, and they make the preliminary incision in the skin into which the maxillae are forced to enlarge the wound sufficiently for the labrum-epipharynx to be inserted. Consequently the physical damage inflicted by a biting simuliid is considerably more severe than that of a mosquito.

The proboscis of *Tabanus*, the horse fly, a representative of the Brachycera, is more robust than that of a mosquito, and the stylets are flattened, blade-like structures (Figure 10). The mandibles work like scissors, cutting with their tips

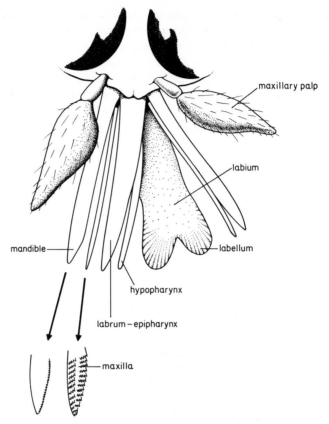

Figure 10. The mouthparts of a female *Tabanus* (Diptera Brachycera) (partly after Wigglesworth 1964).

which are serrated on their inner faces, and the maxillae, moving up and down like a file, are forced into the lesion until the blood flows freely. The labrum-epipharynx and hypopharynx are also in the form of stylets, although more robust than in the mosquito. Much less rigid than these mouthparts is the labium, which again serves as a supporting sheath and is provided with a pair of distal lobes termed labella. During biting, the labium is folded back out of the way, but it is later used to sponge up the exuding blood. Each labellum is provided with numerous fine channels (pseudotracheae), supported by almost complete rings of chitin. Liquid food passes along the pseudotracheae by capillarity to the food canal of the labrum-epipharynx. Male horse flies, like male mosquitoes, do not possess mandibles and do not bite.

  Pangonia is a genus of Tabanidae, the majority of species of which have an extraordinarily elongated proboscis. The proboscis of *P. longirostris* (Figure 11) is twice as long as the rest of the body. Most of the species have been observed feeding at flowers, sucking up nectar whilst hovering in front of a bloom in the manner of a bombyliid or a hawk moth. Several species also attack man and cattle, and the feeding tactics which they adopt have been the subject of some controversy. It was at one time thought that the flies did not alight on the host to feed, but instead pierced its skin with their proboscides by making sudden darts towards it and then fed upon exuding blood whilst hovering.

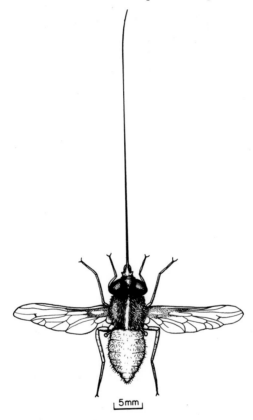

Figure 11. *Pangonia longirostris* male (drawn from a photograph in Tetley 1918).

The labium of Pangoniinae, however, is long and flexible, unlikely to be usable as a stabbing instrument, and this view was not unanimously subscribed to. Recently Goodier (1964) has reviewed accounts of blood-feeding in Pangoniinae and concludes that it is very unlikely that feeding ever occurs whilst hovering. The flies may hover close to a host and test its skin with the tips of their labella, perhaps giving the illusion of feeding. Penetration of the skin by the piercing stylets (mandibles and maxillae) occurs, however, only after settling. The labium is folded back during penetration. In species with a long proboscis, the angle between skin surface and proboscis must be very acute. Such species, after penetration of the stylets has been effected, may from time to time partially hover with only their front legs maintaining contact with the host. This position presumably allows for a greater angle of attack and probably gave birth to the idea that penetration and feeding could be conducted whilst in flight.

The Cyclorrhapha have lost both mandibles and maxillae (except for the palps) as independent entities, and their mouthparts consist of a highly developed labium, labrum-epipharynx, and hypopharynx. Males and females

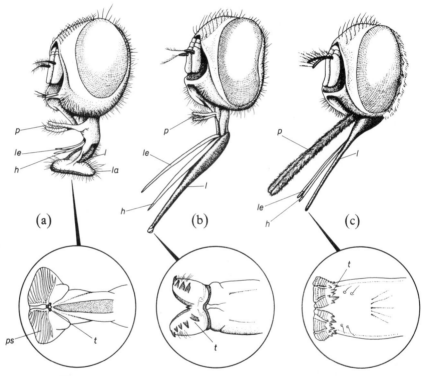

Figure 12. The heads of cyclorrhaphous flies showing the different forms of the proboscis and modifications of the labella (partly after Wigglesworth 1964). (a) *Musca autumnalis* male with soft labium, and pseudotracheae and minute prestomal teeth on labella. (b) *Stomoxys calcitrans* female with hardened labium and well-developed prestomal teeth. The labrum-epipharynx and hypopharynx have been separated from the labium. (c) *Glossina palpalis* female with hardened, slender labium, supporting maxillary palps, and large prestomal teeth and rasps. *h*=hypopharynx; *l*=labium; *la*=labellum; *le*=labrum-epipharynx; *p*=maxillary palp; *ps*=pseudotracheae; *t*=prestomal teeth.

have similar mouthparts. Most species of Cyclorrhapha feed at exposed liquid surfaces, passing saliva onto the food and drawing it up again, together with dissolved food, by means of pseudotracheae on the two large labial labella (Figure 12a). The pseudotracheae have a diameter of about 0.01 to 0.02 millimetres. The labium in these species is soft and well-provided with muscles. It has a dorsal groove to accommodate the saliva-carrying hypopharynx and the labrum-epipharynx. A ventral groove on the labrum-epipharynx serves, as in flies of the other suborders, as a food canal, being almost closed below by the hypopharynx (Figure 14). In addition to pseudotracheae, the labella of a

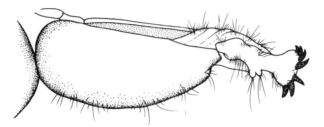

Figure 13. The proboscis of *Musca crassirostris* (after Austen 1909). The labium is soft but there are large teeth on the labella. This species represents a condition between *M. autumnalis* (Figure 12a) and *Stomoxys* (Figure 12b).

typical cyclorrhaphan, a blowfly for example, also bear series of minute, rasping teeth (prestomal teeth) which break up the substrate into particles small enough to pass into the pseudotracheae. These flies may feed on blood oozing from an already formed wound and sometimes take what is left over after a tabanid has had a meal. Some species, for example *Musca (Philaematomyia) crassirostris* (Figure 13), have larger rasping teeth on their labella with which they can scratch the skin surface sufficiently deeply to draw blood which they mop up with their pseudotracheae. The labium remains soft and retractile. A further development in this series of blood-feeding Cyclorrhapha is exemplified by the stable fly, *Stomoxys calcitrans* (Figure 12b). The prestomal teeth are large and the labium itself is hardened and pointed, capable of being jabbed into vertebrate skin by a thrust of the body generated by the legs. Thus, the function performed in mosquitoes and horse flies by mandibles and maxillae is here, in the absence of these organs, performed by the labium. Similar mouthparts are found in hippoboscids, although in these flies the proboscis is withdrawn inside the head capsule when not in use. The culmination of this line of development is seen in tsetse flies (*Glossina*) (Figure 12c). They have more numerous, although smaller, prestomal teeth than *Stomoxys*. The labium is very slender, so fine in fact that it requires support, and this function is performed by the enlarged maxillary palpi. Thus the labium, which in mosquitoes and horse flies is the organ supporting the mandibles and

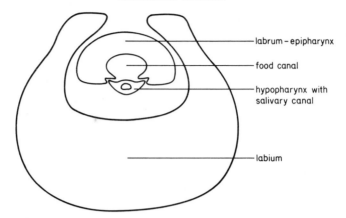

labrum – epipharynx

food canal

hypopharynx with salivary canal

labium

Figure 14. Transverse section through the mouthparts of a cyclorrhaphous dipteran.

maxillae, in *Glossina* needs itself to be supported by a part of these very structures whose function it has taken over. This could hardly be surpassed as an example of the versatility of insect mouthparts. For a more detailed account of the structure and operation of dipteran mouthparts, reference should be made to the descriptions of Patton and Cragg (1913).

Fleas (Siphonaptera) are allied to Diptera, but their mouthparts (Figure 15) are peculiar and the homology of their components is not generally agreed upon. Mandibles are absent, and the wound is made by a pair of blade-like structures which are usually interpreted as maxillary laciniae. These enclose a small salivary canal ventrally. Between the laciniae is a long, thin epipharynx (the labrum is a short lobe situated in front of the base of the epipharynx) which is grooved ventrally to form a food canal (Figure 16). The epipharynx and laciniae are protected and guided by the labium and its palps. The hypopharynx is rudimentary. The maxillary palps are quite long and carried on a basal lobe which is probably the stipes.

Lavoipierre (1965) reviewed the feeding methods of blood-sucking insects and concluded that 'blood sucking arthropods feed, in the main, either from the lumen of a blood vessel or from a blood pool resulting from laceration of blood vessels.' Mosquitoes will do both but this is unusual; they are normally vessel feeders. Other vessel feeders are bugs, fleas, and lice whilst blood-feeding Brachycera and Cyclorrhapha are pool feeders, as also are simuliids. Vessel feeders do not usually take blood from capillaries but from venules or small veins.

Both pool feeders and vessel feeders must overcome the problem of blood clotting. The channels up which blood is drawn are of very small diameter and easily blocked. Anticoagulin is present in the saliva of several blood-sucking insects. When the salivary glands of *Glossina* are excised from a living fly, feeding continues for a while but is eventually terminated by clots of blood

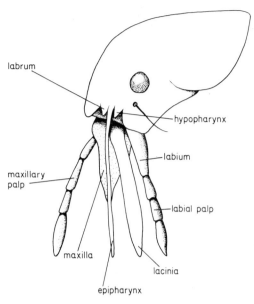

Figure 15. Head and mouthparts of a flea (Siphonaptera).

blocking the proboscis (Lester and Lloyd 1928). Anticoagulin has been found in the saliva of the bugs *Rhodnius, Triatoma,* and *Cimex,* and in the fly *Musca crassirostris,* as well as in *Glossina.* However it has not been found, although looked for, in *Stomoxys,* and among mosquitoes it is present only in some of the species of *Anopheles* and is absent from the saliva of *Aedes* and *Culex. Stomoxys* is such a vicious biter that it causes blood to flow very freely and by feeding rapidly evidently dispenses with the need to prevent clotting. It is

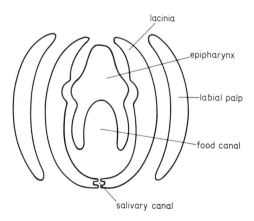

Figure 16. Transverse section through the mouthparts of a flea (Siphonaptera).

more difficult to explain the absence of anticoagulin from most mosquitoes, although it may be connected with the fact that blood passing up the proboscis is much diluted with saliva.

2. Lice

OF ALL INSECTS, lice are the most completely committed to parasitism. In fleas the larval stages are free-living, but for lice the bird or mammal host is the only environment and all life stages from egg to adult are spent on the host's body. Movement of lice between individual hosts is restricted largely to times when the host is attending its young, and it is probably exceptional for most lice to be divorced, even momentarily, from physical contact with their hosts. Indeed, lice very soon die if deprived of the living body of their host.

The parasitic lice are assigned to the orders Mallophaga and Anoplura. They most resemble, and probably share a common ancestor with, bark and book lice (Psocoptera). These latter are small, free-living insects, sometimes winged, which live on the bark of trees, in old birds' nests, amid dusty books and papers, stored cereals, and in other places that are rich in the organic debris, moulds, or lichens on which they feed. It is not difficult to imagine the free-living ancestor (or ancestors) of the parasitic lice, probably already living in nests of vertebrates and aided by small size, taking to living entirely upon the larger animals' bodies where food and shelter were adequate. Adaptations to an ectoparasitic existence followed, the grosser morphological changes being loss of wings (if present), reduction of eyes, and development of dorso-ventral flattening, a tough integument, and often enlarged tarsal claws (Figure 21). Hopkins (1949), in a detailed study of the host associations of mammalian lice, considers all lice (Mallophaga and Anoplura) to be derived from a single stock, most probably parasitic upon birds. We have, however, only living forms to work on, fossils of lice being unknown, and a final decision on this point is unlikely to be reached.

Mallophaga and Anoplura are similar in many ways, and the main difference between them, the difference that prevents many entomologists from uniting them in a single order, is in the structure of the mouthparts. In Mallophaga these are of a modified biting type and the insects feed mostly by chewing feathers and hair. However some Mallophaga (e.g. Gyropidae) are believed to eat the secretion of sebaceous glands, and most will take blood if this is accessible. Scratching by the host in response to the presence of lice may cause small wounds which provide lice with blood. A chicken and turkey louse (*Menacanthus stramineus*) chews holes in the developing quill and takes blood, and mem-

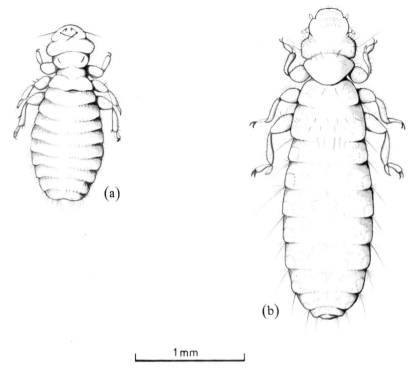

1 mm

Figure 17. Mallophaga Amblycera. (a) *Austromenopon lutescens* female from ringed plover. (b) *Actornithophilus patellatus* female from inside a curlew's quill.

bers of the Ricinidae probably feed entirely on blood from holes pierced in the skin of the hosts (small, passerine birds) by their mandibles (Clay, pers. comm.). *Piagetiella* lives in the pouches of pelicans, attached to the mucous membrane by its mandibles. Here blood and mucus must be the diet since no feathers are present. *Piagetiella* moves onto the feathers only to lay its eggs. There are also records of biting lice eating other lice, and Waterston (1926) identified a moth's scale, feather fibres (all cut to the same length), part of a seed coat, and mineral grains in the crop of *Falcolipeurus* (=*Esthiopterum*). He suggests that the mineral particles may serve a triturating function, analogous to the grit in the gizzard of a bird.

In Anoplura, on the other hand, the mouthparts (Figure 5) are highly modified for piercing the skin of the host and drawing up blood which is the only food taken. Clay (1949b) describes a form of piercing mouthparts in the mallophagan *Trochiloecetes* which further emphasises the similarity between the orders, as well as indicating that piercing mouthparts have evolved more than once among lice.

The Mallophaga are divided into three suborders: Amblycera, Ischnocera,

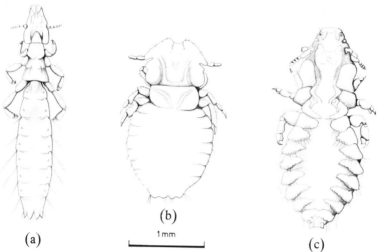

Figure 18. Mallophaga Ischnocera. (a) *Fulicoffula lurida* female from waterhen. (b) *Trichodectes melis* female from badger. (c) *Philopterus ocellatus* female from hooded crow.

and Rhyncophthirina. The first two are large groups and can be distinguished by the fact that the antennae of Ischnocera are filiform and easily seen whilst those of Amblycera are concealed in pockets in the head. The Amblycera (Figures 17, 19) most resemble the free-living Psocoptera. They have retained their maxillary palps, lost in the other two suborders and the Anoplura, and, as a group, their diet tends to be more generalised. Also, they move about more on the host's body, and species of the genera *Actornithophilus* and *Austromenopon* parasitising waders have been found (Thompson 1957) on the eggs of their hosts, especially when these are near to hatching. Amblycera and Ischnocera are parasites of both birds and mammals, but in both groups the majority of species are bird parasites. Ischnocera (Figure 18) are more diversified in structure than Amblycera, and this reflects their greater specialisation to different environments. Not only are they usually host-specific, but each species is often confined, on birds at any rate, to a particular region of the host's body from which it seldom wanders. The diet is more exclusively one of hairs and feathers (keratin) than in Amblycera.

The third suborder of Mallophaga, the Rhyncophthirina, includes only two species, *Haematomyzus elephantis* (Ferris 1931) a louse of African and Indian elephants, and *H. hopkinsi*, a parasite of warthogs. The mouthparts of *Haematomyzus* are rather curious, being of the biting type but situated at the tip of a rostrum-like structure, and it is thought to feed upon blood. Mukerji and Sen-Sarma (1955) describe the anatomy of *H. elephantis* and suggest that *Haematomyzus* should be placed in a separate order Rhyncophthiraptera.

Symmons (1952) examined the heads of a number of lice and found a

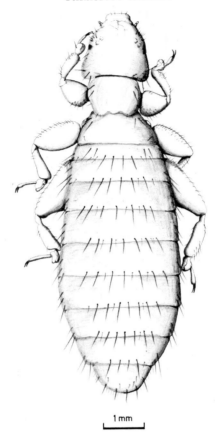

1 mm

Figure 19. *Laemobothrion circi* female (Mallophaga Amblycera), an enormous louse measuring almost a centimetre in
length. The drawing was made from a specimen collected in
Cheshire, 1947, from a juvenile marsh-harrier.

progressive reduction in the tentorium, the internal skeleton of the head, in
the series Psocoptera, Amblycera, Ischnocera, Rhyncophthirina, and Anoplura. It was concluded that this was the phylogenetic order of appearance.

The Anoplura (Figure 20) are found only on mammals, and their known
distribution in the mammalian orders, together with that of the mallophagan
suborders, is shown in Table 1.

The fact that Amblycera and Ischnocera are parasites of both birds and
mammals indicates that life on these two host groups demands no great
diversification of structure. It is interesting to note, however, that the avian
Mallophaga have paired tarsal claws, whilst those on mammals tend to have
only one on at least some of the legs. Anoplura also have only one tarsal claw
(Figure 21) on each leg. This tendency is found among louse flies (Hippo-

Mammal orders	Mallophaga			Anoplura
	Amblycera	Ischnocera	Rhyncophthirina	
Monotremata	–	–	–	–
Marsupialia	x	–	–	–
Insectivora	–	–	–	x
Dermoptera	–	–	–	x
Chiroptera	–	–	–	–
Primata	x	x	–	x
Edentata	–	x	–	–
Pholidota	–	–	–	–
Lagomorpha	–	–	–	x
Rodentia	x	x	–	x
Cetacea	–	–	–	–
Carnivora Pinnipedia	–	–	–	x
Carnivora Fissipedia	x	x	–	x
Tubulidentata	–	–	–	x
Proboscidea	–	–	x	–
Hyracoidea	–	x	–	x
Sirenia	–	–	–	–
Perissodactyla	–	x	–	x
Artiodactyla	x	x	x	x

Table 1. The host associations of lice parasitic on mammals, x denotes presence (mostly from Hopkins 1949).

boscidae) where the mammal-infesting genera often have a reduced number of tarsal claws.

The life history of lice is geared to the body of the host. The eggs are cemented singly or in groups onto hairs or feathers and they require the warmth of the host's body for development. There are three nymphal instars in those species that have been studied, and the duration of the life cycle is only about one month. The constant conditions on the host's body allow for continuous breeding throughout the year, and so a louse population could build up very rapidly to a high figure. Nuttall (1917) records 10 428 individuals of human louse, *Pediculus humanus* (Figure 20e), on a single shirt. On wild mammals, Hopkins (1949) finds the louse population to be of very variable size but often small. There is some evidence that young birds and mammals, and also sickly individuals, often carry the heaviest populations. Buxton (1941) has shown that some men cannot support a louse population; also that man's resistance to the head louse (*Pediculus humanus capitis*) increases with age. Host immunity is possibly an important check on the growth of a louse population. Another very important check is preening or grooming by the host. Murray (1961) has shown that the number of *Polyplax serrata* (**Figure** 20d) on a mouse (*Mus musculus*) depends very largely on the efficiency of the host's oral self-grooming. Buxton (1939) has described an intraspecific control in the human louse. When the louse population is very high, males have

abundant and frequent opportunities of copulating, and in these circumstances
the females are often injured and their reproduction is impaired. Usually the
number of females tends to exceed that of males in a population, and in some
species, for instance *Damalinia americana* (Ischnocera), males are unknown.
Matthysse (1944) has shown *D. bovis* to be capable of parthenogenetic reproduction. With an equal sex ratio the population on a host would rapidly tend
to become genetically homogeneous if movement of the lice between hosts
was slow. This situation could be disadvantageous, and its development is
retarded if the sex ratio favours the females and some females reproduce
parthenogenetically.

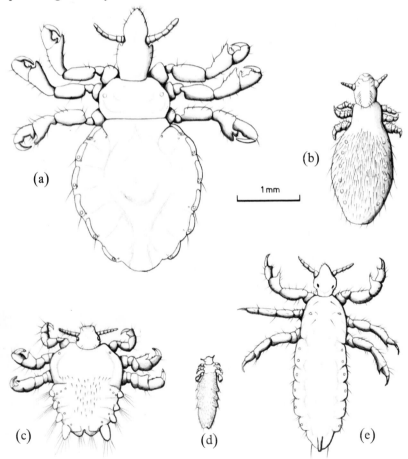

Figure 20. Anoplura. (a) *Haematopinus suis* male from pig.
(b) *Linognathus ovillus* female from sheep. (c) *Pthirus pubis*
female, the crab louse of man. (d) *Polyplax serrata* female
from wood mouse. (e) *Pediculus humanus humanus* male, the
human body louse.

Murray and his co-workers have recently (1965) published the results of their researches into the lice of antarctic seals. The southern elephant seal, which is infested by the louse *Lepidophthirus macrorhini*, comes ashore only twice a year for periods of from three to five weeks. It is only at these times that the lice can reproduce, and they make the most of the short time available, each female laying up to nine eggs a day. The warmest external parts of the seal's body are the hind flippers, both at sea and on land, and it is here that the lice tend to congregate. The Weddell seal inhabits colder water than the elephant seal, but

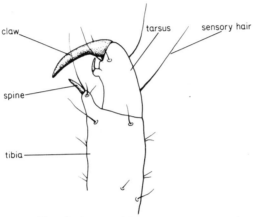

Figure 21. Terminal part of front leg of *Pediculus humanus* female (after Keilin and Nuttall 1930).

it comes out of the sea onto the ice almost daily. Its louse, *Antarctophthirus ogmorhini*, is extraordinarily well adapted to enduring cold, surviving a temperature of −20°C for 36 hours and reproducing rapidly at temperatures between 5°C and 15°C. *L. macrorhini* on the elephant seal requires a temperature of at least 25°C for rapid multiplication. Lice generally are very sensitive to temperature and most soon die when their host's body cools after death. High temperatures also may cause mortality by preventing the eggs from hatching, and high summer temperatures can be lethal even to adult lice. Lice are negatively phototropic and so avoid sunlight. Brinck (1948) has shown that in summer in Sweden horse lice migrate to the shaded parts of the host, but even so, at this time the total louse population drops to a minimum.

The responses of *Pediculus humanus* to varying physical conditions were studied by Wigglesworth (1941), who showed that an air temperature of 29°C to 30°C is preferred and that high humidities are avoided. This might explain the disappearance of lice from people suffering from a fever. *P. humanus* was not attracted by any stimulus studied, but it makes avoiding movements when a favourable stimulus ceases to operate or when a less favourable stimulus is encountered. Other preferences detected in the human louse are for rough surfaces rather than smooth surfaces, and for cloth smelling of man or of lice and their excreta.

The sensitivity of lice to low temperatures has led Eichler (1936) to suggest this as the reason for their absence from animals such as bats which go into a state of true hibernation with a greatly reduced body temperature. However some hibernating mammals, for example marmots, are lice infested; also tropical bats which do not hibernate are as louse-free as their temperate allies.

Other attempts at explaining the absence of lice from certain groups of birds or mammals have been made. Aquatic mammals are often free of lice, probably because few have emulated the seal lice and become adapted to withstand prolonged periods of submergence. *Lepidophthirus* and *Antarctophthirus*, in addition to being able to survive low temperatures, have their bodies covered by small flat scales which retain a surface film of air when under water. Certainly no lice have been found on whales or sea-cows, but aquatic birds support many Mallophaga. Association with ants, because the formic acid which many produce kills lice, has been suggested as the reason for the absence of lice from ant-eaters and pangolins, although the scaly armour of pangolins would seem in any case an unsuitable environment for lice.

Among birds, over 200 species of passerines have been seen to indulge in a curious habit termed 'anting'. In this the bird may squat on the ground near to an ants' nest and allow the ants to crawl in its plumage, or it may actually pick up ants and place them amongst its feathers. Only formic acid-producing formicine ants are used. One of the reasons for this behaviour may be louse eradication, but ornithologists differ on the subject, and it may be noted that the myrmecophagous green woodpecker is by no means spared the attentions of lice. Simmons (1966) notes that birds examined immediately after 'anting' had numbers of dead and dying feather-mites in their plumage.

Host-Specificity

The high degree of host-specificity in lice is summed up by Hopkins and Clay (1952) in the following words: 'The host is to a louse what the locality is to a free-living insect, and lice are so extremely specific in their host-associations that it is rare in many groups to find indistinguishable lice on specifically distinct hosts, and common in a few groups to find that hosts which differ only subspecifically possess quite different lice.' The fact that lice seldom leave the body of their host is responsible for their host-specificity. Since the entire life cycle can be spent on one animal, there is no advantage in the louse habitually leaving it in search of new hosts in a hostile environment. Limited movement is thus the rule, and transfer from host to host occurs almost always at times of close proximity between individual hosts of the same species, as when feeding young or copulating. The latter is the only occasion when cuckoo lice can transfer, unless they do so by phoresy (see below). The European cuckoo has at least three species of host-specific Mallophaga. Opportunities for transference between different species of host are probably rare, although a predator-prey relationship makes one-way traffic to the predator possible. Stragglers, the name given to lice on the 'wrong' host, are uncommon and the recorded instances are mostly from predatory birds.

Hopkins has observed that Anoplura on strange hosts have difficulty in walking and fixing their eggs to the hair, and they usually die within a very short time. An exception is the human louse, *Pediculus humanus*, which can live and breed on the pig. Other parasites of the pig and man, such as the fleas *Pulex irritans* and *Tunga penetrans*, and some ticks and nematodes, are interchangeable. Fortuitous physiological similarities between the two hosts appear to be the only explanation.

A very high degree of host-specificity among lice is to be found, as previously indicated, among some Ischnocera. For example, the distribution of species of *Philopterus* (Figure 18c) on British crows (*Corvus* species) is as follows:

P. atratus on *C. frugilegus* (rook)

P. corvi on *C. corax* (raven)

P. ocellatus on *C. corone corone* (carrion crow) and *C. corone cornix* (hooded crow)

P. guttatus on *C. monedula* (jackdaw)

Only the two subspecies of *C. corone* share the same species of *Philopterus*. In South Africa, different subspecies of the hyraxes *Procavia capensis* and *Heterohyrax syriacus* each supports a different species of *Procavicola* (Ischnocera) (Hopkins 1949).

In Amblycera and Anoplura host-specificity is rather less rigid, although instances of closely allied lice being found on quite dissimilar hosts are unusual. The genus *Colpocephalum* (Amblycera), however, has a wide distribution throughout the Aves, and *C. turbinatum* is found both on pigeons and on species belonging to many genera of the Falconiformes (Clay *pers. comm.*).

Whilst it is usual for a genus of lice to be restricted to an order of birds or mammals, several other exceptions can be cited. *Polyplax* (Anoplura) is found on both rodents and shrews, *Heterodoxus* (Amblycera) on dogs and marsupials, and even among Ischnocera, *Saemundssonia* infests waders, cranes, and tubenoses. But these are exceptions.

Many of the apparent anomalies in host associations may be the result of artificial contact between hosts, as in zoological gardens, menageries, or gamebags, or, more rarely, natural straggling. Straggling may well account for the distribution of species of *Saemundssonia* (= *Docophorus*) on auks in British waters recorded by Waterston (1914) (Table 2).

Each species of auk has its own species of louse, found on all individuals examined, but in addition three of the bird species harboured lice 'belonging' to other auks. The birds concerned, puffin, razorbill, and guillemot, are the three species that most frequently consort with one another on cliffs or at sea.

Host species	Black guillemot	Guillemot	Razorbill	Puffin	Little auk
Number of birds examined:	11	11	6	6	10
Number of birds with:					
S. fraterculae	–	–	–	6	–
S. calva	–	11	–	2	–
S. celidoxa	–	2	6	2	–
S. grylle	11	–	1	–	–
S. merguli	–	1	–	–	10

Table 2. The distribution of feather lice of the genus *Saemundssonia* in a collection of British auks (from Waterston 1914).

The parasitisation of falcons and pigeons by *Colpocephalum turbinatum* probably derives from the predator-prey relationship of these birds, and a similar secondary infestation in past ancestry must account for the presence of a species of *Rallicola* (Ischnocera), a genus normally found on rails, coots, crakes, and similar birds, on the non-parasitic cuckoo *Centropus*. *Centropus*, like rallids, nests among grasses and reeds near to water. That straggling does not very frequently lead to permanent establishment is, however, emphasised by the fact that the European cuckoo is not habitually parasitised by any of the genera of lice associated with its foster parents.

Clay (1949a) makes the point that the host distribution of lice is equivalent in many ways to the geographical distribution of free-living animals. The separation of the host population into two geographically isolated parts also isolates the lice. Thus speciation may proceed simultaneously in the two groups. The high degree of host-specificity in lice points to a very long established relationship, and it would seem that lice were present on the distant ancestors of the groups of hosts on which they are now found. As the hosts evolved, so too did their lice, although usually less conspicuously, and from this it follows that lice may be useful supplementary indicators of the affinities of their hosts. The ostrich and rhea, though native to different continents, are infested by closely related Ischnocera. The louse infesting the kiwi (*Apteryx*), a bird that in the past has been thought to be allied to the ratites, is however quite different and related instead to a louse on rails. The affinity between *Apteryx* and rails had independently been realised on morphological grounds. Similarly, flamingoes are considered, on the basis of their lice and other ectoparasites, to have greater affinity with ducks than with storks, even though several of their anatomical features suggest a stork-like ancestor. Flamingoes are parasitised by four genera of lice and three of these are otherwise restricted to the ducks (Anatidae). The anopluran genus *Pediculus* is common to man and the chimpanzee, and allied to *Pedicinus* from cercopithecoid monkeys; a situation that reflects the accepted relationships of the mammals.

Some so-called rules pertaining to the host relations of ectoparasites have been formulated (Eichler 1948), and lice illustrate these very well. Examples to which Fahrenholtz' Rule is applicable have already been given: 'In groups of permanent parasites the classification of the parasites usually corresponds directly with the natural relationships of the hosts.' An extension of this is Szidat's Rule: 'Primitive hosts are parasitised by primitive parasites, specialised hosts by specialised parasites.'

Examples of louse distribution based strictly upon geography rather than host phylogeny are scarce. However two grebes, *Podiceps auritus* and *P. caspicus*, are infested by *Aquanirmus bucomfistii* and *A. americanus* respectively in America, but in Europe both species carry only *A. colymbinus* (Edwards 1965). Also Clay (1964) cites the case of two species of gannet (*Sula*) that share three species of *Pectinopygus* with different geographical ranges.

Having emphasised at some length the normal host relationships of lice, some exceptions to the general rule must now be considered. Firstly, there are cases of 'discontinuous distribution'. The amblyceran genus *Laemobothrion* (Figure 19) is found today on birds of prey, storks, and rails, and it is thought that this is the remains of a once more widespread and continuous distribution that has become broken up by extinction of the lice on many groups of hosts. Similarly Anoplura are absent from all Carnivora except dogs and seals. The ecological gaps left by extinctions may be filled by other lice. Straggling has already been discussed and it seems that it now rarely leads to the establishment of a permanent association (secondary infestation). If this had always been the case, however, no host would be parasitised by dissimilar louse genera (unless lice are very polyphyletic which is excessively unlikely). Instead, we find a species of tinamou supporting representatives of no fewer than twelve lice genera. Clay (1949a) explains secondary infestations by presuming 'that establishment on a new host was more possible at a time in the evolution of the louse before it had developed extreme host-specificity and when the hosts themselves, less divergent during the earlier stages of their evolution, offered a more uniform environment'.

When species belonging to different genera of Mallophaga coexist on a bird host, each is often restricted to a localised part of the host's body. As already remarked, this is especially evident among Ischnocera, and it adds a further dimension to specificity, allowing more than one louse species to live without interspecific interference on the same host (Figure 22). The Ischnocera of the head and neck region are often not greatly flattened, rather slow-moving, and they have short, more rounded bodies and large, broad heads to accommodate the large mandibular muscles. If this type strayed to other regions of the body, it could be readily removed during preening, but the head and much of the neck are out of reach of the bird's beak. In contrast, the Ischnocera of the back and wings are greatly flattened, elongated lice with much narrower heads and smaller mandibles, and they are capable of rapid movement across the broad feathers of their domain. Besides these two major types there are several minor intermediate forms and one outstanding one. This latter is typified by *Actornithophilus patellatus* (Figure 17b), one of the Amblycera, which has adopted the very specialised habitat of the inside of the quill or shaft of the

Figure 22. Distribution of Mallophaga on the glossy ibis (from Dubinin in Dogiel 1964). (1) *Ibidoecus bisignatus* on head and neck. (2) *Menopon plegadis* on breast, abdomen, and flanks. (3) *Colpocephalum* and *Ferribia* species on wings and tail. (4) *Esthiopterum raphidium* on back.

primary and secondary feathers of the curlew (Waterston 1922). This louse is only very rarely found outside feathers. It eats the pith of the shaft, and all stages from egg to adult may be present in large numbers inside a single feather.

Among Anoplura, specialisation to different body regions of the host occurs more at the specific than the generic level. On sheep, for instance, *Linognathus pedalis* is normally found on the legs whilst *L. ovillus* (Figure 20b) predominates on the head. Murray (1963) has studied and described ecological and behavioural differences between these two species. Man is parasitised by two quite closely related species of lice, *Pediculus humanus* and *Pthirus pubis* (Figure 20c). *Pthirus*, with very large tarsal claws, is usually found associated with the strong pubic hairs, whilst *Pediculus* is found on the head and those regions of the body covered by clothing. Further, *P. humanus* is differentiated into two subspecies (or possibly sibling species), *P. humanus humanus* (= *corporis*, *vestimenti*) (Figure 20e) being on the body and *P. humanus capitis* on the head. *P. h. humanus* is the larger form with relatively shorter antennal segments, and it lays its eggs on clothing. *P. h. capitis* has a darker and tougher integument than *P. h. humanus* and it lays its eggs on the host's hairs. Neither form is entirely restricted to its principal region of the host, but their differentiation is maintained in spite of some gene-flow because they are only slightly interfertile. Crosses between the two subspecies frequently result in offspring that are intersexes (Keilin and Nuttall 1919).

Phoresy

Phoresy is a special type of interrelationship between two species in which one attaches itself to, and is carried by, the other, usually for purposes of transport only. It is 'free passage without food'. The habit is practised by several groups of parasitic insects, but Mallophaga are one of the chief exponents of it, and Clay and Meinertzhagen (1943) review instances of phoresy involving these insects. Nearly all the records concern species of *Brueelia*, *Sturnidoecus*, and *Philopterus* (Ischnocera) and the usual transporter is *Ornithomya*, a winged hippoboscid fly (Figure 23). The lice attach themselves by their mandibles to the wing veins or to the bases of hairs. Lice have also been found, although very infrequently, attached to mosquitoes and other blood-sucking flies, fleas, dragonflies, bees, and butterflies. In the last three instances, the lice must have been doomed, with no chance of regaining a host. However, in the other cases it would be possible, in varying degrees, for the carrying insects to transport the lice to a new host. Rothschild and Clay (1952) describe the case of a starling examined immediately after death and found to be infested by fifteen lice and one *Ornithomya*. Host and parasites were wrapped in a cloth and on examination two hours later it was found that seven of the lice had attached themselves to the abdomen of the fly. Some lice may regularly use this as a method of escape from dead hosts. It is possible that on a living host the body of a fly holds little attraction for lice, but as the host's body cools after death the fly becomes relatively more attractive or less repellent than the cooling flesh. The general rarity of phoresy, however, seems to indicate that very often it is merely an accidental association, and as the carriers are only rarely specific parasites of the same host as the lice, it offers only a slim chance of survival to the lice.

Exceptionally, a relatively high incidence of phoresy has been reported. Corbet (1956a) made an intensive study of *Ornithomya fringillina* on starlings on Fair Isle. From sixty-eight birds, 156 flies were taken and sixty-eight of these (43.5%) had lice attached to them. Probably all of the lice were *Sturni-doecus sturni*, a specific parasite of starlings, and most of them were found on the abdomens of the flies, firmly attached to cuticular folds by their mandibles. One fly was carrying as many as twenty-two lice. In this case it is unlikely that

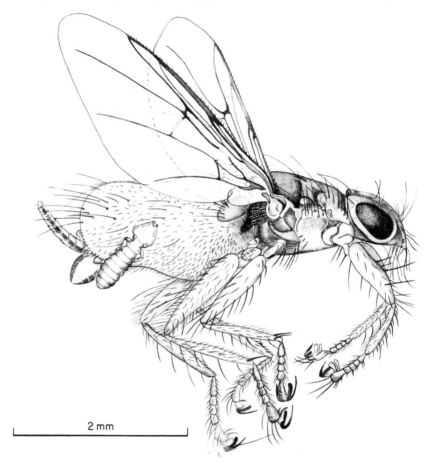

2 mm

Figure 23. Feather lice (Mallophaga) being carried by hippo-boscid fly. This is an example of the phenomenon known as phoresy (redrawn from Rothschild and Clay 1952).

phoresy is an accidental phenomenon, and in *S. sturni* it may play a significant role in transferring lice from one host to another. Although the carrier, *O. fringillina*, is by no means restricted to starlings, it is likely, because starlings are gregarious, that most of the movements of the flies are between starlings.

The Anoplura lack mandibles with which they could grasp a carrier, and as might be expected, there are few records of phoresy that concern them. *Pediculus humanus*, however, has been found attached to house-flies, and *Haematopinus tuberculatus* to *Lyperosia*, carriers in both cases belonging to the dipterous family Muscidae.

3. *Fleas*

FLEAS ARE ENDOPTERYGOTE INSECTS and they have, in contrast to lice, a larval stage that is totally different in form from the adult. The eggs are laid loosely on the body of the host, which may be either a mammal or a bird, and they drop off to hatch upon the floor of the host's burrow or nest. The larvae (Figure 24) are white, legless, eyeless creatures, resembling in some ways the maggots of Diptera, and they are provided with stout body hairs and biting mouthparts. They feed on organic debris, particularly the faeces of adult fleas. When fully grown, which is after two larval moults in most species, the larva pupates in a silken cocoon spun from a secretion of its salivary glands. The cocoon is oval in shape and the outside is frequently festooned with any available organic fragments.

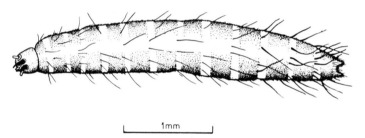

1mm

Figure 24. A fully grown flea larva (*Spilopsyllus cuniculi*) showing mostly the ventral surface.

The adult flea feeds only upon its host, from which it extracts blood by means of its piercing and sucking mouthparts (Figure 15). It has some of the same modifications to an ectoparasitic life as a louse: a tough integument which is difficult for the host to rupture, tarsi adapted for clinging, small size, loss of wings (although Sharif in 1935 demonstrated that the pupa retains remnants of wing-cases), and the eyes, if present at all, are small and simple. The body is also strongly flattened, but whereas in lice the flattening is dorso-ventral, in fleas it is lateral. This is probably associated primarily with the flea's ability to slip through the host's fur, but it contributes also to the notorious attribute of fleas for leaping as a method of escape and reaching the host, which demands that the back legs operate in the vertical plane. Over thirty centimetres may be covered in a single leap.

Neville and Rothschild (1967) investigated the jumping mechanism of fleas and found that it is probably a modified flight mechanism. The pleural arch of the flea metanotum, a rounded depression above the insertion of the hind legs, is homologous with the wing-hinge ligament of some flying insects, and like the latter it contains a mass of resilin. Resilin is a colourless protein that, like rubber, can store and release energy. In fact, in its properties it approaches an ideal rubber more closely than any known rubber (Weis-Fogh 1960). Bennet-Clark and Lucey (1967) studied the mechanics and energetics of a jump by the rabbit flea, *Spilopsyllus cuniculi*, and showed that the energy required in a jump is far in excess of that which could be supplied by direct muscle action. The main propulsion in a jump comes from the downwards swing of the hind femora from a near-vertical to a horizontal position. However, the depressor tendon to the hind femur is so arranged that contraction of the femoral

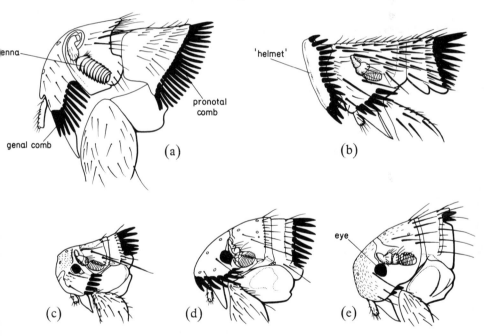

Figure 25. Head and prothorax of various fleas to illustrate variations in comb development. (a) *Histrichopsylla talpae* (×80). (b) *Stephanocircus dasyuri* (×90), a 'helmet flea', so-called because of the curious comb-bearing development on the front of the head. It belongs to the small family Stephanocircidae, parasites of marsupials. (c) *Spilopsyllus cuniculi* (×90). (d) *Ctenocephalides felis* (×90). (e) *Archaeopsylla erinacei* (×80). (b,c,d,e, after Smart 1965.)

depressor muscle causes pressure to be exerted on the resilin pad; it does not, at first, swing the femur downwards. The energy stored in the resilin as a result of the pressure upon it is suddenly released when a small muscle pulls the depressor tendon to a position that allows the depressor muscle to pull the femur down. The surge of energy released by the resilin propels *S. cuniculi* into the air with an acceleration of 135 gravities!

Because the rate of release of energy stored in resilin is largely independent of temperature, the flea can jump equally well at low and high temperatures, something it would be incapable of if it relied only upon muscle contraction. In those fleas that are poor jumpers, for example permanent nest-dwelling forms such as *Rhadinopsylla* and *Tarsopsylla*, the pleural arch is very much reduced.

Another striking feature of adult fleas is the presence on certain sclerites of a series of tooth-like spines forming the comb or ctenidium. These spines are in reality highly modified setae (Hopkins and Rothschild with Traub, in press). Many fleas have a genal comb on the head and a pronotal comb on the thorax (Figure 25), although one or both of these may be reduced or absent in particular families. The combs help to retain the flea amongst the fur or feathers of its host, and Humphries (1967a) finds a positive correlation between the width of the space between adjacent spines of the comb of a flea species and the diameter of the hairs of its usual host (Table 3). The distance between

Flea species	Average spacing of pronotal comb tips (μ)	Range of width of 50 host hairs (μ)	Host
Ctenocephalides felis (Figure 25d)	52	24–44	Cat
Spilopsyllus cuniculi (Figure 25c)	35	20–39	Rabbit
Palaeopsylla minor	28	11–18	Mole
P. soricis	20	3–15	Shrew
Ischnopsyllus octatenus	13	5–9	Pipistrelle bat

Table 3. The relationship between the form of the comb of some fleas and the width of the hairs of their usual hosts (from Humphries 1967a).

spine tips is not merely a function of the size of the flea. It is slightly wider than the maximum diameter of the appropriate host's hair. Since each spine narrows apically, the space between spines decreases towards the base of the comb. The combs therefore lock onto the hairs if the flea is dragged backwards through the fur. When a flea is resting it adopts a vertical or oblique posture with respect to the surface of the host, its head close to the host's skin and its hindparts raised. A host attempting to remove its fleas is most likely to seize the posterior part of a flea and attempt to drag it out of its fur or plumage backwards. The combs make such removal during grooming or preening more difficult. Removal by the host is one of the chief hazards confronting a flea. It is significant that the hedgehog, which because of its spines is unable to effectively groom itself, frequently supports very large populations of *Archaeopsylla erinacei*, and this flea has only vestigial combs (Figure 25e).

Traub (1966) cites other examples of correlation between the structure of the comb and the type of host. For instance, nocturnal and fossorial mammalian hosts support fleas whose pronotal combs tend to be composed of narrower, straighter, and more pointed spines than the spines in the pronotal combs of fleas on diurnal hosts. The mechanical significance of this is unexplained. Analogous structures to the combs of fleas are found in several other ectoparasitic insects, emphasising that their most dangerous enemies are frequently the hosts themselves.

Fleas, although usually slightly larger than lice, are still relatively small insects. One of the biggest is the mole flea, *Histrichopsylla talpae* (Figure 25a), which measures almost six millimetres in length.

Many adult fleas can survive for periods of weeks or even months without feeding and away from a host. This is in very marked contrast to the ability of

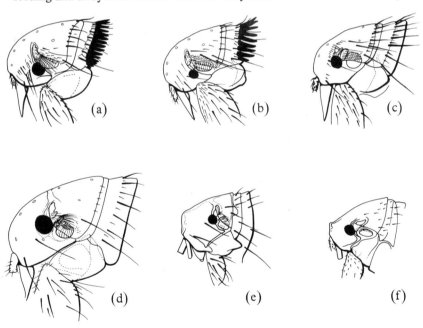

Figure 26. Head and prothorax of various fleas. (a) *Cerato-phyllus gallinae* (×90). (b) *Nosopsyllus fasciatus* (×80). (c) *Xenopsylla cheopis* (×90). (d) *Pulex irritans* (×90). (e) *Echidnophaga gallinacea* (×110). (f) *Tunga penetrans* (×100). (All after Smart 1965.)

lice whose continual dependence on their hosts has resulted in a high degree
of host specificity. Accordingly, we find that fleas are much more catholic in
their tastes than lice. For instance *Ceratophyllus gallinae* (Figure 26a), the
most abundant of British bird fleas, has been found on well over one hundred
species of birds and it has turned up as a straggler on several mammals.

The American cottontail rabbit is infested by two species of flea, *Cediopsylla
simplex*, which is found on the head, and *Odontopsyllus multispinosus*, which is
found on the back and flanks (Haas and Dickie 1959). Usually, however, fleas
have not become morphologically adapted to living on particular regions of
the host in the manner of many lice. As a consequence there is not a great deal
of gross variation in the structure of fleas. However 'stick tight' fleas, such as
the poultry flea, *Echidnophaga gallinacea* (Figure 26e), and to a lesser extent
the rabbit flea, *Spilopsyllus cuniculi* (Figures 25c, 27c), have adopted a rather
characteristic form. The female stick-tight flea is more or less permanently
attached by its enlarged mouthparts to the host, nearly always on the head, and
to ensure close contact with the host's skin, the front of the head is flattened
and the frons angulate. Also, in accordance with its sedentary existence, the

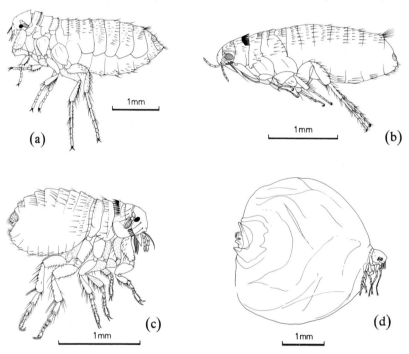

Figure 27. (a) *Ceratophyllus styx* female, the sand-martin
flea. (b) *Ctenophthalmus nobilis* female from wood mouse.
(c) *Spilopsyllus cuniculi* female, the rabbit flea. (d) *Tunga
penetrans* female, the 'jigger'.

leg muscles are reduced, the thoracic segments narrow, and the thoracic spiracles rudimentary (Jordan and Rothschild 1906). Female stick tight fleas very rarely change hosts, but the males lead a 'normal' active existence. This habit is carried to its extreme by fleas of the *Tunga*-type (Suter 1965). *Tunga penetrans* (Figures 26f, 27d) is the jigger or chigger of South America, from whence it was introduced to Africa, and in both continents it is an insect of medical significance. The female jigger burrows into horny areas of skin, most usually about the nail-bases or between the toes of natives' feet, and there it remains, almost completely sealed off from the outside world, for the rest of its life. Before entering its host it measures barely one millimetre in length, but once established its abdomen becomes greatly distended, up to about five millimetres in diameter, and its legs slowly degenerate. It is responsible for much irritation (causing the sufferer to 'jig' about) and secondary infections may develop. The male *Tunga*, which is not sedentary, copulates with the female after the latter has completely penetrated the host.

Mating behaviour has been studied in detail in the hen flea, *Ceratophyllus gallinae* (Humphries 1967c). The behaviour is initiated by apparently fortuitous contact between the abdominal cuticle of the female and the maxillary palps of the male. The cuticle of the female possibly carries a pheromone (or ectohormone) which, on perception, causes the male flea to hold its antennae vertically instead of in the resting downwards position, and to push against the female with its head. The female responds by facing away from the male

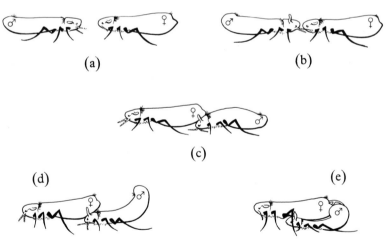

Figure 28. Sequence of behaviour leading to copulation in the hen flea. (a) Partners approach each other and when maxillary palps of male contact female, (b) the male's antennae are erected. (c) Male moves behind female and pushes against her with lowered head, his antennae clasping her abdomen. (d) Male raises apex of abdomen. (e) Copulation. (Redrawn from Humphries 1967c.)

so that the male can push head and thorax beneath her abdomen. The erect antennae of the male grasp the ventral part of the female's abdomen and, apparently, stimulation of their inner surfaces causes the male to curve his abdomen over his back and so bring the genitalia of the two sexes into contact (Figure 28). On the inner surfaces of the antennae of many male fleas there are batteries of suckers which assist the male in holding the female during copulation (Rothschild and Hinton 1968). Males of *Spilopsyllus* and *Echidnophaga* do not have these suckers; their females are semi-sessile. The copulatory apparatus of a male flea is an amazingly complicated organ. Its morphology has been studied by Traub (1950) and Holland (1955), and the act of linkage by Humphries (1967d), again using *C. gallinae*. The male genitalia are guided into the correct copulatory position by a complex arrangement of lobes and setae, eventually enabling the male to insert his penis rods into the female spermathecal duct. Each step in alignment must be successfully completed before the sequence can proceed, and Humphries (following Jordan) suggests that species-specific differences in male flea genitalia have a species isolating function preventing copulation between partners belonging to different species. Some hybridisation between bird fleas has, however, been recorded. This type of isolating mechanism is less efficient than one that operates nearer to the commencement of courtship.

Adult fleas feed only on vertebrate blood. Whilst the blood of several species of host is frequently adequate for survival, more precise requirements seem to be necessary, in some species at least, for breeding to proceed. It has been found, for example, that reproduction of the rabbit flea, *Spilopsyllus cuniculi*, is under the control of the sexual cycle of its host.

Mead-Briggs and Rudge (1960) showed that the female rabbit flea's ovaries do not develop until the flea feeds on a pregnant rabbit or its newborn young. Rothschild and Ford (1966) demonstrated experimentally that ovarian maturation, egg-laying, and regression of the rabbit flea are under the control of the hormones secreted by the anterior lobe of the pituitary gland of the host. During the last ten days of the rabbit's pregnancy the blood concentrations of corticosteroids and oestrogens rise steeply, thus stimulating maturation of the flea and synchronising both reproductive cycles. A few hours after the young rabbits are born the fleas leave the doe and feed on her litter, and eggs are laid in the nest which provides a suitable environment for larval development. The ovaries of any female flea remaining on the rabbit after parturition, or returning to the doe when egg-laying is completed, undergo regression probably under the influence of the rising level of progestins in the host's blood (Exley, Ford, and Rothschild 1965; Rothschild and Ford 1966). The period of egg-laying in the nest does not exceed two to three weeks (average ten days) after which the fleas abandon the young and return to the lactating doe to await another pregnancy. It is believed that the combination of high corticosteroid and growth hormone levels in the blood of the nestling rabbits is one of the principal factors governing copulation of the fleas on the newborn young, but there are certainly others. Thus fleas feeding on baby rabbits in large litters mature more rapidly and mate earlier than those feeding on isolated young. Rothschild and Ford (1969) postulate a pheromone-like factor emanating from the nestlings which stimulates both copulation and maturation in the rabbit flea. Fleas on adult rabbits make no attempt at copulation.

Just how the host's hormones act upon *S. cuniculi* is not yet known; they may act directly, but it seems more likely that their effect is mediated through the flea's own hormone system.

There seems to be no comparable intimate relationship between host and parasite in the case of the rat flea, *Xenopsylla cheopis*, which will mature on castrated and adrenalectomised rats (Rothschild and Ford 1966). It could be that *S. cuniculi* is almost unique in its dependence upon host hormones. The name of the human flea, *Pulex*, however, is said to be derived from the Latin word *puella*, meaning a girl, as a reflection of a preference it is alleged to show for young females. Perhaps *Pulex*, like *Spilopsyllus*, is in some way governed by the host's hormones.

Flea larvae may be reared upon a variety of organic materials, provided that these are dry and fairly finely divided, although for the larvae of many species it seems that the host's blood constitutes a most important item in their diet, and this they obtain by eating the faeces of their parents. Adult fleas frequently appear to feed to excess, blood in some cases being passed out of the anus almost totally undigested, and it is likely that this is to ensure a full supply of food for the larvae. The rate of defaecation of *Spilopsyllus cuniculi* is, like ovarian maturation, under the control of the corticosteroid level in the blood of its host (Rothschild and Ford 1966), for the rate rises to the deposition of one blood-saturated faecal pellet every few minutes when adrenocorticotrophic hormone is injected into the host rabbit. This means that, in nature, fleas will drop large numbers of faecal pellets into the nest just before they lay their eggs, so that the larvae are provided with an abundant food supply as soon as they hatch. Larvae of a rat flea, *Nosopsyllus fasciatus*, ingest blood as it is actually passing out of the anus of the adult. This flea is not sedentary like *S. cuniculi* and it spends a lot of time in the host's nest where it is accessible to its larvae. The larvae attach themselves by their mandibles to the adult's pygidial region, and this apparently stimulates the adult to defaecate (Molyneux 1967). *Pulex irritans* and *Xenopsylla cheopis* are somewhat exceptional in not requiring blood as larvae. *P. irritans*, the human flea, is able to complete its larval development on a variety of organic materials, and this enables it to survive in the abode of its host which is usually cleaner than that of other flea hosts.

Because lice are very little affected by seasonal conditions, they breed continually. Fleas however, being apart from the host for a considerable portion of their life cycle, are not so protected from the climate, and they do often undergo seasonal cycles. In temperate regions, adult fleas usually reach maximum abundance during the warmer months. A seasonal rhythm is especially marked in such fleas as *Ceratophyllus styx* (Figure 27a), the sand-martin flea, which inhabit perennial birds' nests, occupied by the host for only a short period during each year. Fleas, especially the females, may have long adult lives, and a specimen of *Pulex irritans* has been kept in captivity for nearly three years.

The eggs and larvae of fleas require rather closely defined ranges of temperature and humidity for their development, and the necessary conditions vary from species to species. Bacot (1914b), in an extensive investigation of fleas on rats, discovered that *Nosopsyllus fasciatus* could tolerate a lower range of temperatures than *Pulex irritans*, which in turn was able to withstand lower temperatures than *Xenopsylla cheopis*. The tolerance ranges for the egg stages of these three species are:

Xenopsylla cheopis   13°C to 34+°C
Pulex irritans   8°C to 34°C
Nosopsyllus fasciatus   5−°C to 29°C

These ranges can be correlated to some extent with the geographical distribution of the species, *Xenopsylla* being unable to maintain itself in Britain except in heated buildings, while *Nosopsyllus* is common and widespread in the British Isles. Similar, although narrower, ranges of tolerance are shown by the active larvae of these three fleas, but as soon as the cocoon is spun the larva is shielded from conditions that, earlier in its life, would have proved fatal. *Spilopsyllus cuniculi* is well-adapted to cool climates like its host, the rabbit, and adults of this species can be stored for nine months at −1°C and survive (Rothschild 1965).

The extent of the effect of climatic conditions upon adult fleas has been investigated by Cole (1945) who studied *Xenopsylla cheopis*, a flea of great medical importance as the vector of bubonic and septicaemic plague. It was found that temperature had a differential effect upon the sexes of *Xenopsylla*. On days with a high mean temperature (21°C to 24°C) males outnumbered females on rats, but on cooler days the position was reversed. Both sexes need to feed more frequently at high temperatures to survive, but the effect on males is the greater.

In most field studies of fleas in temperate climates (e.g. Evans and Freeman 1950) it has been found that female fleas outnumber males. This has usually been considered to be a result of the longevity of the female. However, it must be said that there is little information on the sex ratio on emergence from the pupae, and it could be that in some species this always favours the female. One male flea is able to fertilise a number of females.

Geographical distribution

A few fleas, such as the hen flea (*Ceratophyllus gallinae*), have been carried all over the world on their hosts, with the assistance of man, but the distribution of many fleas, in contrast to that of lice, may fall well short of the distribution of the host. This is largely because climatic conditions are not everywhere suitable for the immature stages. Specialisation of the immature stages to a limited range of conditions allows for the parasitism of a single widespread host species by a number of different, allied fleas, each of the fleas being limited in its distribution to a different climatic zone. These zones may overlap, and to prevent interspecific competition resulting in the total replacement of one species by another, each species must, in its own particular zone, be better able to survive than any of the other species. An example of this is provided by rat fleas, some of which have already been mentioned. Even within the relatively narrow range of climatic variation experienced in the British Isles, other examples may be found. *Ctenophthalmus nobilis* (Figure 27b), a parasite of several small mammals and especially voles, occurs in Britain as two subspecies with different distributions, *C. n. nobilis* being almost exclusively confined to the east and south, and *C. n. vulgaris* predominating in the west and north. Among bird fleas, *Ceratophyllus garei* is quite widespread in Britain, but the very similar *C. borealis* is found in abundance only on the islands off the northern and western coasts from which *C. garei* is absent or, at most, very scarce. Both species are parasites of ground-nesting birds.

The development of these situations is dependent on the limited dispersal of fleas, which allows a species to evolve in comparative isolation and without interference from competitors. Once a new species has evolved, a spreading of its geographical range, and possibly of its host range also, may occur and bring it into competition with other species. The result of such competition may be that the new species replaces its competitors; alternatively the territory of the new species may be invaded by a competitor which exterminates it. However the extermination of one species by another is an extreme and probably rare event. During speciation, an emergent species is very closely adapted to its limited environment, so that the microhabitat of its formation represents, as long as it remains unchanged, a stronghold for the species in which it is unlikely to be bettered by an invading competitor. Holland (1958) cites the case of two allied species of *Atyphloceras*. *A. bishopi* is a parasite of a widespread nearctic vole (*Microtus pennsylvanicus*), but it occurs in only a small part of the range of its host in eastern North America. The only closely related species to *A. bishopi* is *A. nuperus*, a flea found only on voles in the mountains of central and southern Europe. Holland believes this to be the relic of an ancient and continuous holarctic distribution. The factor preventing reinvasion of the old range is probably effective competition from other fleas.

Host Associations

The ecological requirements of flea larvae, referred to previously, have a profound effect on the host relations of the adult. Smit (1957b) expresses the situation precisely as follows: '. . . the immature stages of each species of flea are adapted to a certain environment. Most fleas are not strictly host-specific, but are nest-specific, since they must to a large extent be dependant on ecological factors governing their development . . .; the adult's choice of blood is often rather varied and seems of secondary importance in determining host-specificity.' Nevertheless, laboratory experiments have frequently demonstrated (e.g. Samarina *et al.* 1968) that polyphagous fleas are more fertile on some of their hosts than on others, and this must be an effect of the hosts' blood.

Ash (1952) studied the flea fauna of birds' nests in Britain, and some of his results are reproduced in Table 4. It may be presumed that physical conditions in birds' nests are largely determined by their situation, and the most variable condition is probably humidity. Fleas in the cocoon are usually very susceptible to desiccation, water loss causing body shrinkage and rendering ineffectual the movements necessary for a successful ecdysis and emergence from the pupa.

Ash's results confirm the observation of Rothschild (1952) that the three most common British bird fleas, *Ceratophyllus gallinae*, *C. garei*, and *Dasypsyllus gallinulae*, are zoned according to the preferred nesting sites of their hosts, thereby minimising the possibility of interspecific competition. Thus *C. garei* is almost confined to ground-nesting birds, *D. gallinulae* prefers nests in bushes, tall reeds, or scrub, whereas *C. gallinae* favours nests in higher situations, particularly holes in trees or walls. These three species have broad host ranges, but other bird fleas are often more restricted. The sand-martin flea, *Ceratophyllus styx*, is host-specific, and this, no doubt, is a result of the high degree of adaptation demanded by the special physical conditions found in the nesting burrow, and by the need for synchronisation with its periodic occupation by birds. Two fleas of the house martin, *Callopsylla waterstoni* and *Frontopsylla laeta*, provide convincing evidence as to the importance of the hosts' nest in flea distribution, for these two species are found only in those nests built on cliffs, never in nests beneath the eaves of houses. Likewise nests of the rockhopper penguin on Macquairie Island are infested by *Parapsyllus magellanicus* only when they are in a sheltered position (Murray and Vestjens 1967). *Ceratophyllus rossittensis* parasitises only hooded and carrion crows. The dry, arboreal nests of these birds again offer a specialised environment, although other factors must operate in this association since other birds, for instance rooks and magpies, build apparently similar nests yet escape infestation by *C. rossittensis*.

Ash failed to find fleas in any of the birds' nests built among the branches of trees which he examined, and in general it may be said that the drier conditions prevailing in most birds' nests, in comparison with mammalian nests, constitute a more difficult environment, and this no doubt partly explains why fleas are most successful as parasites of mammals. The major reason for the latter, however, must be the impermanent nature of most birds' nests, together with their only temporary occupation.

Number of nests infested by each flea species, with the number of fleas in parentheses.

Host species	Nest situation	No. examined	No. infested	Ceratophyllus gallinae	C. garei	C. fringillae	Dasypsyllus gallinulae
Blackbird	bush	24	10	2 (25)	0	1 (1)	10 (325)
Song Thrush	,,	17	3	2 (15)	0	0	0
Chaffinch	,,	15	11	7 (27)	0	0	9 (198)
Linnet	,,	2	1	1 (1)	0	0	0
Spotted Flycatcher	,,	4	1	1 (7)	0	0	3 (179)
Hedge Sparrow	,,	5	3	1 (1)	0	0	1 (2)
Wren	,,	4	2	1 (2)	0	0	3 (10)
Long-tailed Tit	,,	3	3	1 (1)	0	1 (1)	0
Skylark	ground	1	1	0	1 (7)	0	0
Meadow Pipit	,,	1	1	0	1 (505)	0	0
Tree Pipit	,,	2	0	0	0	0	0
Woodlark	,,	5	2	1 (1)	0	0	2 (64)
Willow Warbler	,,	1	1	0	0	0	1 (6)
Robin	,,	7	4	0	1 (71)	0	4 (22)
Reed Bunting	,,	1	0	0	0	0	0
Coot	,,	1	0	0	0	0	0
Lapwing	,,	1	0	1 (1)	0	0	0
Common Partridge	,,	1	1	0	1 (290)	0	0
Montagu's Harrier	,,	1	1	0	0	0	0
Blue Tit	hole	2	2	2 (1305)	0	1 (2)	1 (6)
Great Tit	,,	2	2	2 (683)	0	1 (1)	2 (7)
Tree Creeper	,,	2	1	1 (26)	0	0	0
Nuthatch	,,	1	1	1 (78)	0	0	1 (2)
Starling	,,	1	1	1 (5)	0	1 (3)	0
Little Owl	,,	1	1	1 (2)	0	0	0
Pied Woodpecker	,,	1	0	0	0	0	0
Green Woodpecker	,,	1	0	0	0	0	0
Wood Pigeon	tree	2	0	0	0	0	0
Jay	,,	1	0	0	0	0	0
Magpie	,,	1	0	0	0	0	0
Carrion Crow	,,	1	0	0	0	0	0

Table 4. The occurrence of four species of fleas in birds' nests in Britain (from Ash 1952).

Mammal fleas, like bird fleas, are usually polyphagous. Gabbutt (1961) gives an account of the distribution of small mammals and their associated fleas in Labrador, and concludes that the fleas show both a host and a habitat preference. The flea found most abundantly was *Megabothris quirini*, and it was taken from species of the allied rodent genera *Clethrionomys*, *Microtus*, and *Phenacomys*. In black spruce muskeg *M. quirini* preferred *Clethrionomys* to *Microtus*, and in conifer-lichen woodland *Phenacomys* was preferred to *Clethrionomys*. *Phenacomys* might therefore be said to be the preferred host, but *Phenacomys* is primarily an inhabitant of semi-tundra, and is much less frequent in lichen woodland which is where *M. quirini* was caught most frequently. There would therefore appear to be operating on the distribution of *M. quirini* conflicting preferences for host and habitat.

More research is required to establish the relative extents to which hosts' blood, pelage or plumage, habitat, nest characteristics, and other factors contribute to the host ranges and preferences of fleas.

The British small mammals are parasitised by a number of fleas (Smit 1957a) of which only a few are host-specific. Of the commoner species, *Archeopsylla erinacei* is found only (except for stragglers) on hedgehogs, *Typhlocerus poppei* on wood-mice, *Spilopsyllus cuniculi* on rabbits and hares, *Ctenophthalmus bisoctodentatus* on moles, and *Leptopsylla segnis* on house-mice. Of the species with a range of hosts, some are confined to certain discrete host groups. Bat fleas (*Ischnopsyllus*, *Nycteridopsylla*), for instance, are not found on other mammals, *Malaraeus* and *Megabothris* are typically parasites

Flea species	% host individuals infested			
	Apodemus sylvaticus (*wood mouse*)	Clethrionomys glareolus (*bank vole*)	Microtus agrestis (*field vole*)	Sorex araneus (*common shrew*)
Nosopsyllus fasciatus	1	0	0	0
Malaraeus penicilliger	1	17	2	1
Megabothris walkeri	1	3	17	1
M. turbidus	4	12	2	0
Ctenophthalmus nobilis	38	57	34	0
C. bisoctodentatus	0	0	0·5	0
Rhadinopsylla pentacantha	1	3	0·5	0
Doratopsylla dasycnema	1	1	1	26
Palaeopsylla soricis	1	2	1	40
P. minor	0	0	0	0·5
Peromyscopsylla silvatica	0	0	20	0
Histrichopsylla talpae	1	6	12	12
Number of each host examined:	788	281	368	292

Table 5. The distribution of flea species on small mammals in Bagley Wood, Oxford (from Elton, Ford, Baker, and Gardner 1931).

of microtine rodents (voles), and *Nosopsyllus* is characteristic of murine rodents. Outside Britain, the Pygiopsyllidae are essentially fleas of Australasian marsupials. However, some species are extraordinarily catholic in their choice of hosts, as for instance *Ctenophthalmus nobilis* which is common on voles and wood-mice and not infrequent on moles and shrews.

Elton *et al.* (1931) record the distribution of fleas on wood-mice, voles, and shrews in Bagley Wood near Oxford (Table 5). These figures emphasise the general lack of host-specificity among fleas. Even though the hosts examined represent two mammalian orders, only four out of the twelve flea species found were restricted to one host species, and three of these four species were present in very low numbers. Many of these small mammals share the same runways and so exchange of their fleas is facilitated.

No doubt, as with bird fleas, the conditions of the host's dwelling place are of great importance in determining the extent to which different fleas parasitise mammals, but little information appears to be available regarding these conditions. It is interesting that monkeys and wild ungulates, which do not resort to a habitual sleeping place, are usually free of fleas.

Dunnet (1950) suggests that hosts with similar pelage share the same or allied fleas, and instances some fleas common to shrews and moles (e.g. *Histrichopsylla talpae*). The structure and function of the combs, described above, makes this likely.

Sometimes physical association between quite unrelated hosts has led to the evolution of closely allied species of fleas on each. *Nearctopsylla brooksi* is a parasite of New World mustellids (weasels, mink) which prey upon shrews, the hosts of other *Nearctopsylla* species. Similarly, *Ceratophyllus*, primarily a genus of bird fleas, also includes a weasel flea, *C. tundrensis*. Association of hosts through cohabitation perhaps explains why *Ornithopsylla laetitiae*, a flea allied to the rabbit flea *Spilopsyllus cuniculi*, has become a parasite of manx shearwaters and puffins which nest in old rabbit burrows. *S. cuniculi* has sometimes been taken on puffins; presumably a common ancestor of *Ornithopsylla* and *Spilopsyllus*, some time in the past, also transferred its allegiance.

Host location

Observations suggest that fleas are capable of detecting their hosts from a range not exceeding a few centimetres, and they usually do not stray far from the nest, burrow, or run of the host. Fleas are drawn towards their host mainly, it seems, by the sense of smell which is acute enough to enable them to distinguish between different species of vertebrate. The odour of the 'wrong' host may in some cases have a repellent effect. Horses are not attacked by fleas, and neither, it is said, are those people who work with them. Some fleas are also attracted by warmth and carbon dioxide. Regarding the former, Bates (1962) was able to collect numbers of fleas by dragging a hot-water bottle covered by a bird skin through undergrowth.

Fleas are very sensitive to air currents and vibrations, to which they often react by leaping. These leaps seem to be random in direction, and they serve either as a means of escape or of gaining the body of a host. During a jump a flea cartwheels in the air, and it holds either its first or second pair of legs upwards over its back with the tarsal claws directed forwards (Figure 29). This is evidently a grappling device to assist the insect in retaining contact with the fur of a host should it strike the host with its back first (Rothschild

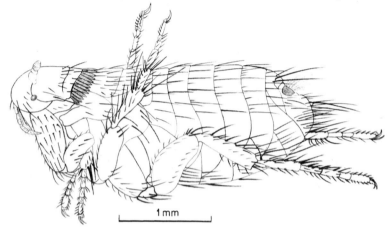

Figure 29. A rat flea (*Nosopsyllus fasciatus*) in the middle of a jump. The middle legs are raised above the back to act as grappling irons as the insect spins through the air (redrawn from Rothschild 1965).

1965). The receptor organ for air currents is the very specialised sensilium or pygidium (Figure 30); destruction of this organ renders the flea insensitive to air currents. Vibrations can initiate emergence from the cocoon, and this obviously increases the chances of a newly hatched flea finding a host.

Fleas have a poorly developed visual sense, having no compound eyes and only a single pair of ocelli. Nevertheless, it has been shown (Humphries 1968) to be of importance in the host-finding behaviour of *Ceratophyllus gallinae*. When the young birds leave the host's nest in spring most of the flea population is left behind, and the fleas overwinter as adults in their cocoons in the nest. They emerge the following spring and at first are negatively phototactic and remain in the old nest, possibly for mating purposes. After a few days, however, the fleas become negatively geotactic and positively photo-tactic so that they climb vegetation surrounding the nest. Here they orientate themselves towards the light, and jumping is elicited when there is a sudden fall in light intensity, as when a bird passes close. *Ceratophyllus styx*, studied by Bates (1962), also overwinters as an adult in its cocoon, in old nests of sand martins in burrows in the sides of steep sandy banks and cliffs. Emergence of the adults is stimulated by warmth in spring, and it often precedes the return of the hosts to their burrows. The fleas may disperse over the cliff face to other burrows which they apparently recognise by their horizontal floors. They congregate on horizontal surfaces, but somehow seem able to avoid gathering on the cliff top; perhaps they require a shaded situation. New hosts are infested when they come to nest.

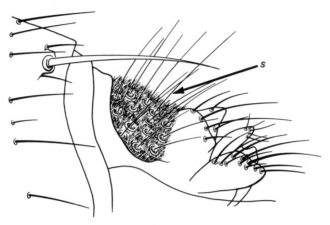

Figure 30. The elaborate sensilium *s* on the ninth abdominal tergum of a flea (*Xenopsylla cheopis*) (after Smit 1957b).

Information on the efficiency of host-finding is supplied by the work of Evans and Freeman (1950). Using the mouse *Apodemus* and the vole *Clethrionomys* they found that within twenty-four hours of removing all fleas, hosts in the field sometimes acquired as many, or more, fleas than they were originally carrying. For instance, seventeen fleas were taken from a vole the day after it had had five fleas removed. Migrating birds on Fair Isle became reinfested with local fleas a few hours after their release from traps in which they had had their fleas removed (Rothschild 1958). Impressive data are also provided by an experiment by Mead-Briggs. Two hundred and seventy marked *Spilopsyllus cuniculi* were liberated in a meadow with an area of 2000 square yards. Three rabbits liberated into this meadow were, within a few days, infested by nearly half of the fleas. Whilst away from a host, fleas may prolong their survival by drinking water from damp earth (Humphries 1967b).

4. Blood-sucking Flies

THE MOUTHPARTS OF ADULT Diptera show a wider range of form than those of any other order of insects, and their feeding habits are accordingly varied. The habit of penetrating the skin of warm-blooded vertebrates and drawing up a blood meal has evolved independently in all three of the dipterous suborders, and the associated modifications of the mouthparts to form a piercing proboscis have been discussed in Chapter 1. Blood-sucking species occur rather sporadically in a systematic list of Diptera. They are found in the following families:

Suborder Nematocera
    Culicidae (mosquitoes)
    Ceratopogonidae (biting midges)
    Simuliidae (black flies)
    Psychodidae (sand flies)
Suborder Brachycera
    Tabanidae (horse flies, clegs)
    Rhagionidae (snipe flies)
Suborder Cyclorrhapha
    Muscidae (Stomoxydinae) (stable flies, horn flies)
    Glossinidae (tsetse flies)
    Hippoboscidae (louse flies)
    Nycteribiidae (bat flies)
    Streblidae (bat flies)

The last three families constitute a group termed the Diptera Pupipara which is the subject of the next chapter.

Nematocera

In Nematocera, blood-sucking is confined to the females, they alone being equipped with mandibles and maxillae which can pierce vertebrate skin. Males often feed at flowers, and the proboscis was probably originally an organ adapted for the taking of nectar. It is still used exclusively for this purpose by both sexes of the Bibionidae, a family that has several primitive morphological features.

By no means all of the species in the four nematocerous families listed above are blood-suckers. Among Culicidae, members of two of the three subfamilies, the Dixinae and the Chaoborinae, are non-biters, whilst one tribe (Megarhinini) of the third subfamily, the Culicinae or mosquitoes, has reverted to nectar feeding. Relatively few species of Ceratopogonidae take vertebrate blood. Most Simuliidae attack warm-blooded animals, but only one of the two subfamilies of Psychodidae, the Phlebotominae, contains biting species. Even amongst those species that regularly feed on blood, nectar-feeding is probably more widespread and of greater physiological importance as a source of energy than has hitherto been realised (Lewis and Domoney 1966).

In Simuliidae blood has been shown to be necessary for egg development in a number of species. The genus *Simulium* (Figure 31a) is a large one, and world-wide in distribution. The species are generally daytime biters and attack both birds and mammals. Adults tend to move from the breeding areas in search of hosts, and in 1923 swarms of *Simulium columbaschense* spread from the banks of the Danube and descended upon domestic animals in such numbers that nearly 20 000 horses, cattle, sheep, goats, and pigs are reported to have died as a result. Toxaemia and anaphylactic shock, rather than loss of blood, probably accounted for most of the fatalities. The aptly named *S. damnosum* similarly inflicts much suffering on man and cattle in Africa.

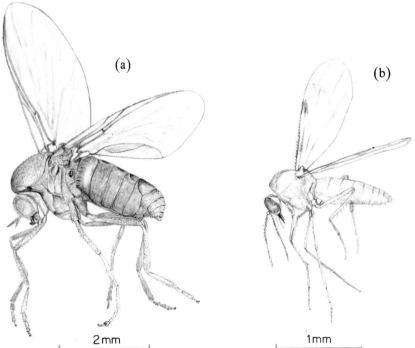

(a)

(b)

2mm

1mm

Figure 31. Blood-sucking Diptera Nematocera. (a) *Simulium equinum* female (Simuliidae). (b) *Culicoides pulicaris* female (Ceratopogonidae).

The Ceratopogonidae as adults are interestingly diverse in their feeding habits. Most either suck plant juices or pounce on and kill small insects about their own size. Some, however, fall into the small category of ectoparasites of invertebrates. Thus some *Forcipomyia* species cling temporarily to the wings of such insects as lace-wings and butterflies while they pierce the integument of the wing veins and feed on the internal fluid. Other species in the genus feed on caterpillars. Likewise, species of *Atrichopogon* feed through the arthrodial membranes of oil beetles and mealy bugs, *Pterobosca* species attack dragonflies and other large insects, and *Culicoides anophelis*, in feeding on gorged female mosquitoes, may perhaps be described as a hyperparasite. However only species of *Culicoides*, *Lasiohelea*, and *Leptoconops* take blood from warm-blooded vertebrates. Members of the genus *Culicoides* (Figure 31b) are very small midges, mostly with spotted wings and exceedingly blood-thirsty, which occur in hordes on mosses and marshy places in Scotland and elsewhere and make human outdoor activity very nearly intolerable when they are at the peak of their activity. One of the most abundant British species is *C. impunctatus*, the Scottish midge.

Blood-sucking Psychodidae are, with few exceptions, confined to the genus *Phlebotomus* (Figure 32a). The non-biting psychodids are easily recognised, hairy little flies resembling in appearance minute moths, especially when resting with wings folded in a roof-like fashion (tectiform) over the back. The blood-sucking species differ in having longer legs and in resting with their wings separated. Species of *Phlebotomus* are frequently referred to as sand flies. They are nocturnal insects, resting during the day in shady places including houses. Although very small they can inflict a painful bite, and *P. papataci* is a very serious pest in parts of eastern Europe and the Middle East. It bites man, particularly about the ankles and wrists, and also domestic animals, frogs, and even caterpillars.

The larvae of all of these nematocerans are more or less aquatic. Those of the Simuliidae are characteristically found in turbulent, well-oxygenated water, culicid larvae inhabit ponds and small, often temporary, collections of standing water, larvae of ceratopogonids occur in streams, ponds or in very wet mud, while the larvae of phlebotomine psychodids may be described as semi-terrestrial, surviving on damp surfaces beneath stones and in rock crevices in regions which are predominantly hot and dry.

Any insect that is of medical importance naturally attracts a great deal of research, and it is probably true to say that more has been written about mosquitoes than about any other group of insects. The yellow fever mosquito, *Aedes aegypti* (Figure 32b), has recently been comprehensively monographed by Christophers (1960).

Whilst female mosquitoes are most renowned for their biting, the most striking feature about the males is their habit of swarming. Many kinds of flies, and also other winged insects, form aerial swarms which consist of from

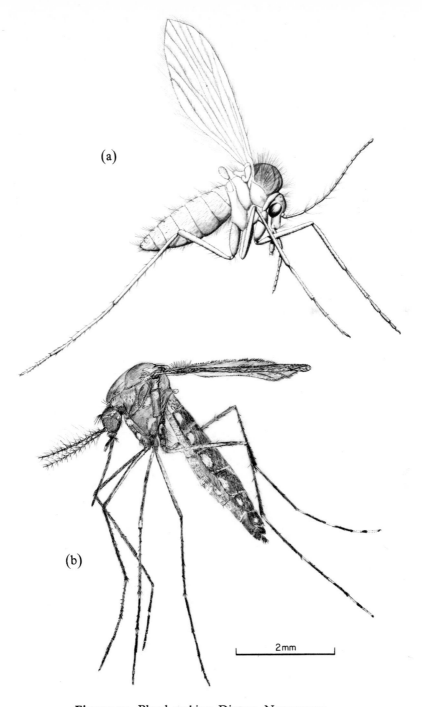

(a)

(b)

2mm

Figure 32. Blood-sucking Diptera Nematocera.
(a) *Phlebotomus papataci* female (Psychodidae).
(b) *Aedes aegypti* female (Culicidae) (from Smart 1965).

two or three to a large number of individuals, usually all of the same species, and all flying within a limited space and centred above, below, or beside a particular landmark or marker. In the case of mosquitoes, the marker is often a small bush or post; each individual in the swarm centres its flight about this (Downes 1958b) and the flight incorporates a large vertical movement, so that the swarm appears as a more or less attenuated column of insects. Different species of mosquitoes often use rather different markers, and they may fly at different distances from the marker. This helps to keep the swarms unispecific. Swarms are usually almost entirely composed of males, and they occur mostly at dawn or dusk. The purpose or biological function of the mosquito swarm is not at all clear, although it is generally supposed to serve a sexual function. A male mosquito by itself is not a conspicuous animal, but a swarm of males is much more obvious, and it may be that female mosquitoes are visually or acoustically attracted to the swarm and enter it in order to find a mate. It is reported (Shannon 1931) that some of the most brightly-coloured mosquitoes do not swarm. Haddow and Corbet (1961), however, observing swarming mosquitoes from the top of a tower in a forest in Uganda, failed to see mating, and other workers have usually found that the amount of mating resulting from swarming must be only a fraction of the total matings.

Male mosquitoes have the ability to detect and locate females for themselves from a range (in *Aedes*) of about ten inches. The antenna of the male is equipped with many whorls of very long hairs, and its sub-basal segment is enlarged to house a battery of sense cells (Johnston's organ) which is stimulated by movements of the antenna. In flight, the beating of the wings of the female mosquito makes a whining noise which causes the hairs on the male's antenna, and also the shaft of the antenna itself, to vibrate, and the sense organ is thus stimulated. Females of different species emit rather different notes; *Anopheles maculipennis* buzzes between middle C and E flat, whilst *Theobaldia annulata* varies from A to B. Although the males respond to a rather broad range of frequencies, they may be partially selective in seeking out females of their own species. Haskell (1961) reports that the response of the male to the female is lost if more than one source of attractive sound is presented at the same time. It follows, therefore, that males can locate only isolated females, and further doubt is thrown on the function of the swarm as a device for bringing the sexes together. It has even been suggested that swarming is a vestigial activity, but this idea can scarcely be maintained in view of the time and energy expended by male mosquitoes in its performance. A quite different explanation of the function of an aerial swarm is suggested by Wynne-Edwards (1962). He considers it to be an epideictic demonstration designed to provide 'information' to the participating individuals on the population density, this information being used in a postulated physiological or behavioural feed-back system leading to the adjustment of the rate of reproduction, so as to maintain or restore the population's optimum density.

Having mated, the female mosquito enters a daily rhythm of activity which varies from species to species. Although this is not rigidly adhered to by all the females, it is usually possible to describe a species, on the basis of the females' times of activity, as nocturnal or, less commonly, diurnal, with one or two peaks of activity a day. During the active period the females seek out hosts to obtain a blood meal.

Female mosquitoes respond to a variety of stimuli in locating their meal. Laarman (1959) working with *Anopheles maculipennis atroparvus* found that smell is important in locating the warm-blooded host, and that heat and moisture together are probably releasing stimuli for alighting upon the host. Heat also assists in the localisation of the host, providing a gradient that the mosquito can follow up. Further, this thermotaxis is activated by carbon dioxide. Similarly Wright and Kellogg (1962) write '. . . it appears that *Aedes aegypti* are stimulated to fly in search of a warm-blooded host by an alteration in the ambient carbon-dioxide level, and that they are guided to the appropriate surface partly by colour, but mainly by warm, moist convection currents rising from it.' *Aedes aegypti* is most attracted by dark colours, and it usually approaches the shaded parts of a host. Host movement is also believed to be of importance in enabling mosquitoes to find their food. As far as man is concerned, mosquitoes, like other groups of ectoparasites, do not attack all people with equal vigour. Some people are scarcely ever bitten by them, and the entire Esquimaux race is said to be immune.

A blood meal is known to be needed by many species of mosquitoes in order for them to mature their eggs; in other species a blood meal may increase the rate of egg production. *Culex pipiens pipiens*, which usually bites birds, is unable to produce eggs until it has fed, but *C. pipiens molestus*, which attacks man and seems to be merely a biological race of *C. pipiens*, will lay eggs before feeding (i.e. it is autogenic). This difference may be a result of *C. p. molestus* acquiring necessary nutriments during its larval life. Both forms occupy similar larval habitats but the pupa of *C. p. molestus* contains more protein than that of *C. p. pipiens*. How this is achieved is unknown.

Many mosquitoes are polymorphic and names such as *Aedes aegypti*, *Culex pipiens*, and *Anopheles maculipennis* each cover a host of forms that are distinguishable only by minute morphological details, and also often by different behavioural and physiological traits. The two forms of *C. pipiens* just mentioned provide one of the better known examples. It seems very likely that intensive studies of other insect groups will reveal many similar instances in which a widespread and apparently variable species is found to be, in fact, an aggregate of biological races or sibling species.

Brachycera

There are a few reports in the literature of rhagionid flies, especially of *Symphoromyia* in North America and *Spaniopsis* in Australia, attacking mam-

mals, but the habit is evidently abnormal in the family. The only large group
of brachycerous Diptera whose species habitually suck blood is the family
Tabanidae.

The Tabanidae are robust flies of powerful flight with large heads and large
eyes. In the males the eyes often meet in the dorsal mid-line of the head, and
the facets in the upper part of the eye are often larger and demarcated from
those of the lower part, both features probably associated with swarming.
In life the eyes of both sexes are usually beautifully striped with gold, green,
or purple. As in the Nematocera it is only the females that take blood, although
carbohydrate food in the form of nectar or honeydew is also often taken, and
the significance of blood in their physiology is not clear. Wilson (1967) found
that in *Tabanus lineola* and *T. fuscicostatus* a blood meal was a necessary
prelude to oviposition, and that carbohydrates were required to prolong the
adult life beyond four or five days. The front tarsi are provided with contact
chemoreceptors which are sensitive to sugar solutions. Tabanids are attracted
to hosts by carbon dioxide, and they may be captured on sticky traps baited
with dry ice. There seems little doubt, however, that vision is very important
in host location.

One group of Tabanidae from the warmer parts of the world, the Pangoniinae (Figure 11), are characterised by a greatly elongated labial proboscis
which is used to withdraw nectar from deep inside flowers, although with their
mandibles and maxillae the females also manage to take blood meals (page 9).
However a few species have lost their mandibles and subsist entirely on nectar.
It is argued by Downes (1958a) that nectar feeding is a secondary development
in Tabanidae and that the family was originally dependent upon blood. The

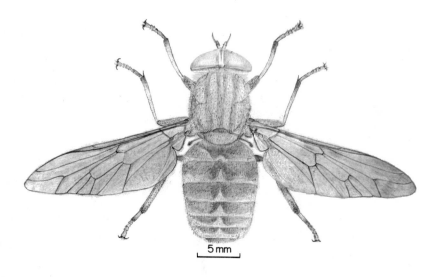

5 mm

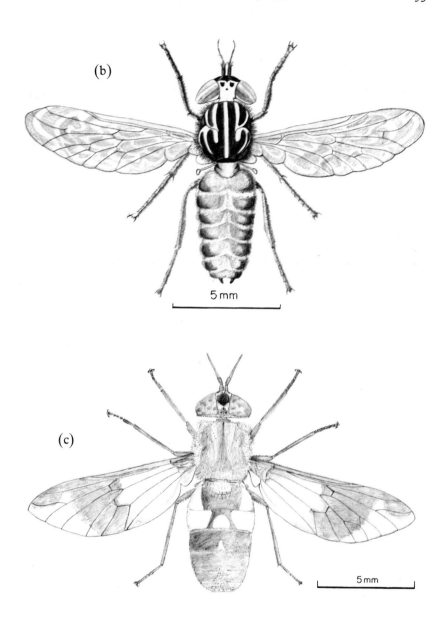

Figure 33. Blood-sucking Diptera Brachycera of the family
Tabanidae. (a) *Tabanus sudeticus* female. (b) *Haematopota
pluvialis* female (modified from Patton and Cragg 1913).
(c) *Chrysops caecutiens* female.

eggs of tabanids are laid in masses, usually upon vegetation beside standing water, and the larvae are carnivorous.

Three genera of tabanids (Figure 33) are represented in the British fauna, *Tabanus*, *Haematopota*, and *Chrysops*. *Tabanus* and *Haematopota* usually attack cattle and horses, and *Chrysops* is particularly attached to deer, although they are not specific in their choice of host and representatives of all three genera readily attack man. Species of *Haematopota*, or clegs, are especially bothersome in this respect, and their silent flight and stealthy, persistent approach often allow them to take a human meal without penalty. The front tarsi are provided with numerous sensory hairs which may play some part in the cleg's ability to alight undetected. Clegs usually attack man about the arms and body, even penetrating clothing, whereas *Chrysops* nearly always makes for uncovered parts of the head and neck. This leaves the legs vacant for those species of *Tabanus* or horse-flies (e.g. *T. micans*, *T. bromius*) that attack man. This approximate sharing or 'parcelling up' of the environmental resources is a feature which constantly presents itself in some form or another in the study of parasites. It restricts interspecific competition and is a corollary of the rule, sometimes referred to as Gause's Theorem, which states that two or more species occupying the same ecological niche cannot for long coexist.

The species of *Tabanus*, unlike those of *Haematopota* and to a lesser degree *Chrysops*, produce a loud buzz when flying, and this has a very aggressive sound when made by the larger species. However, it does not cause the panic or 'gadding' in cattle that the buzz of a warble-fly (*Hypoderma*) induces. This is most interesting, since immediate pain is inflicted only by the tabanid, the warble-fly merely laying eggs on the animal's coat. The vernacular name of 'gad-fly' when applied to *Tabanus* is unwarranted.

In general, tabanids are much less important as vectors of disease than are mosquitoes, but the severity of their biting is such that they have influenced the behaviour of both man and animals. Where horse-flies are abundant in parts of the Middle East, Asia, and Africa, pastoral tribes are sometimes forced to migrate to less infested regions during the weeks when the flies are active, or alternatively work in the fields may have to be done during the night. It has been estimated that a grazing animal may lose something in the order of 100 millilitres (one-sixth of a pint) of its blood to tabanids in a day in areas that are heavily infested with flies. Even the hippopotamus with its tough hide is plagued by tabanids. Tinley (1964) reports that these large mammals spend much of their time during the summer more or less submerged in water as much to avoid the attention of tabanids as to escape from the heat of the sun. He describes how a young hippopotamus, on being approached by a boat, broke away from the herd and ran out of the water but was at once so set upon by *Tabanus* that it promptly turned round and plunged back into the water, evidently preferring to face a potential human enemy rather than suffer the onslaught of these vicious flies.

Cyclorrhapha

As mentioned in the introduction to this chapter, there are two distinct groups of cyclorrhaphous Diptera which suck vertebrate blood. Firstly there is a group of flies allied to the ordinary house-fly, *Musca*, that spend most of their time away from the host and only visit the host for short feeding periods, and then there is a group of highly specialised ectoparasites, the louse flies and bat flies, which spend the greater part of their adult life, indeed sometimes all of it, on the body of the host. This latter group merits a chapter on its own so that the present section concerns only the blood-sucking Muscidae and their close allies, the Glossinidae.

Typically, cyclorrhaphous Diptera are equipped with a soft and elongated proboscis or labium which forms a sucking tube up which liquid food may be drawn (Figure 12a). The labium has at its apex two small lobes or labella. These bear on their ventral surfaces numerous fine channels (pseudotracheae) and also some minute rasping teeth (prestomal teeth). The labella are applied to the substrate on which the insect is feeding, saliva is poured onto the food, and dissolved food is drawn up the pseudotracheae and into the sucking tube of the labium. Solution of food is assisted by the rasping action of the prestomal teeth. With this equipment these flies can feed at the nectar of flowers or on other fluid vegetable products, and on the exudations of dead or excreted animal matter, and a few can even kill other insects and feed upon their body juices. In the family Chloropidae, tropical species of *Hippelates* and *Siphunculina* feed on liquid at the eyes, mouths, and noses of mammals. *Chlorichaeta tuberculosa* in another cyclorrhaphous family, the Ephydridae, is another such 'eye fly', and in the Muscidae, species of *Hydrotaea* habitually take sweat from humans. A few species of *Musca* have even developed the habit of attending feeding *Tabanus* and taking blood issuing from punctures made by the horse-flies. These muscids may be described as commensals, and the next step into parasitism is taken by *Musca (Philaematomyia) crassirostris*. This fly actually creates a wound for itself by rasping the skin of mammals with its prestomal teeth.

Stomoxys calcitrans (Figure 34a), the cosmopolitan stable fly, has gone a stage further. In this and allied flies that constitute the subfamily Stomoxydinae, the proboscis itself is hardened into a stabbing organ with large prestomal teeth for penetrating vertebrate skin (Figure 12b). Both man and his domestic animals are attacked by *Stomoxys*. In Africa the species is said to be especially predisposed to biting dogs' ears. *S. calcitrans* is, at first sight, easily mistaken for a house-fly but its bite at once reveals its true identity. It is very persistent, concentrating in attacking the legs and ankles of man, but it is an agile fly and disconcertingly difficult to swat.

In Britain there are two other biting muscids. These are *Haematobia stimulans*, which also feeds on man and livestock, and *Lyperosia irritans*, the horn-fly, which avoids man and has a predilection for the heads of cattle.

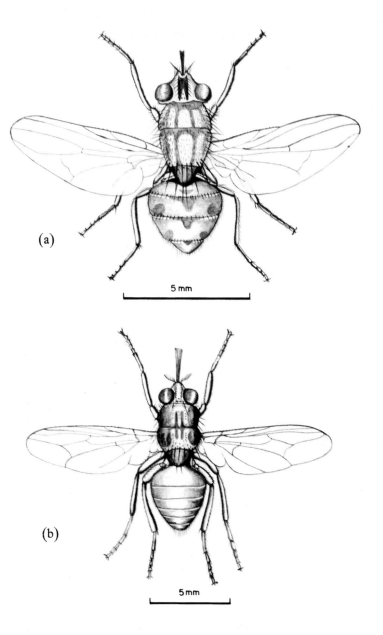

Figure 34. Blood-sucking Diptera Cyclorrhapha. (a) *Stomoxys calcitrans* female (Muscidae), the stable fly. (b) *Glossina palpalis* male (Glossinidae), a tsetse fly (both modified from Patton and Cragg 1913).

A few other related genera and species are found in other regions.

Both sexes of these parasitic cyclorrhaphous flies have similar mouthparts and thus, in contrast to the Nematocera and Brachycera, the males as well as the females can take blood.

The tsetse flies are blood-sucking insects of immense medical and economic importance. They are all contained within the genus *Glossina* (Figure 34b), of which there are about twenty-two species confined to 'fly belts' in Africa and Arabia south of the Tropic of Cancer. Although allied to the Muscidae, a separate family Glossinidae has been erected for them. Glossinids differ from muscids in that the female lays not eggs but large larvae; they are larviparous. The eggs, which contain very little yolk, hatch whilst still in the female and the larvae are nurtured inside the uterus of the female parent by a secretion from so-called milk glands. Larvae are laid singly at intervals of about ten days, and it has been estimated that one fly will produce about twelve larvae during its life, an exceptionally low fecundity for an insect. Blood taken by the adult fly is the sole food source during the life cycle, and at least three blood meals are necessary in the production of each larva. A fully grown larva weighs almost as much as its mother. A recent claim that flies which were fed on pregnant hosts produced larvae at a faster rate was later retracted. In some muscids small larvae at an early stage of development are deposited by the female, but the glossinid larva is fully grown when deposited, and has only a brief active period of about an hour in which it burrows into loosely packed, moist soil before pupating. This habit of laying fully developed larvae is shared with the Diptera Pupipara in which the duration of free larval life is even further shortened.

Both sexes of *Glossina* are biters and they feed predominantly upon the wild game of Africa. Unfortunately from a human standpoint, some species also bite man and domestic mammals. Sleeping sickness in man, and nagana and other diseases of cattle, horses, and camels, are forms of trypanosomiasis caused by flagellate Protozoa of the genus *Trypanosoma*. These diseases, unless treated at an early stage, are usually fatal. The natural reservoirs of the try-panosomes are the blood of certain wild mammals, especially various ante-lopes, which are immune to their effects, and tsetse flies are the vectors or distributive agents. The fly takes in the trypanosomes when it bites a diseased animal. In the insect's body the trypanosomes pass through certain develop-mental stages (page 100) so that the fly eventually becomes highly infective, passing on the protozoan whenever it bites a fresh host.

The genus *Glossina* has been the subject of intensive biological study primarily aimed at finding ways by which its very serious effects upon man in Africa may be reduced, and many interesting facts have come to light. Tsetse flies are not at all restricted in their choice of hosts but different species have different habitat preferences, and therefore their hosts do tend to differ. There are three major ecological groups. The group centred upon *G. morsitans* includes flies of open savanna country with sparse shade, and they feed upon the wild ungulates that abound, or used to abound, in such regions. During the dry season, however, *G. morsitans* migrates to moister, more wooded areas, and some flies may travel distances of up to eight miles (Nash 1948). The *G. fusca* group, on the contrary, contains forest-inhabiting flies which are larger than the other species, but of very much less importance to man since they bite mostly bush pigs and other animals of the forest, avoiding open grazing land. Species of the *G. palpalis* group are also found, typically, in dense rain forests, usually about rivers and lakes, where the blood of large reptiles such as monitor lizards and crocodiles forms a major item in their diet. This grouping is based largely upon the monumental work of Swynnerton (1936). In recent years, serological testing of the blood meals of *Glossina* has contributed to more precise information on the feeding habits of tsetse, and Weitz (1963) distinguishes five groups:

> GROUP 1. Feed mainly on suids.
> *G. swynnertoni, G. austeni, G. fuscipleuris, G. tabaniformis*
> GROUP 2. Feed mainly on suids and bovids.
> *G. morsitans*
> GROUP 3. Feed mainly on bovids.
> *G. fusca, G. pallidipes, G. longipalpis*
> GROUP 4. Feed mainly on other mammals.
> > *G. brevipalpis* (mostly hippopotamus), ?*G. longipennis* (mostly rhinoceros)
> GROUP 5. Feed on most available hosts including man and reptiles.
> *G. palpalis, G. tachinoides*

Groups 1 and 2 comprise the *morsitans* group, groups 3 and 4 the *fusca* group, and group 5 the *palpalis* group.

The *G. palpalis* group of species have been described as opportunist feeders, and their wide range of potential hosts is evident from a serological analysis of the blood meals of a sample of *G. tachinoides* collected in Northern Nigeria (Jordan, Lee-Jones, and Weitz 1962). In order of magnitude of contribution to the meals, blood from the following hosts was detected: primates (mostly man), bovines, unidentified mammals, reptiles, various mammals, porcupines, and birds. Nevertheless, even amongst these species with a vast number of possible host species, different feeding habits have been detected. *G. tachinoides* hunts at no more than two feet above the ground, and since it is unable to penetrate undergrowth, it is restricted to clearings and the open spaces at the edges of lakes and rivers. *G. palpalis*, on the other hand, will fly considerably higher than *G. tachinoides* and hunt in areas of quite dense undergrowth. Nash (1948) published results that highlight this rather striking difference in the habits of the two species. A record was made of the numbers of *G. palpalis* and *G. tachinoides* settling at different heights on a man standing erect in a

stream bed in Nigeria, in both the dry season and in the wet season. The
following data are condensed from Nash (1948).

| | % settling above knees | | % settling below knees | |
	Dry season	Wet season	Dry season	Wet season
G. tachinoides	21	34	79	66
G. palpalis	85	75	15	25

Tsetse flies are only locally very common, even when wild game, and therefore possible food, is plentiful. This has given rise to the suggestion that their
numbers are checked by density-independent factors such as climate, rather
than by limitations in the food supply. Certainly there is probably heavy pupal
mortality due to desiccation and drowning in the dry and wet seasons respectively. Food, however, might not be so abundant as at first sight appears, since
the flies have preferred hosts which are not always the most abundant species.
This emerges very clearly from data (Glasgow 1963) comparing the frequency
of potential hosts and the actual sources of blood in the guts of specimens of
G. swynnertoni from the same area of East Africa (Table 6). Warthog provided

Mammal	% frequency of mammal	% blood in G. swynnertoni
Impala	70	1
Dikdik	8	0
Giraffe	7	0
Waterbuck	4	0
Hartebeest	4	0
Grant's gazelle	3	0
Warthog	3	77
Lesser kudu	1	0
Rhinoceros	0.2	2
Buffalo	0.02	14

Table 6. The frequency of possible hosts and the origin of blood actually taken
by *Glossina swynnertoni* in east Africa (modified from Glasgow 1963),

most food for G. swynnertoni although it is well down the table of abundance,
and the most plentiful possible host, the impala, contributed only 1% of the
blood in the tsetse sample. Zebra are not known to be attacked by any species
of Glossina, and hartebeest by one subspecies only, G. morsitans submorsitans
(Glasgow 1967).

Most species of tsetse fly are day-biters and vision seems to play a large
part in host finding. A black screen, 4 feet long and 3 feet high, moved on
wheels, elicits a response from G. morsitans at a distance of 50 yards (Chapman 1961). G. morsitans is a fly of open country and, as may be expected, its
distance vision is superior to that of forest-dwelling species such as G. medi-

corum. A slowly moving truck has been used as an attractant in sampling flies of the *morsitans* group. However, other factors besides movement must operate in enabling *Glossina* to locate a host. Bursell (1961) found that mammals that spend a great part of their day lying up in shade are more regularly bitten than those that do not. Most tsetse flies also spend the hottest part of the day resting on leaves of trees and shrubs.

Environmental factors, physiological condition, and behaviour are intimately related in *Glossina*, and much work is still needed to clarify the situation. Working with *G. morsitans*, Pilson and Pilson (1967) showed that the species will feed at temperatures between 18°C and 32°c. Below 18°C flies are inactive, and above 32°C they become photonegative. Whatever the temperature, feeding stops abruptly at the onset of darkness. During daylight the flies spend a good deal of time, like most Cyclorrhapha, resting on exposed surfaces, especially tree trunks and branches, but when the temperature rises above 32°C they retire to shady holes and crevices. The nocturnal resting places of the species were not located, although it is known that *G. swynnertoni* and some other species pass the night on the foliage of bushes and trees.

5. *Diptera Pupipara:*
Louse Flies and Bat Flies

THE DIPTERA PUPIPARA is a group comprising three small families of 'higher' Diptera (suborder Cyclorrhapha). These families are the Hippoboscidae, which are ectoparasites of birds and large mammals, and the Nycteribiidae and Streblidae, which are ectoparasites only of bats. Relationships within the group are obscure and it is thought by many dipterists to be polyphyletic, although in all three families there is a characteristic slit on the dorsal surface of the second antennal segment (Figure 41b). Theodor (1964) finds great similarity between the genitalia of Hippoboscidae and Streblidae whilst considering the similarity between Streblidae and Nycteribiidae in proboscis structure and several other characters to be largely an outcome of convergence. Theodor concludes that the three families share a common ancestry and that the Nycteribiidae were the first family to become distinct. Similarities have been found by others between Hippoboscidae and tsetse flies, especially regarding mouthpart structure and in the habit of larviposition, and though affinities are now rather tenuous, the two groups may have had a common muscoid-like ancestor. Whatever might have been the origins of the families of the Pupipara, they now share several biological characteristics. The relationship with the host is always a very close one and some species, such as *Melophagus ovinus* (sheep ked), never leave its body. Vertebrate blood is the only food taken, as it is in *Glossina*, and all Pupipara share with the tsetse flies the curious habit of larviposition. The female fly does not lay eggs but instead retains them in her uterus until they hatch, when she nourishes the larva on secretions of the uterine 'milk' glands until it is fully grown. Only one larva is carried at a time so that the rate of reproduction is low, one female producing perhaps between ten and twenty larvae during its life. The fully grown larva is ultimately extruded through the vagina, after which it quickly forms a puparium without feeding further. This means that blood is the sole source of food of an individual during its life, and these flies have been able to suppress the larval stage which in typical insects obtains the greater part, if not all, of an individual's nutriment. Micro-organisms in mycetomes (page **xvi**) supply the Pupipara with essential accessory food factors that are lacking in vertebrate blood.

Morphologically, the Pupipara bear the unmistakeable stamp of vertebrate

ectoparasites, already seen in lice and fleas. Their bodies are much flattened
dorso-ventrally and provided with a tough, bristly cuticle, powerfully clawed
legs, and mouthparts in both sexes adapted for piercing the host's skin and
sucking blood. Eyes and antennae are small, the latter sunk into grooves as in
fleas and amblyceran lice. Movement on the host is rapid and effected in a
sideways manner. Some species are devoid of wings while some are fully
winged but with rather feeble powers of flight. In body length they are usually
between five and ten millimetres, considerably larger than most lice and fleas,
and large numbers are not therefore found upon a single host unless it be a
large mammal. In cases of unusually heavy infestations, some hippoboscids
have occasionally been seen to stab their mouthparts into the abdomen of a
gorged neighbour, drawing up the contained blood as fast as the victim could
withdraw it from the host.

Hippoboscidae

Hippoboscids have a variety of common names amongst which the most
familiar are louse flies and keds. They have been the subject of a world revision
by Maa (1963) who lists about one hundred species.

Only a few species are host-specific, most utilising a range of hosts that
appears to be determined as much by ecological considerations as by the
phylogenetic relationships of the hosts. In this respect they resemble fleas.
Bequaert (1930) has reviewed the host associations of species of *Hippobosca*,
and his findings with some additions may be summarised as follows:

Species of Hippobosca	Normal hosts
equina (Figure 35)	domestic horses (Europe)
capensis (*longipennis*)	dog, hyena, fox, cheetah, lion, serval, civet, leopard
fulva	hartebeest
maculata (*variegata*)	domestic horses and cattle
rufipes	domestic horses (Africa)
hirsuta	waterbuck and allied antelopes
struthiornis	ostrich
camelina	camel and dromedary
martinaglia	impala

Thus ungulates are the principal hosts of the genus, but the occurrence of one
species on ostriches causes Oldroyd (1964) to remark that apparently 'to
Hippobosca the ostrich is not a bird, . . . but is simply a gazelle with feathers!'

Species of *Hippobosca* (Figure 35) are fully winged and can fly quite well.
However *Melophagus ovinus* (Figure 36) has completely lost not only its wings
but its halteres as well. This is the sheep ked, and its entire life cycle is spent
sheltered in the fleece of its host. A related and similar species, *M. rupicaprinus*,
is found on the chamois.

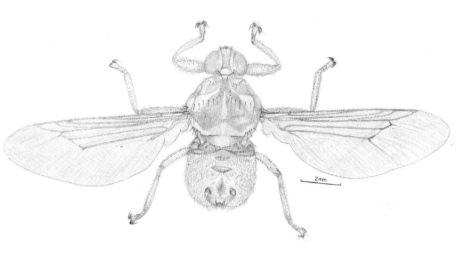

Figure 35. *Hippobosca equina* (Diptera Pupipara), a fully-winged hippoboscid ectoparasitic upon horses.

Lipoptena cervi (Figure 37) is the deer ked. It may be regarded as somewhat intermediate in the state of its wings to *Hippobosca* and *Melophagus*, for it starts its adult life in a fully winged condition, but loses its wings as soon as a deer is reached. This species was studied in Denmark by Haarløv (1964) who describes wing loss in a male fly: 'The process started with a crack on its rear margin, just distal to the level where the basal ribs suddenly get feeble, and where the wing widens rather abruptly. The severance lasted 24 hours. A few days later the other wing was lost in the same way. In both cases it was evident that the legs did not participate directly in the process, but that it seemed mainly to be provoked by the constant rubbing against the hairs when the animal was moving in the coat.'

The female *L. cervi* deposits either third-stage larvae or sometimes puparia containing fourth instar larvae or pupae. Usually the larva or puparium drops to the ground, but instances were noted (Haarløv 1964) of puparia remaining attached to the coat of the deer, reminiscent of the usual state of affairs in *Melophagus*. Adults do not emerge from the puparium for a considerable time. Cowan (1943) found that *L. depressa* spent between 43 and 214 days in the

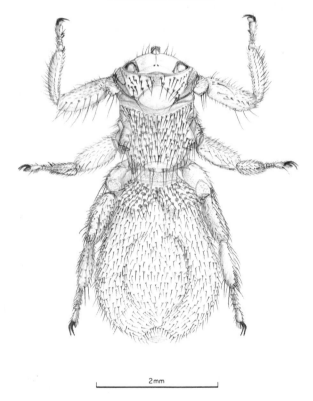

2mm

Figure 36. *Melophagus ovinus* (Diptera Pupipara), the sheep ked, an apterous species of hippoboscid which does not leave its host.

puparium. Emergence occurs towards the year's end. The adult *L. cervi* are fairly good fliers over short distances. They prefer wooded areas, and female deer-keds are apparently not inclined to fly around in search of hosts but prefer to wait in trees for a suitable host to pass by. I have several times had winged *Lipoptena* drop onto me when walking under trees in the Scottish Highlands; perhaps movement attracts them.

L. cervi has been recorded from several species of deer, and stray individuals have even been found on birds. Reindeer introduced to Rothiemurchus, Scotland, were quickly parasitised by it (Kettle and Utsi 1955). The reindeer is a new host for *L. cervi* since the fly does not extend northwards into the natural range of the reindeer. However the red deer is its principal host; Haarløv (1964) found that only one of twenty-two fallow deer examined was infested by it, whilst at the same time eleven out of twelve red deer running in the same park were parasitised. This may be related to the fact that the fallow deer tended to keep to open areas away from trees. The highest number

2 mm

Figure 37. *Lipoptena cervi* (Diptera Pupipara), the deer ked,
a hippoboscid which sheds its wings after reaching a host.
The drawing depicts a specimen that has lost its right wing.

of keds on one deer recorded by Haarløv was sixty-two, and this particular
animal was an old stag in poor condition. The low state of health of the stag
was related to the high infestation in that a healthy host is the main agent
controlling the number of its parasites. Kettle and Utsi (1955) likewise note
that the most heavily parasitised Scottish reindeer were animals that died soon
afterwards.

Hardenberg (1929) found that healthy sheep harboured a maximum of
about ten *Melophagus ovinus*, whereas up to four hundred keds might be found
on a sick sheep. When additional keds were added to the population on a
healthy host they were soon eliminated. *M. ovinus* moves around a great deal
on the host's body. Individuals do not remain for long at the skin surface
because they seem unable to tolerate prolonged exposure to near blood
temperature. They frequently move to the surface of the fleece and it is at

these times that they can readily pass to a fresh host when sheep come into brief bodily contact. Tetley (1958) found that dissemination onto new hosts increased in sunshine, and he considered this to be due to the increased temperature within the fleece. The 'dissemination potential' varied from sheep to sheep, probably because of differing fleece characteristics with respect to temperature. According to Tetley, a sheep with a high dissemination potential might act to some extent as a disinfector of a flock, since the keds it acquires quickly try to leave and some will perish in making unsuccessful attempts to transfer to a fresh host.

Lipoptena cervi and *Melophagus ovinus* tend to aggregate on certain regions of the bodies of their hosts. This is apparent from the data in Table 7, which give the percentage distribution of the insects on different body regions. The

Region of host	Lipoptena cervi on red deer Adults	Melophagus ovinus on sheep Adults	Puparia
Back	0	20	0
Belly	2	10	5
Flanks	20	70	45
Neck	25	0	50
Head	3	0	0
Groin and axillae	35	–	–
Anal region	15	–	–

Table 7. The percentage distribution of deer and sheep keds on different regions of their hosts' bodies (from Haarløv 1964, and Evans 1950).

figures are from Haarløv (1964) for *Lipoptena*, and Evans (1950) for *Melophagus*. Tetley (1958) also describes the distribution of *Melophagus*, and he found adults to be concentrated beneath the neck and shoulder, on the flanks and in the groin, a distribution resembling that which Haarløv records for *Lipoptena*. Both *Lipoptena* and *Melophagus* tend to avoid the back and belly, and this is probably at least partly because these areas can be reached by the host with its legs or teeth. Remains of *L. cervi* have been found in a deer's stomach. Also birds, especially starlings, frequently perch on the backs of sheep and deer and search them for ectoparasites. However, other factors also contribute to the distribution pattern of keds on their hosts. Haarløv found that the highest skin temperatures of red deer, which are favoured by *Lipoptena*, are at the groin and axillae. Another factor is the microstructure of skin and coat. On the back, belly, and neck of a deer, the contour hairs are too stout for *Lipoptena* to obtain a good grip with its claws, but those of groin and axillae are of more suitable diameter. The neck of a deer would appear to be too vulnerable a site for *Lipoptena*, and also the skin temperature is relatively low here, but nevertheless it is a popular location. This can be explained by the nature of the coat on the neck. There is a dense growth of woollen hairs which provide a suitable 'foothold', and the large contour hairs probably afford protection. Yet another factor to be considered, and an important one, is the availability of blood in different regions. The proboscis of *Lipoptena* can penetrate to a depth of 900μ, and regions having most blood vessels within reach are the groin and axillae, with fewest on the back and neck. However, the neck is more easily penetrated than the back because of fewer elastic fibres in the tissue overlying the blood vessels. In conclusion to his study, Haarløv writes that 'no single factor can be singled out as a key factor in determining the distributional pattern of deer keds. In each niche the ked population presumably depends on a balance of several more or less important factors.'

Turning now from mammal to bird hippoboscids we find again a range of wing development, from the fully winged condition of flies like *Ornithomya* (Figure 38) to flightless forms with considerably reduced wings, exemplified by *Stenepteryx* (Figure 39) and *Crataerina*. *Stenepteryx* is a parasite of martins, *Crataerina* of swifts. They deposit their larvae in the nests of these birds and

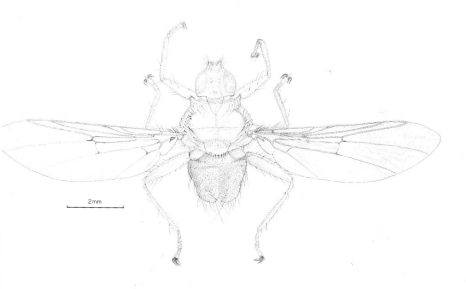

2 mm

Figure 38. *Ornithomya avicularia* (Diptera Pupipara), a fully-winged hippoboscid ectoparasite of several species of woodland birds.

the adult flies seldom leave the host or its nest. The emergence of adult flies from puparia is synchronised with the return in spring of the birds to their nesting sites. Thus loss of flight is no disadvantage to these flies, which have become confined and specially adapted to a particular group of hosts. The fully winged *Ornithomya* species, on the contrary, have a much wider range of hosts. In Britain there are three common species of *Ornithomya*, and their host ranges and distribution have been investigated by Hill (1962). The major hosts of these species are as follows:

O. avicularia: Falcons, hawks, owls, cuckoos, pigeons, and the larger passerines (crows, thrushes, starlings).

O. chloropus: Grouse, curlews, and other waders, plovers, merlins, and (=*lagopodis*) passerines (starlings, wheatears, tits, pipits).

O. fringillina: Small passerines only, particularly warblers, hedge sparrows, robins, and titmice.

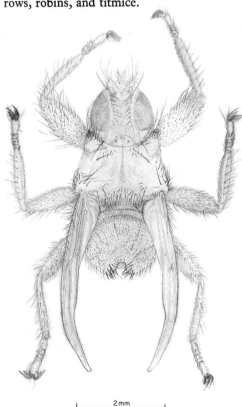

2mm

Figure 39. *Stenepteryx hirundinis* (Diptera Pupipara), a species of hippoboscid which parasitises martins and is incapable of flight.

Thus these three species are by no means host-specific, but the amount of host-sharing between them is slight. They are, in fact, quite clearly separated ecologically, and this is apparent when the usual habitats of the host birds are examined. Hill (1962) obtained 66% of his specimens of *O. avicularia* from woodland birds, 98% of *O. chloropus* from moorland birds, and 90% of *O. fringillina* from hedgerow birds. *O. chloropus* is therefore clearly associated with moorland birds, but there is a fairly extensive overlap between the habitats of the hosts of *O. avicularia* and *O. fringillina*. However, *O. avicularia* was found generally to parasitise birds over eight inches in length, whilst *O. fringillina* was generally restricted to birds less than six inches long. How host size affects the flies is not clear, but it nevertheless enables their host ranges to be distinguished and is the factor which, directly or indirectly, prevents extensive competition between the two species. Hill also studied their geographical distribution, and found that, in Britain, *O. avicularia* and *O. fringillina* did not occur much north of latitude 60°N, whilst *O. chloropus* was not found much to the south of this latitude. Corbet (1956b, 1961) records that juvenile starlings on Fair Isle were parasitised almost exclusively by *O. chloropus*, whilst in Essex the only hippoboscid found on these birds was *O. avicularia*. However, in Scandinavia (Hill, Hackman, and Lyneborg 1964) *O. chloropus* ranges further south than in Britain, and there is considerable geographical overlap between all three species.

Ornithomya species are parasites mostly of juvenile birds, but not of nestlings. The puparia are most probably situated in loose earth or litter, rather than in the nest material of the host as they are in *Stenepteryx* and *Crataerina*. The *Ornithomya* larva forms a puparium promptly on extrusion from the female, and the black puparium is carried about by the female fly on the host for about an hour before being dropped. The fly does not appear able to determine the type of substrate on which the puparium will fall.

In the case of *Hippobosca equina* (Figure 35) however, the female, which is fully winged, has been observed (Thompson 1955) to actually push puparia into the litter of a forest floor. This closely resembles the behaviour of *Glossina*, which seek out and sometimes congregate in large numbers on moist, loosely-packed earth or litter on which to lay their larvae.

The greater infestation of juvenile birds by *Ornithomya* is attributed by Hill to their ineptitude in ridding themselves of ectoparasites. Increasing numbers of damaged *Ornithomya* are found on young birds as the year advances and the birds gain experience. Corbet (1961) found a maximum of 62% *O. chloropus* with damaged wings on starlings on Fair Isle, and a maximum of 25% *O. avicularia* with wing damage on the same host species in Essex. Corbet suggests that the larger size of *O. avicularia* may render it more vulnerable to damage during the host's preening activities.

In Britain all of the bird hippoboscids have one generation during the year which is synchronised with the nesting behaviour of the hosts, although

Corbet (1961) found indications that *O. avicularia* might produce a partial second generation in south-east England. The *Ornithomya* species emerge from their puparia about mid-summer when there are many juvenile birds about (Hill 1963). Females mate soon after emergence and a single mating serves the female with sufficient sperm for the rest of her life. This contrasts with *Melophagus*, which mates repeatedly and in which it is thought (Bequaert 1953) that the spermatozoa and secretions of the male may help to nourish larvae developing in the female following earlier matings.

Nycteribiidae

The 195 described species of Nycteribiidae are all ectoparasites of bats. Awesome in appearance (Figure 40), they are perhaps the most specialised of all flies. They have become so modified for their peculiar mode of life that they no longer resemble flies, at least superficially, and they are so completely committed in their structure and biology that it is difficult to conceive of them taking any further major evolutionary steps.

Nycteribiids have completely lost their wings, and the wing muscles have dwindled so that the thorax is dwarfed by the rounded abdomen. This, and the long, bristly legs with prodigious claws give the insect a spiderish look. The insertions of the legs, and also the small head, have been displaced to a dorsal position. Well-developed combs are present and these function as in fleas (page 32), inhibiting backwards removal of the insect from its head-downwards resting attitude in the host's fur.

Predominantly an Old World group centred perhaps in the lands of the Malaysian subregion (Theodor 1967), species of Nycteribiidae are remarkable for their wide geographical distribution. *Penicillidia dufouri* has been found

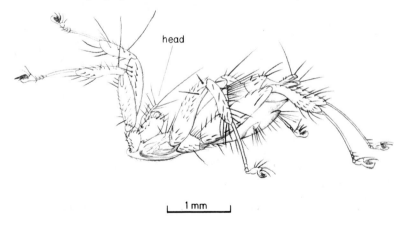

head

1 mm

Figure 40. A nycteribiid (Diptera Pupipara), *Stylidia biarticulata* female, in side view (after Theodor 1967).

from France and Spain east to Israel and Iran, in North Africa, the Himalayas and Kashmir, and, as a different subspecies, in China, Japan, and Formosa. It has been taken from bats collected at altitudes of 3000 feet and 8000 feet. Presumably the body of the host affords protection from climatic extremes, although several species have a more limited distribution than that of their hosts. Wide distributions are often correlated with a lack of host-specificity, the parasites attacking different hosts in different regions, but even so many species are associated with particular groups of bats. Of the three species that have been recorded in Britain, *Stylidia biarticulata* has been found only on horseshoe bats (*Rhinolophus*) and *Nycteribia kolenatii* has seldom been taken from a host other than Daubenton's bat, even though there are allied species of bat which are much commoner. Of the two subfamilies (Theodor 1967), the Nycteribiinae are parasites of the lesser or insectivorous bats (Microchiroptera) and the Cyclopodiinae are parasites of the larger fruit-eating bats or flying foxes (Megachiroptera). One species of bat may be infested by several species of Nycteribiidae (*Miniopterus schreibersi* has twelve species of *Nycteribia*, three of *Stylidia*, and nine of *Penicillidia* recorded from it), and one species of nycteribiid may attack several species of bat, but in the latter case the bats may be congeneric or at least belonging to the same family. Only rarely do the bats belong to more than one family.

Detailed observations of a nycteribiid in nature have yet to be made, but Rodhain and Bequaert (1916) studied *Cyclopodia greeffi* on captive individuals of *Cynonycteris straminea*, a fruit-eating bat. The parasites very seldom left their hosts, resting motionless for long periods with only the apex of the abdomen protruding through the host's fur. They avoided the head and seemed to prefer the neck and beneath the wings. Female *Cyclopodia* left the bats only to deposit their larvae. Males passed occasionally from host to host, probably in search of mates. Deprived of a host, the flies survived only about twelve hours. The females lay their larvae on dry surfaces in the roosting places of the bats. After expulsion from the female, the larva of *C. greeffi* adheres to the substrate and the female stands astride it and presses down upon it with her thorax, thereby securely fixing it. Some other Nycteribiidae have been said to deposit their larvae on the fur of the host, as in the hippoboscid *Melophagus*, but this behaviour is probably not typical of the group. The rate of reproduction of Nycteribiidae is apparently higher than that of other flies which lay large larvae, for the female *C. greeffi* produces larvae at intervals of two to six days, probably for most of the year.

Streblidae

Like nycteribiids, all Streblidae are ectoparasites of bats, though there are records of stray individuals being found on doves, parrots, and an opossum (Kessel 1924). Structurally they are not quite so modified as Nycteribiidae and most species are winged (Figure 41c). When on the host, the wings are neatly folded so as not to obstruct the fly's passage through the bat's fur. Head combs have been developed (Figure 41a). While the Nycteribiidae are predominantly an Old World group, the Streblidae are numerous in the New World as well, although everywhere they are confined to those lands lying approximately between latitudes 40°N and 40°S (Jobling 1951). In most of this region the temperature is sufficiently high all the year round to permit bats to remain active and avoid the necessity of hibernation. It may be that the Streblidae are restricted in their distribution by an inability to adapt, either in their life cycles or physiology, to periods of hibernation in the host when the

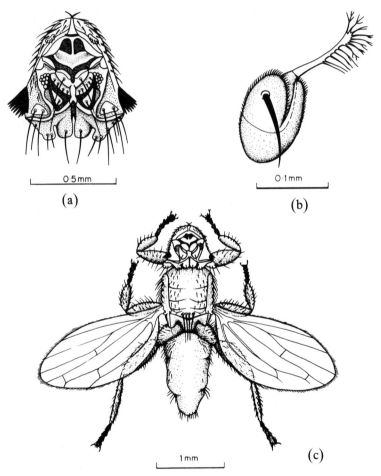

Figure 41. A streblid (Diptera Pupipara), *Euctenodes mirabilis*. (a) Dorsal view of head. (b) Antenna showing the slit in the second segment characteristic of Pupipara. (c) Dorsal view of adult. (All from Kessel 1924.)

body temperature falls drastically. Nycteribiidae are evidently not limited in this way.

As with other Pupipara, streblids are not usually host-specific and their host ranges are determined largely by the habitat of the bats. Thus different species of bats sharing the same roost are likely to be infested by the same species of Streblidae.

Most Streblidae deposit their fully developed larvae upon surfaces in the bat roost, like nycteribiids. However, larvae of *Ascodipteron* fall to the ground and pupate there, recalling the typical habit of Hippoboscidae. *Ascodipteron* is a most peculiar insect (Figure 42a), and it used to be considered worthy of a family on its own. The newly emerged flies have well developed wings, but when the female gets onto a bat (flying fox) it bores into the skin by means of its proboscis, and sheds not only its wings but its legs as well (Muir 1912). The abdomen becomes bloated (physogastric) as the fly takes in blood from the host, and only its posterior portion remains unenclosed by host tissue. Fully developed larvae are pushed out of this opening. This behaviour is very similar to that of fleas like *Tunga* (page 35). The male *Ascodipteron* is free-living and retains its legs and wings.

Other Flies that have Parasitic Adults

There are two other flies which do not belong to the Pupipara but which are in many respects biologically comparable with them. The first of these is *Carnus hemapterus* (Figure 42b), the best known representative of the family

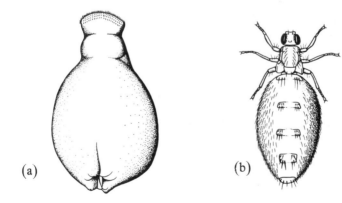

(a)   (b)

Figure 42. Two parasitic adult-flies showing marked disten-sion of the abdomen (physogastry). Both are apterous. (a) *Ascodipteron africanum* (Pupipara, Streblidae), the much-modified, sessile female parasitic upon bats (after Jobling 1939). (b) *Carnus hemapterus* (Cyclorrhapha, Carnidae), a female that has lost its wings after becoming established on a nestling bird (from Séguy 1950).

Carnidae (Cyclorrhapha: Acalypterata). It is a small fly, about two millimetres long, and its larva feeds upon organic debris in birds' nests, as do the larvae of many flies. *C. hemapterus* is unusual, however, in that the adult flies have adopted a parasitic mode of life, living amongst the developing feathers of the nestling birds. Once on its host, *C. hemapterus* loses its wings in the manner of *Lipoptena*, and its abdomen becomes much distended. Its diet is not certainly known, but since its proboscis appears to be too weak to penetrate skin in order to obtain blood, it most probably feeds upon moist skin exudations. The parasitic habit of *C. hemapterus* most likely evolved after the habit of egg-laying in the nutritionally rich medium of birds' nests was developed, parasitism by the adult flies being taken up as a consequence of their close physical contact with nestlings while engaged in laying eggs. This contrasts with the way in which parasitism by *Glossina* and the Pupipara is generally thought to have evolved, that is by the adult flies gradually acquiring a taste for blood and seeking hosts only for the purpose of feeding. However, it is possible that the ancestors of *Glossina* and the Pupipara were first led into the proximity of vertebrates by selecting sites for laying their eggs or larvae which were frequented by vertebrates. Such sites, for instance the ground beneath a bat or bird roost, are rich in organic detritus, and they provide a substrate so nutritionally rich that the free, feeding life of a fly larva could be shortened as a step towards the habit of larviposition.

The family Braulidae is another acalypterate family, and it is apparently

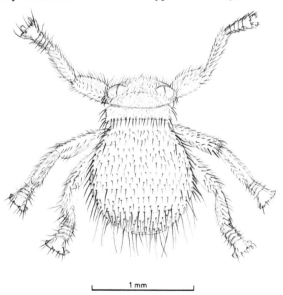

1 mm

Figure 43. The bee louse, *Braula coeca* (Diptera Cyclorrhapha).

represented by only one species, the bee louse *Braula coeca*. *Braula* is a very aberrant fly, highly adapted for an ectoparasitic existence, and it used to be placed with the Pupipara until it was discovered (Skaife 1921) that it lays eggs, not fully grown larvae. Little over one millimetre in length, *B. coeca* closely resembles a mite with its rounded body and total absence of either wings or halteres (Figure 43). Its eyes are vestigial and it is provided with long legs and strong claws. *B. coeca* is an ectoparasite of honey-bees (*Apis*), living in the dense growth of plumose hairs on the bee's thorax and elsewhere. Queen bees are especially infested by it, and over one hundred have been reported from a single bee. For food, *Braula* moves to the proboscis of the bee and takes regurgitated nectar and saliva. Eggs are laid in the hive, and the larvae eat wax and pollen.

Several species of *Braula* have been described but it is now thought that they are all forms of the one rather variable species. *B. coeca* has a very wide distribution, corresponding with that of its host, being found even in countries such as Britain to which the honey-bee has been introduced by man. This is typical of many adult ectoparasitic insects, their microhabitats varying little in physical characteristics wherever they are found. One of the few exceptions is the family Streblidae which has already been discussed. *Braula* has been found in both North and South America, most of Europe, South Africa, Japan, and Tasmania. It possibly originated in the mediterranean region, where it is now most abundant.

6. Bugs, Earwigs, Beetles, and Moths that are Parasitic as Adults

THE GROUPING TOGETHER of the Hemiptera, Dermaptera, Coleoptera, and Lepidoptera in one chapter is very largely a matter of convenience, for the orders have little in common save that each contains a few parasitic species amongst a majority of non-parasitic species. The Hemiptera and Dermaptera are exopterygotes, and their nymphs as well as their adults are parasitic. The other two orders belong to the Endopterygota. The Dermaptera and Coleoptera have biting mouthparts, the Hemiptera have piercing and sucking mouthparts, and the Lepidoptera have sucking mouthparts.

Hemiptera

The order Hemiptera (true bugs) is one of the largest groups of insects, and is divided into two suborders, Homoptera and Heteroptera. The Homoptera, which includes aphids, scale insects, cicadas, leaf-hoppers, and similar types, differs from the Heteroptera in having forewings of uniform texture throughout, that is the wings are not partly membranous and partly sclerotised, and in holding the wings at rest in a roof-like or tectiform manner instead of flat with the forewings of the two sides overlapping. Being exopterygotes, the same general form and habits are shared by adults and young stages (nymphs) alike. Plant sap is the diet of the majority of species, and it is extracted by means of a piercing and sucking proboscis (page 4). With mouthparts of this form, it is not surprising that some species have deviated from vegetarianism towards a diet of animal body fluids. Even in characteristically phytophagous families, blood-sucking has occasionally been reported (Usinger and Myers 1929). Obligatory zoophagous forms are also found in plenty amongst the Heteroptera, several families of which are partly or entirely predatory in habit, overwhelming smaller insects and other arthropods and sucking out their body fluid. A few of the Heteroptera have turned to larger game, birds and mammals, and it is these blood-sucking parasites that concern us here. The most familiar parasitic bug is unquestionably the bed-bug, *Cimex lectularius* (Figure 1a), and the family Cimicidae will be considered first.

Cimicidae (*Hemiptera*)

Some authors (e.g. Southwood and Leston 1959) include the Anthocoridae in the family Cimicidae. Anthocorids and cimicids are basically similar, but anthocorids are mostly predatory bugs, not parasitic, and may be distinguished from cimicids in the strict sense by the possession of three well-developed ocelli on the vertex of the head. Among anthocorids, members of the sub-family Lyctocorinae are most closely allied to the Cimicidae. These are bugs which, unlike most other anthocorids, do not live on growing plants, but inhabit barns, compost heaps, birds' nests, and other situations where dead vegetable matter is accumulated, and here they feed upon small insects. *Lyctocoris campestris* is a cosmopolitan species closely associated with man. It is frequently found in hen-houses where it may feed upon small caterpillars. However, it is also recorded (e.g. Woodward 1951) as occasionally living amongst human clothing and bedding, and biting man, thus demonstrating how the full-time parasitism of Cimicidae could have arisen.

Cimicids are very flattened bugs, often of rounded outline, as a result of their method of feeding and also, more especially, their habit of hiding away in cracks and crevices for prolonged periods. No doubt due to their passing the daylight hours in dark retreats, their colouration is a sombre and uniform brown. Cimicids can run rapidly but they have dispensed with wings. Being only temporary ectoparasites they are not adapted for clinging to their hosts, and they lack the modified tarsi and the combs characteristic of many permanent ectoparasites of birds and mammals. There are very few records of Cimicidae being captured on hosts away from the nest, roosting place, or bed; nevertheless some such journeys must occasionally be made so as to infest isolated host colonies.

Seventy-four species of Cimicidae, distributed amongst twenty-two genera, have been described, and their physiology, biology, and taxonomy are the subject of a recent monograph (Usinger 1966). There is a high degree of specificity in the host relations of Cimicidae.

Host families	Genera of Cimicid parasites
Vespertilionidae (Typical Bats)	*Bucimex, Propicimex, Cimex, Cacodmus, Aphrania, Stricticimex*
Molossidae (Free-tailed Bats)	*Primicimex, Propicimex, Stricticimex, Loxaspis, Crassicimex*
Emballonuridae (Sac-winged Bats)	*Loxaspis, Leptocimex*
Pteropodidae (Fruit Bats)	*Aphrania, Afrocimex*
Noctilionidae (Fish-eating Bats)	*Latrocimex*
Hipposideridae (Old World Leaf-nosed Bats)	*Stricticimex*
Rhinolophidae (Horse-shoe Bats)	*Cimex*
Cynocephalidae (Flying Lemurs)	*Cacodmus*
Hominidae (*Homo*) and domestic animals	*Cimex, Leptocimex*
Hirundinidae (Swallows, Martins)	*Oeciacus, Paracimex, Ornithocoris, Hesperocimex*

Host families	Genera of Cimicid parasites
Apodidae (Swifts)	*Paracimex, Ornithocoris, Cimexopsis, Synxenoderus*
Phasianidae (Domestic Fowl)	*Cimex, Ornithocoris, Haematosiphon*
Columbidae (Domestic Pigeon)	*Cimex*
Muscicapidae (Flycatchers)	*Cimex*
Psittacidae (Parrots)	*Psitticimex*
Cathartidae (New World Vultures)	*Haematosiphon*
Strigidae (Typical Owls)	*Haematosiphon*
Tytonidae (Barn Owls)	*Haematosiphon*
Falconidae (Eagles, Falcons)	*Haematosiphon*
Furnariidae (Oven-bird)	*Caminicimex*

Twelve genera are bat parasites, nine are bird parasites, and only one genus, *Cimex*, contains species which attack birds and bats in addition to man and domestic animals. The genus *Cacodmus* has been associated with the flying lemur as well as with bats.

Three species attack man: *Cimex lectularius*, *C. hemipterus*, and *Leptocimex boueti*. *C. lectularius* is the bed-bug parasitising man, bats, rats, chickens, and occasionally other domestic animals, over the greater part of the world but avoiding the tropics. *C. hemipterus*, a closely allied species, attacks man, chickens and, rarely, bats in tropical and subtropical regions in both Old and New World. The third species, *L. boueti*, is known only from West Africa where it parasitises bats as well as man. No disease is known to be transmitted by cimicids to man.

The birds attacked by cimicids are most commonly martins, swallows, and swifts, which are birds that frequently nest in caves. Some species of *Leptocimex* which are bat parasites exhibit the typical troglobiont features of pale colouration and long legs and antennae. Caves are almost certainly the original home of Cimicidae and it is not difficult to imagine some ancestral species adopting the habit of feeding on a warm-blooded animal, probably a bat, instead of hunting in the guano (accumulated faeces) for arthropodan prey. A hibernating bat will offer little resistance to a small blood-sucker. It is tempting to assume that man also acquired his association with these insects as a cave-dweller. This, however, is not certain. Pigeons, such as the rock dove, which nest and roost in caves, are not parasitised. It was apparently only after the rock dove was domesticated and kept in dovecots that it developed its association with *Cimex columbarius*, a sibling species of *C. lectularius*.

It is believed that *C. lectularius* originated in warm-temperate regions of the palaearctic, but as urban areas expanded and dwellings became better heated, it was able to extend its range northwards. It is known to have been a British insect in the sixteenth century, and it soon became common in seaports and other large towns. In 1939 it is estimated that four million people in greater London were affected by it, and in the eighteenth century it gave employment to several firms of professional bug exterminators. John Southall, author of *A Treatise of Buggs*, enjoyed a flourishing business, as did Messrs. Tiffin and Son who proclaimed 'May the Destroyers of Peace be destroyed by us, Tiffin and Son, Bug Destroyers to Her Majesty and the Royal Family'. However, today bed-bugs are rare in Britain, relegated to this status by insecticides, particularly DDT, and by improved standards of domestic hygiene.

Temperature is an important factor in the life of the bed-bug (Johnson 1940, 1942). The number of generations passed during a year is directly dependent upon the ambient temperature. In tropical climates *C. lectularius* may have twelve generations a year whereas in London, in an unheated room, the most that it can achieve is one complete generation and a partial second. The lower temperature threshold for egg-hatching and adult activity is between 13°C and 15°C for *C. lectularius*, somewhat higher for *C. hemipterus*. The

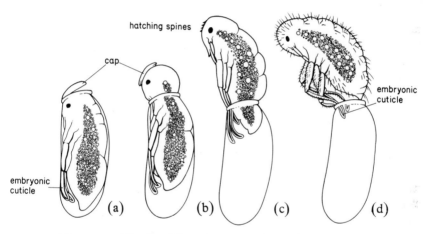

Figure 44. The hatching of a bed bug (*Cimex lectularius*). (a) and (b) The cap of the egg is lifted as peristaltic movements force fluid into the head of the nymph. (c) Nymph almost free of egg but still invested by embryonic cuticle which, (d), is eventually ruptured and slipped off posteriorly after the insect swallows air (from Sikes and Wigglesworth 1931).

optimum temperature for *C. lectularius* development is about 28°C, and for *C. hemipterus*, the tropical bed-bug, 32°C. *C. lectularius* is correspondingly more tolerant than *C. hemipterus* of short periods of low temperatures. Both species can tolerate a wide range of humidity.

C. lectularius rests during the daytime behind skirting-boards, loose wallpaper, in cracks in walls, or narrow crevices in wooden bedsteads. Several are usually found together in such harborages, and it is here that the white, reticulately-sculptured eggs are laid on rough surfaces. The young nymph, when about to hatch, rhythmically contracts its abdomen, causing the fluid pressure to increase in its head until the egg cap is pushed off (Figure 44). Small teeth on the embryonic cuticle of the labrum and vertex (Figure 45) may help the young bug to rupture investing membranes and escape from the egg (Sikes and Wigglesworth 1931). The bug, during development, must have at least one blood meal in each of its five nymphal instars. If undisturbed, up to twice its own weight of blood will be consumed in one meal. After feeding, the bug retires to its resting place for several days whilst the blood is digested.

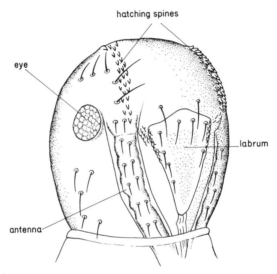

Figure 45. Enlarged view of head of hatching nymph of *Cimex lectularius* still enclosed in embryonic cuticle. The hatching spines assist the insect to free itself of other investing membranes (from Sikes and Wigglesworth 1931).

Symbiotic bacteria probably aid digestion; two disc-shaped, white mycetomes lie laterally in the abdomen beside the gonads. Neither does the male bug mate nor the female produce eggs until after feeding. Egg-laying commences five or six days after a meal and may, under optimum conditions, continue for about six days (Johnson 1942). Bed-bugs are able to withstand long periods of starvation. At 22°C an adult male can survive 143 days, a female 131 days, and a first instar nymph 84 days without feeding. Bacot (1914a) kept once-fed bugs in an unheated outhouse for eighteen months, and found at the end of this period that several adults and fifth instar nymphs were still alive. However, if starved bugs are compelled into even short periods of activity, their life span is reduced (Mellanby 1939).

Most of the activity of bed-bugs is directed towards searching for a host. They avoid light and are most active just before dawn (Figure 46) (Mellanby

1939). The antennae are held out straight in front of the head and evidence suggests that they are guided towards a suitable host by following up a temperature gradient. Carbon dioxide also attracts bed-bugs, but it seems that host odour plays little part in host location. Usinger (1966) writes 'it is clear that the bed-bug cannot detect a host beyond 5 feet and probably not beyond a few inches.'

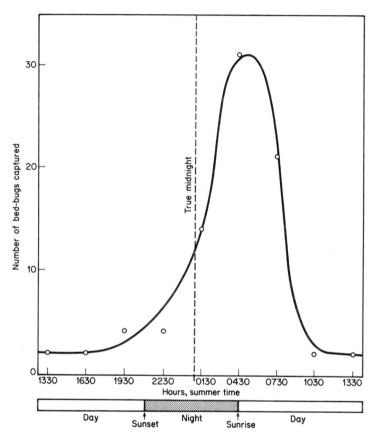

Figure 46. The numbers of *Cimex lectularius* captured in three-hour periods during twenty-four hours. There is an activity peak about dawn (from Mellanby 1939).

It has been suggested that bed-bugs follow scent trails between their own retreats and the sleeping place of their host, but this has not been experimentally substantiated. They are well endowed with scent glands but these are most likely to be for rendering them distasteful to would-be predators. The odoriferous substances are carbonyl compounds, and dwellings infested by bugs have a distinctive and unpleasant odour.

Perhaps the most remarkable and unusual feature of cimicid biology is the method of copulation and insemination, a full account of which is given by Carayon in Usinger (1966). The male bed-bug recognises his partner by sight, but his sight is weak and mistakes are made. Homosexual behaviour is common and in the laboratory interspecific matings are frequent. The normal insectan method of insemination is not practiced by Cimicidae. The genitalia of the male are asymmetrical, the left paramere alone being developed; the male pierces the female's abdominal integument with this paramere and injects into her a large quantity of spermatozoa. The female has a paragenital system (Figure 47) to cope with this extraordinary male behaviour. In *C. lectularius* the fifth abdominal segment of the female is notched ventrally on its posterior border. This notch (paragenital sinus) opens into an integumental pocket (ectospermalege) which is associated with a mesodermal pouch (mesospermalege). During copulation the male paramere is stabbed into the ectospermalege and sperm injected in a compact mass into the mesospermalege. The position of the female paragenital system varies between cimicid genera, and is associated with different copulatory positions. The mesospermalege contains amoebocytes which ingest some of the spermatozoa and sperm plasma. Surviving sperm pass out of the mesospermalege into the haemocoel of the female

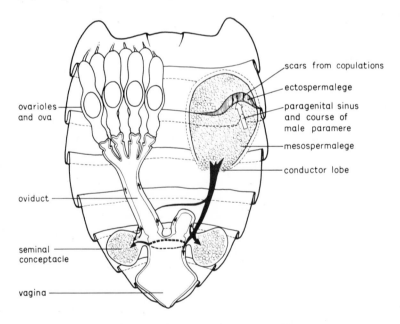

Figure 47. Diagram of insemination and the paragenital system in female *Cimex*. The right ovary and oviduct have been omitted. Arrows indicate the route followed by the spermatozoa (after Usinger 1966).

by way of a 'conductor lobe'. This apparently serves to orientate the sperm so that they swim towards the bases of the oviducts. It is possible that the sperm follow an increasing oxygen gradient in this migration. Each oviduct has a well-tracheated basal swelling, the seminal conceptacle; the sperm penetrate these sacs and may be stored there for a time. Here, as in the mesospermalege, some sperm may be ingested by amoebocytes. In function the seminal conceptacles are partly analogous to spermathecae in other insects, but they are of mesodermal origin and not homologous. The walls of the oviducts are perforated by a network of minute tubes (spermodes), and the sperm travel by way of these to the ovarioles and ova.

The ectospermalege evidently has the function of restricting damage to the female from the rapacious onslaughts of the male. It is reinforced by a thick layer of endocuticle, and may be required to withstand several penetrations by male parameres because bed-bugs mate repeatedly and frequently. *Afrocimex* is exceptional in that the male as well as the female has a functional ectospermalege. Perhaps homosexual unions are more frequent in *Afrocimex* than in other genera.

Primicimex is an unspecialised form with no spermalege at all. The male paramere is thrust through the almost unmodified dorsal abdominal integument of the female, and after multiple copulations scars form in the regions most accessible to the male. There is some hypertrophy of the integument in these regions, probably indicating an incipient spermalege. At the other extreme of spermalege development stands *Stricticimex*. In this genus the mesospermalege has an elongated conductor lobe which carries the sperm right to the genital tract so that they do not have to cross the haemocoel. Traumatic insemination is not restricted among insects to the Cimicidae, being found also in the families Anthocoridae and Polyctenidae.

As a result of their repeated matings, female cimicids may accumulate very large quantities of spermatozoa. Those sperm that are engulfed by amoebocytes may provide food for the bug, and Hinton (1964) considers that traumatic insemination enables females to survive when hosts are unavailable. However, Davis (in Usinger 1966) records that starved females avoid mating, and goes on to suggest that hypergamesis is related to the very small chance a sperm has of reaching an ovum. This, however, does not explain why some sperm are engulfed by amoebocytes.

The male cimicid's enthusiasm for mating is apparently unimpeded by any need to perform a courtship display, and the stabbing home of a paramere cannot demand either the time or the co-operation of the female that is required by, for example, a male flea. This suggests that in nature there is little likelihood of different species meeting, and therefore interspecific reproductive barriers at the pre-insemination level have been dispensed with. In the laboratory, however, cross-species matings have been observed frequently. Male *C. hemipterus* will mate readily with female *C. lectularius* but such unions are not fruitful; the female *C. lectularius* are poisoned by the foreign seminal fluid and die. The severity of this physiological barrier to interspecific mating is a further indication of the infrequency of the event under natural conditions.

Polyctenidae (Hemiptera)

Polyctenidae (Figure 48) are allied to Cimicidae. They are found in tropical and subtropical regions of both Old and New Worlds, but they are rare bugs. Like so many insect ectoparasites they attack bats, and it is to this group of mammal hosts that they are exclusively confined. Bats are unable to groom themselves with their limbs, and this may account for their susceptibility to infection by insects the size of bugs, as well as flies like Nycteribiidae and Streblidae. Polyctenids measure three to five millimetres in length and, bearing in mind the small size of most bats, they must impose a significant burden of irritation and loss of blood on their hosts. Maa (1959) records that 17 bats (*Cynopterus* and *Megaderma*) from a cave in Malaya yielded 22 specimens of Polyctenidae, 59 Streblidae, and 222 Nycteribiidae. All of the Polyctenidae were on the *Megaderma* and one bat was infested by fourteen of the bugs.

Ferris and Usinger reviewed the family in 1939 and recognised 6 genera and

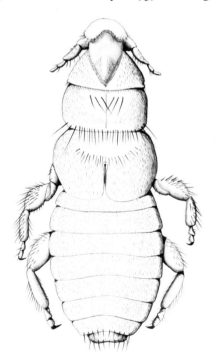

Figure 48. Female *Adroctenes horvathi* (Hemiptera, Polyctenidae) (from Weber 1930 after Jordan).

18 species. However, only 19 specimens from the Old World were available for study, compared to 228 specimens which Maa (1964) was able to examine. Maa recognises 14 Old World species and summarises the host relationships in the family.

Polyctenid species	Bat host	
Eoctenes ferrisi	Emballonura	⎫
E. coleurae	Coleura	⎬ Emballonuridae
E. intermedius	Taphozous	⎭
E. nycteridis	Nycteris	Nycterididae
E. spasmae	Megaderma (Megaderma)	⎫ Megadermatidae
Polyctenes molossus	Megaderma (Lyroderma)	⎭
Adroctenes horvarthi	Rhinolophus	Rhinolophidae
A. jordani, magnus	Hipposideros	Hipposideridae
Hypoctenes petiolatus, quadratus, clarus, faini	Tardarida (Old World)	⎫
Hesperoctenes cartus, hermsi, setosus	Tardarida (New World)	⎬ Molossidae
H. abalosi, angustatus, eumops, giganteus, fumarius, limai, longiceps, vicinus	Molossus	⎭

Closely related bats have similar parasites and the family conforms well to Fahrenholtz' rule (page 25); no bat species has so far been recorded as harbouring more than one polyctenid species (Maa 1964). The hosts all belong to six of the total of sixteen extant families of Microchiroptera. The Polyctenidae are divided into two subfamilies, Hesperocteninae and Polycteninae. The Hesperocteninae are represented in both Old and New Worlds and attack bats of the families Rhinolophidae, Molossidae, and Hipposideridae. Polycteninae are restricted to the Old World, where they parasitise Megadermatidae, Emballonuridae, and Nycterididae.

The family name is descriptive of the many combs that adorn these bugs. Combs may be present on the basal antennal segments, gena or other part of the head, pronotum and prosternum, mesonotum, ventral surface of the first abdominal segment, and the last tarsal segments. Polyctenids lack compound eyes and ocelli, and their wings are greatly reduced. Their two posterior pairs of legs are long and slender, but their front legs are short and stout. The entire body is very hairy. Unlike Cimicidae, Polyctenidae are viviparous, an exceptional faculty amongst Heteroptera, and this, together with their prolific comb development, indicates that they live almost entirely on the body of the host. Little is known, however, about their biology.

Reduviidae (Hemiptera)

This is a very large family of about 4000 described species. Most of them are predators of other insects. *Reduvius personatus*, a black bug fifteen millimetres or so long, has a holarctic distribution and not infrequently flies into lighted rooms at night in southern England. One of its chief prey species in certain areas is said to be the bed-bug. *R. personatus* will, in self-defence, bite man.

In America there are several species of the genus *Triatoma* which frequently bite vertebrates and withdraw blood. They are commonly called cone-nose bugs or, from their habit of biting the lips of sleeping people, kissing bugs. A well-known species is *Triatoma sanguisuga* which lives in rodent nests but occasionally invades houses. Its bite is said to be excruciatingly and immediately painful, the effect of a toxin which is injected into the wound with the saliva. Burr (1954) recounts how an Asian species of reduviid was once used as an instrument of torture by the Emir of Bokhara, the bugs being kept in a pit into which the emir's prisoners were thrown. When the bugs had no prisoners to feed on, they were given pieces of raw meat. *Panstrongylus* (= *Triatoma*)

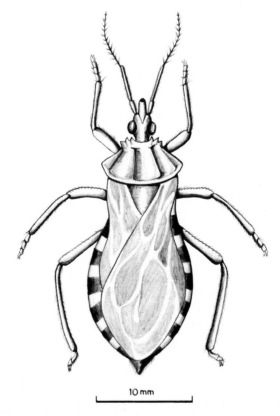

10 mm

Figure 49. *Panstrongylus megistus*, a cone-nose bug which transmits Chagas' disease to man (modified from Patton and Cragg 1913).

megistus (Figure 49) is a regular biter of man in America, but its bite is much less painful than is usual for the family. This is what one would expect of an insect which feeds exclusively on vertebrate blood.

Because cone-nose bugs are mobile animals, not confined to a single host species, they are important vectors of disease (page 104). *T. sanguisuga* and *P. megistus* are two of the commonest vectors of the protozoan that causes Chagas' disease in Central and South America. Another vector of this disease is *Rhodnius prolixus*, a cone-nose bug that has been extensively used in laboratory experiments on the hormonal control of insect development.

Cone-nose bugs can stridulate. They have a series of ridges on the ventral surface of the thorax against which the rigid tip of the rostrum can be rubbed. The squeaking sound produced may serve to warn off predators. In the British press in 1966 the United States Pentagon was reported to be 'planning to send bed-bugs [sic] to help to win the war in Vietnam . . . Their plans are based on the fact that bed-bugs scream with excitement at the prospect of feasting on human flesh [sic!] . . . a sound amplification system would enable the GI, sweating through the jungles of South Vietnam, to hear the anticipatory squeals of a captive bed-bug as it detects the Vietcong lying in ambush ahead. Tests have apparently shown that a large and hungry bed-bug will appropriately register the presence of a man some two hundred yards to its front or side, while ignoring the person carrying it in a specially devised capsule' (*The Guardian*, 7th June 1966). This is intriguing but somewhat improbable. Why, for one thing, should bugs stridulate on detecting a food source?

Dermaptera

The more specialised exopterygote orders, the hemipteroid orders, include among their number the Mallophaga and Anoplura, all species of which are permanent ectoparasites, and the Hemiptera, which includes two families of obligatory ectoparasites. In contrast the less specialised orthopteroid exopterygote orders have only two genera of parasitic species, *Hemimerus* and *Arixenia*. These are both temporary ectoparasites. The reason for the paucity of parasites amongst the orthopteroid orders is possibly partly related to their large size. Few measure under a centimetre in length as adults, and an insect of these dimensions could hardly remain undetected for any length of time by a vertebrate host.

When Burr (1911) separated the order Dermaptera from other orthopteroid groups to accommodate the earwigs, he recognised three suborders: Forficulina, Arixenina, and Hemimerina. The large suborder Forficulina includes all typical earwigs. The other two suborders, each of a single genus only, contain species which are ectoparasites of mammals. Recently, Popham (1961b) has removed the Hemimerina from the Dermaptera and elevated the group to ordinal status, and the same author (1965) places *Arixenia* in a family of its own within the dermapteran superfamily Labioidea.

Hemimerus (*Dermaptera*)

Hemimerus was scientifically recognised in 1871 by Francis Walker, who described *H. talpoides* from Sierra Leone, placing it as a cricket in the family Gryllidae. *Hemimerus*, when viewed from above (Figure 50a), has something of the appearance of a small cockroach, particularly in the shape of its large pronotum and uniform, light brown colouration. It lacks wings however, and the abdominal segments are all visible as in earwigs. Eyes are absent, and the body, which measures a little over a centimetre in length, terminates posteriorly in two short, thread-like, caudal cerci. Verhoeff (1902) considered the systematic position of *Hemimerus* to be between cockroaches and earwigs. Popham (1962) investigated the feeding of *Hemimerus* and concluded that the mandibles probably collect the food and it is passed to the mouth via the maxillae. In Dermaptera, including *Arixenia*, it is the laciniae of the maxillae that collect food and pass it to the mouth by way of the mandibles. This difference, together with others, leads Popham (1961b) to the conclusion that the Hemimerina represent an order of orthopteroid insects with little affinity to any other living order.

The genus *Hemimerus* is confined to tropical Africa, ranging from Portuguese Guinea south to the Tropic of Capricorn and from west coast to east. This distribution coincides with that of the giant pouched rat, *Cricetomys*. *Cricetomys* is the principal host of *Hemimerus*, and eight of the nine described species are specific parasites of it. The remaining species, *H. morrisi*, has been found on the long-tailed pouched rat, *Beamys major*, a close relative of *Cricetomys*. This extreme specialisation in host associations is remarkable. The taxonomy of *Cricetomys*, like that of many other African genera of rodents, is very confusing, but Walker (1964) refers the numerous named forms to a single species, *C. gambianus*. Different forms of *C. gambianus*, distinguished largely on characteristics of the pelt, occur in different regions, and it is probable that each species of *Hemimerus* occurs on a different form of *C. gambianus*. Rehn and Rehn (1936) observed a partial correlation between the texture of the coat of *Cricetomys* and the structure of the last sternite of the female *Hemimerus* species infesting it. In some species the last sternite can lock onto the ventral surface of the last tergite, thereby closing the anal orifice. Such species usually live on rats with a thin, coarse coat, and populations of *Cricetomys* having this type of coat are found in the more arid regions. It is possible that the locking of the anus is a device to restrict water loss.

Nearly all specimens of *Hemimerus* in collections have been taken from the bodies of *Cricetomys*. They are generally distributed over the host and 'exceedingly active in eluding capture, slipping through the hair in a surprisingly flea-like manner' (Rehn and Rehn 1936). How much time *Hemimerus* spends away from the host in the extensive burrows of *Cricetomys* is not known, but it is probably not very much since the insects are viviparous and all stages have been found on hosts. Dead rats are quickly abandoned. Features of *Hemimerus*

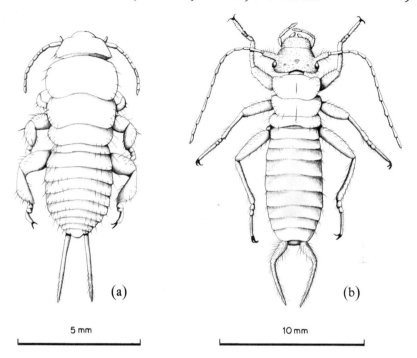

(a)   (b)

5 mm   10 mm

Figure 50. Two parasitic earwigs. (a) *Hemimerus talpoides* from a pouched rat. (b) *Arixenia jacobsoni* associated with bats.

which indicate that the coat of the host is the usual environment are the absence of eyes and wings, the enlarged tarsi which are furnished with patches of small setae to assist clinging, short legs for pushing the flattened body through fur, and viviparity.

The diet of *Hemimerus*, as revealed by the examination of gut contents, consists of small slices of host epidermis but not hair (Popham 1962), and of spores and sporangia of a fungus which grows on the skin of *Cricetomys* (Jordan 1910).

Arixenia (Dermaptera)

Arixenia is a parasite of bats. The species are considerably larger than *Hemimerus*, measuring about two centimetres in length, and they have fewer obvious adaptations to living on the host's body. In fact one species, *A. jacobsoni* (Figure 50b), may not be a parasite at all, but feed only on the guano of bats. Wings are absent, but *Arixenia* retains a pair of small, compound eyes and the position of the three ocelli can be located although they are probably non-functional. The colouration is dark brown, and the body terminates in a pair of cerci which are much more heavily sclerotised than those of *Hemimerus* and approach the condition of the forceps of earwigs. Only two species of

Arixenia are known, and both have a very restricted distribution although they are apparently abundant where they occur.

The first species to be described was *A. esau*. This was discovered in the giant Subis cave on the Niah River, Sarawak, in a colony of the naked bat, *Cheiromeles torquatus*. Lord Medway (1958) describes the locality: 'The colony of these bats at Niah is far out of reach, about 250 feet overhead, high in the semidarkness of the cave roof in the centre of a very tall chamber . . . An estimated 20 000 individuals, closely packed, roost here. Directly below them and nowhere else . . . are large numbers of *Arixenia esau*, most of them gathered into four groups, each clustered at the highest point of one of the odd poles which stick up from the guano, rotten relics of the climbing equipment of birds' nest collectors.' No *Arixenia* were found on *Cheiromeles* captured away from the roost, only on dead or dying bats, lying on the floor of the cave, over which they crawled with 'an excited jerky gait, exploring every fold and pouch, or grazing restlessly on the pitted skin.' *Arixenia* crawling over Lord Medway's hands fed mostly on the sweaty skin between the fingers, scraping the skin surface with the mandibles and moving steadily forwards and upwards the whole time. Some earwigs were seen to feed on living adult tenebrionid beetles inhabiting the Niah cave, and captive individuals readily fed upon a variety of crushed insects.

The only other known species of *Arixenia*, *A. jacobsoni*, was also discovered in a bat-inhabited cave, this time in southern Java. The host bat is *Tardaridus mops*. Its discovery by Jacobson is recounted by Burr and Jordan (1912): 'The cave is called by the natives *Gouwa Lawa*, which means *bat-cave*, on account of the tremendous numbers of bats which frequent it . . . The most conspicuous insects inhabiting the cavern are . . . the Earwigs . . . they crawl in countless numbers on the surface of the guano and everywhere on the rocky walls. Evidently they live on the various larvae feeding on the guano, but besides this, they are constantly waging a terrible war against each other, the victors devouring the bodies of their slain mates . . . A more loathsome spectacle than these thousands of ugly, hairy creatures, running about hither and thither, fighting and devouring each other, can hardly be imagined.' The only indication that *A. jacobsoni* is a parasite of the bats is by analogy with its congener, *A. esau*. However Popham (1962) describes differences in mouthpart structure between the two species which are consistent with the suggestion that *A. jacobsoni* is a predator of arthropods, like many other species of Dermaptera, whilst *A. esau* is a temporary, facultative ectoparasite of bats.

A. jacobsoni has been found in the Philippines and Malaya as well as in Java, but *A. esau*, apart from its type locality, is known from only one other place. This is in Malaya, where many individuals were obtained from a hollow tree colonised by *Cheiromeles torquatus* (Cloudsley-Thompson 1957). The earwigs were collected both from the bodies of the bats and from the guano. Curiously, this same bat roost also yielded a few specimens of *A. jacobsoni*, all from the guano. This is good circumstantial evidence in support of the different feeding habits of the two species.

Coleoptera

Coleoptera have biting mouthparts and the few species that are parasitic as adults feed, apparently, on epidermal secretions and hair on their vertebrate hosts. Species in the small families Leptinidae and Platypsyllidae, together with a few species of Staphylinidae and Scarabaeidae, probably feed obligatorily in this way.

Leptinidae (Hatch 1958) are allied to Silphidae which, as adults, are mostly scavengers of carrion. They are small beetles, two or three millimetres long, with at most rudimentary eyes. Six species have been described. The sole British representative, *Leptinus testaceus* (Figure 51), has poorly understood food requirements. It has most often been found in small mammal nests, especially mouse nests, but voles and mice themselves sometimes carry it. In addition, it has been recorded from bees' nests. A very closely related American species is *L. americanus* which has been found on small rodents and insectivores.

Two other North American leptinids are *Leptinillus validus*, an ectoparasite

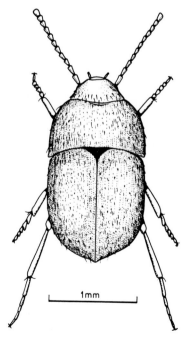

Figure 51. *Leptinus testaceus*, a small beetle sometimes found living on voles and mice.

of the beaver, and *L. aplodontiae,* an ectoparasite of the mountain beaver. The mountain beaver is not at all closely related to the true beaver. A Russian species, *Silphopsyllus desmanae,* is a specific parasite of the desman or water-mole.

Platypsyllus castoris, a curious, totally blind beetle, was in 1869 placed by Westwood in a separate order Acheiroptera. It is now generally regarded as the single known representative of the family Platypsyllidae, although some coleopterists group it with the Leptinidae. Like *Leptinillus validus, P. castoris* is a specific parasite of the beaver. It is found in both nearctic and palaearctic regions and is an obligate parasite both as an adult and as a larva. In this latter respect *P. castoris* provides one of the very few exceptions to the general rule that larval endopterygote insects are unable to adapt to an ectoparasitic existence on a vertebrate host (page xvi). The diet of both larval and adult *P. castoris* is probably skin debris. Eggs are laid in the pelt of the host.

The staphylinid ectoparasites belong to the tribe Amblyopini, which is allied to the Quediini. They have rudimentary eyes and are flightless, both features related to their obligatory, ectoparasitic existence. The most recent review of the tribe is by Seevers (1955). The group occurs predominantly in subtropical and temperate zones of the neotropical region, although a relatively unspecialised form, *Myotyphlus,* was discovered in Tasmania.

Amblyopini are parasites of rats and marsupials. The utilisation of marsupials as hosts is probably a secondary development, the group originally evolving as rodent parasites. A high degree of host-specificity exists.

Amblyopine genus	No. of species	Distribution	Host group
Myotyphlus	I	Tasmania	'rat'
Edrabius	6	S. America	Hystricomorph rodents (*Ctenomys*)
Megamblyopinus	2	S. America	Hystricomorph rodents (*Ctenomys*)
Amblyopinus	27	S. & C. America	Mostly cricetine rodents but 1 sp. on hystri-comorph and 5 spp. on marsupials
Amblyopinodes	6	S. America	Cricetine rodents

The beetles embed their mandibles deeply into the hosts' skin, and they are removed only with difficulty. They tend to congregate either behind the ears or about the base of the tail. The host's skin swells up around the point of attachment. These Staphylinidae, like *Platypsyllus,* live in all their life history stages upon the host.

Parasitic Scarabaeidae are only found on the host as adults, their larvae living in excrement on the ground. Two South American species, *Uroxys gorgon* and *Trichillium brachyporum*, parasitise the three-fingered sloth, *Bradypus*. *Uroxys* infests the neck, *Trichillium* the anal region. The Australian *Macropocopris* lives in the fur of kangaroos.

Lepidoptera

Lepidoptera are almost exclusively phytophagous insects and as adults they have a long, sucking proboscis for withdrawing nectar from flowers. Nevertheless, characteristic of the adaptability of insects, a few adult Lepidoptera have become vertebrate parasites. Best known of these are the so-called eye moths. These habitually feed about the eyes of domesticated ungulates. Like the chloropid eye flies (page 55), they generally suck lachrymal secretions and pus, but occasionally blood is taken. Eye moths are known from Africa and south-east Asia, the genus *Arcyophora* featuring prominently in the records from both areas. *Arcyophora sylvatica* and *Lobocraspis griseifusa*, both of the family Noctuidae, are the principal species in Thailand and Cambodia where they have been studied by Büttiker (1967). Both are nocturnal and males and females have been seen feeding at the eyes of cattle, domestic water buffaloes, and sambar. Büttiker lists twenty-four species of eye-frequenting Lepidoptera from south-east Asia and recognises three categories:

(i) Visitors only to the eyelids, taking lachrymation, pus, and sometimes blood from the conjunctiva and cornea.
Noctuidae (*Arcyophora*, *Lobocraspis*), Pyralidae (*Botyodes*, *Filodes*, *Margaronia*, *Pagyda*), Geometridae (*Hypochrosis*, *Somatina*).

(ii) Fairly frequent eye visitors where they take lachrymation, but also feed on fluid running down the hosts' cheeks.
Pyralidae (*Pionea*), Geometridae (*Semiothisa*).

(iii) Only occasional visitors to eyes, where they take lachrymation.
Noctuidae (*Blasticorhinus*, *Nanaguna*, *Hypena*, *Mocis*), Pyralidae (*Bradina*, *Typsanodes*), Geometridae (*Peratophyga*, *Pingasa*, *Scopula*). In this category there is also a single record for the long-tailed blue butterfly, *Lampides boeticus*.

Several Lepidoptera drink at muddy pools, or on juices exuding from decaying fruit or even animal carcases. The behaviour of the eye moths is merely an extension of this general habit.

A few adult moths feed on fresh fruit and they have a strengthened proboscis for penetrating the rind. One such species, the noctuid *Calyptra eustrigata*, can apparently also pierce vertebrate skin for, at the time of writing, reports have appeared of this moth sucking mammalian blood in south-east Asia (Bänziger 1968).

Quite different from the above species is the genus *Bradypodicola* which is the most remarkable of ectoparasitic Lepidoptera. Unfortunately, very little seems to be known about its biology. Two species are known and both live upon the three-fingered sloth, *Bradypus*, in South America. Both larvae and adults live amongst the dense hairs of the host. The adult moths probably feed upon skin secretions, whilst the larvae may eat the hairs. The hairs of the sloth are unique in being covered in minute pits in which grow green algae, the possible diet of *Bradypodicola* caterpillars.

7. Blood-sucking Insects as Vectors of Human Disease

DISEASES THAT ARE TRANSMITTED by blood-sucking insect vectors have had an enormous impact upon mankind. Just how great has been their effect it is almost impossible to estimate. The notorious black death of the middle ages, transmitted in part by the rat flea, eliminated at least one quarter of the population of western Europe. There cannot be many disasters of equal magnitude in the history of mankind. Nearer to the present day, it has been estimated that at the beginning of this century almost a quarter of the world's population had suffered or was suffering from malaria. About half a million United States servicemen contracted malaria during the second world war. The examples could be multiplied, but it is probably more meaningful to say simply that the present global pattern of economic prosperity has been very strongly influenced by the distribution of insect-borne pathogens. And one thinks particularly of the fly-borne diseases; malaria, sleeping sickness, yellow fever, and forms of filariasis.

It now seems almost incredible that it was not until nearly the end of the last century that blood-sucking insects were conclusively shown to be vectors of disease, although for many hundreds of years previously they had been viewed with suspicion. Malaria, for instance, was believed in both Africa and India to be mosquito-borne at a time when in Europe it was confidently attributed to the miasma, bad air or 'mal air', arising from marshes.

The first proof of insect transmission of human disease was not produced until 1877 when Patrick Manson, working in China, discovered the developmental stages of the nematode worm, *Wuchereria bancrofti*, the causative agent of elephantiasis in man, inside a mosquito, *Culex fatigans*. Towards the end of the nineteenth century several more important discoveries were made. Bruce, in 1895, showed that tsetse flies are the vectors of the African cattle disease nagana, and in 1897 Sir Ronald Ross, working in India, finally demonstrated that the life cycle of the malaria parasite involves both man and the mosquito. The discoveries that the bacterial and viral pathogens of bubonic plague and yellow fever are also spread by insects came only during the present century.

The transmission of pathogenic organisms to man is the most significant aspect of blood-sucking insects, but the direct annoyance caused by persistent attacks should not be overlooked. The affect of the irritation of insect bites is incalculable in terms of human discomfort, and the economy of a country can suffer as a result. *Simulium* at the height of activity, for instance, may render certain regions uninhabitable. *S. wellmanni* is 'possibly one of the most successful destroyers of patience and provokers of profanity in the Colony [Angola]. Natives near wet plains sometimes are compelled to move their kraals on account of it' (Wellman 1908). It is not unlikely that Scotland would have a more prosperous tourist industry if *Culicoides* were absent.

Bacterial and Viral Pathogens

The transmission of disease to man may be either direct or cyclical. Bacterial and viral diseases are transmitted directly by insects, the vector serving merely as a vehicle rather than as a host for the pathogen. No stages in the life cycle of the pathogen are dependent upon the insect. An example is *Pasteurella pestis*, the bacterium responsible for bubonic plague (black death). In some wild rodent populations this disease is endemic. That is, it is continually present, usually at a fairly low incidence, in a state of equilibrium with the host animal. The disease is passed from the blood of an infected animal to that of a healthy one by a flea, the flea injecting bacteria as it bites. An epidemic may result when an infected flea gets onto a domestic rat. The rat is not the usual host of the disease and lacks immunity to it; if the rat population is dense the disease will spread, passing between individuals by way of the rats' own fleas, particularly *Xenopsylla cheopis* and *Nosopsyllus fasciatus*. The disease at this stage has spread from its permanent locus in a wild population to a population that is living in association with man, and when a man is bitten by a rat flea that has previously fed upon a diseased rat, he too may be infected and a case of bubonic plague results. Rat fleas initiate human bubonic plague, but the disease can be passed directly from man to man. The bacteria eventually appear in the lungs of a person suffering from plague, and they are exhaled in droplets which may pass directly to the lungs of other people. This phase of the disease is called pneumonic plague. A human epidemic of plague depends upon both a large domestic rodent population and a large vector population. In the absence of these conditions cases of bubonic plague in man are rare and sporadic. In the United States about two cases are reported every year with a total of 111 since 1908.

Viral diseases, yellow fever for example, are also transmitted directly. As in the case of *Pasteurella pestis*, the disease is endemic in wild host populations, this time in monkeys. Yellow fever occurs in monkey populations in both South America and Africa, and transmission is by species of mosquitoes inhabiting the forest canopy. Monkeys do not appear to be affected by the disease. In South America, people working in the forests may be bitten by mosquitoes and inoculated with the virus. The disease is then carried back to villages and towns where domestic species of mosquito, particularly *Aedes aegypti*, may transmit it through the human population and bring about an epidemic. In Africa, epidemics of yellow fever are often started by infected monkeys being bitten by domestic mosquitoes when entering villages to raid crops. Although *Aedes aegypti* is a very widespread mosquito, yellow fever is not present between west Africa eastwards to South America; it is absent from the entire oriental region. The reason for this is probably that there is no reservoir or endemic form of the disease in oriental monkey populations. Whether the yellow fever virus originated in South America or in west Africa is not known.

Other mosquito-borne viral diseases of man are rift valley fever and dengue. The pathogens are related to the yellow fever virus, as also is pappataci fever virus which is transmitted by sand-flies. Pappataci fever does not have a reservoir in a wild mammal population; the disease is peculiar to man, who has developed considerably immunity to it and is only rarely more than mildly affected by it.

Rickettsiae and Spirochaete Pathogens

Rickettsiae are micro-organisms that resemble viruses. They live inside the cells of arthropods, but a few can be transmitted to man in whom they may be pathogenic. Typhus rickettsiae probably originated in ticks, and were passed to the wild rodent hosts of ticks. From wild rodents they may have been picked up by fleas and transported by them to domestic rodents and to man (murine typhus). Epidemic typhus occurs when lice transmit the disease from man to man (Figure 52). Curiously, infection of man by a louse is not by way of the louse's saliva but through its excreta or its body being scratched into the skin. In evolutionary terms, epidemic typhus is probably of fairly recent origin, since neither man nor the human louse can accommodate the rickettsiae without rapidly suffering harmful effects. Infected human lice, unlike infected fleas or rat lice, quickly die. Epidemics of typhus are most prevalent where standards of hygiene are low, allowing lice to multiply, and in the past the disease has been especially frequent under wartime conditions.

Spirochaetes are a group of organisms related to bacteria. Many are free-living, but a few are obligate parasites. Relapsing fever in man results from infection of the blood by a spirochaete. This pathogen, *Borrelia recurrentis*, is a rodent parasite, and infestation of man has followed a rather similar route to that of typhus rickettsiae. It is transmitted between rodents by ticks, and occasionally an infected tick will bite man. The disease may be spread through a population by the human louse *Pediculus*, but not by *Pthirus*. As in typhus, a man is usually infected through crushing a louse between the fingers and then introducing the micro-organisms into his body as a result of scratching.

Protozoan Pathogens

Protozoan pathogens are usually cyclically transmitted. The pathogen has an obligatory alternation of hosts. An essential part of the life cycle proceeds in

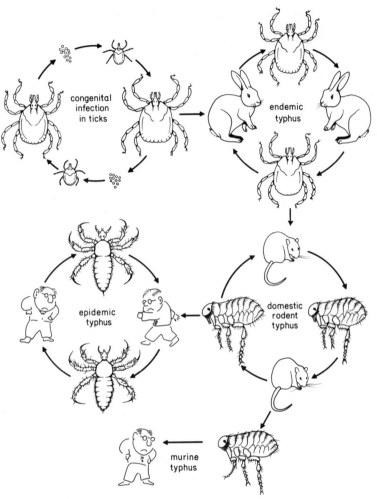

Figure 52. The natural history of typhus rickettsiae. See text for explanation. (Modified after Cameron 1965.)

the vector, which is thus a host rather than merely a carrier, and in consequence there is a delay between the vector becoming infected and being capable of spreading the pathogen. Since the pathogen must be closely adapted to living in its vector host, the vectors nearly always belong to a group of closely allied species.

The genus *Plasmodium* belongs to the class Sporozoa. The life cycle involves an insect vector, a mosquito, in which a sexual cycle (gametogony followed by sporogony) occurs, and a secondary vertebrate host in which asexual reproduction (schizogony) takes place. *Plasmodium vivax* (Figure 53) is the cause of benign tertiary (tertian) malaria in man. It is passed, as a sporozoite, into the blood of man with the saliva of a biting anopheline mosquito. The sporozoite is amoeboid, and it enters a liver cell where it divides by multiple fission to produce a number of small, amoeboid individuals termed cryptozoites. These infect other liver cells and the process, termed schizogony, is repeated. Sometimes a cryptozoite enters a circulating red blood corpuscle and here schizogony again takes place, resulting in the production of small merozoites. When

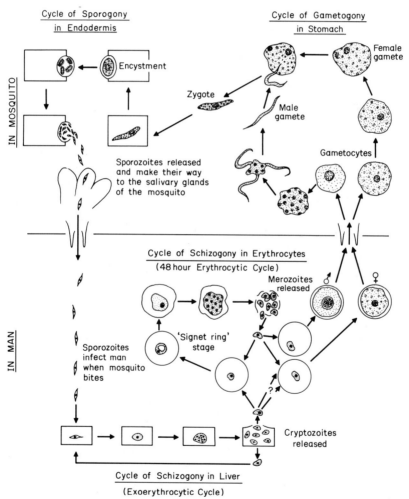

The Life Cycle of Plasmodium vivax

Figure 53. Diagram illustrating the life cycle of the protozoan *Plasmodium vivax* which is responsible for benign tertian malaria in man (after several authors).

the red corpuscle disintegrates, the merozoites are released into the blood plasma, and they enter other corpuscles to start the erythrocytic cycle anew. Erythrocytic cycles continue for a few asexual generations only, whereas the cycles in the liver (exoerythrocytic) may persist for several years. It is the erythrocytic cycle that is responsible for malarial fever. When the erythrocyte disintegrates, toxic products of the metabolism of the parasites are released into the host's blood, together with the merozoites. It is these toxins that cause the fever. One cycle of schizogony, from the entry of a merozoite or crypto-zoite into an erythrocyte, to the subsequent release of the next generation of merozoites, takes forty-eight hours. Fever therefore recurs on every third day and is termed tertiary malaria. After several asexual generations, gametocytes are produced, probably from the erythrocytic cycle although this is not certain. These enter red blood corpuscles but do not form gametes until they are taken into a mosquito. In the insect's stomach, the male gametocyte divides to pro-duce a few elongated male gametes. The female gametes are large, globular cells. Fusion between male and female gametes takes place and the resultant zygote is an elongated and active organism which bores its way into an endo-dermal cell of the mosquito's stomach. Here it encysts and divides to produce many sporozoites (sporogony). When the cyst ruptures, the sporozoites are liberated into the mosquito's haemocoel and they move to the salivary glands, from which organs they can be injected into a fresh secondary host. The life cycle of *Plasmodium vivax* is illustrated diagrammatically in Figure 53. Gametogony and sporogony can proceed only in anopheline mosquitoes.

Benign tertiary malaria is essentially a chronic disease and not usually fatal. The pathogen is dependent upon a long period of infestation of the host's liver cells to produce a sufficient number of gametocytes to complete its life cycle and spread to new hosts. The mosquito host is not destroyed either.

Plasmodium malariae is biologically quite similar to *P. vivax*, but its asexual cycle takes three days and it causes quartan malaria.

A third form which causes malaria in man is *P. falciparum*. It differs from the two preceding pathogens in passing its erythrocytic cycle in larger, deep-lying blood vessels instead of in peripheral vessels. Infected erythrocytes become sticky and agglutinate so that the blood vessels may become occluded, with fatal results. The disease is called malignant tertiary malaria.

Trypanosomes are flagellate Protozoa. They probably evolved as parasites of insects, living in the gut, but contact with vertebrates through blood-sucking insects has led some to adapt to living and multiplying in vertebrate blood. The undulating membrane, linking the flagellum to the body along nearly all of its length, is characteristic of trypanosomes and believed to be a device to assist locomotion in viscous blood plasma. Unlike *Plasmodium*, there is no cycle of sexual reproduction in *Trypanosoma*. In the insect vector, how-ever, they may undergo a cycle of development and multiplication which includes the adoption of a different body form described as the crithidial stage.

In the crithidial stage, the undulating membrane is much shorter than it is in a trypanosome. It originates between the nucleus and the flagellated end of the organism.

Some trypanosomes, for instance *Trypanosoma lewisi* which infects rats and is transmitted by fleas, are not pathogenic to the vertebrate. The pathogens of the domestic cattle disease called nagana are trypanosomes that are non-pathogenic in African wild ungulates and other game animals. *T. brucei*, *T. vivax*, and *T. congolense* all have natural reservoirs, especially in antelopes, and they are transmitted to cattle by tsetse flies. The protozoans multiply and pass through a developmental cycle in the alimentary tract of the fly (also in the salivary glands in the case of *T. brucei*). A mammal may, however, be inoculated directly by the bite from a fly recently interrupted whilst feeding on an infected animal and still containing trypanosomes in its proboscis. Of the three species, *T. brucei* is the most frequently fatal to cattle, and its usual vector is *Glossina morsitans*. Two forms or strains of *Trypanosoma*, morphologically identical to *T. brucei*, occur in man. These are *T. rhodesiense* which causes a very virulent form of sleeping sickness, and *T. gambiense* which is responsible for a more chronic form of the disease. *T. gambiense* is transmitted principally by *Glossina palpalis* and *T. rhodesiense* by *G. morsitans*. The *rhodesiense* form of sleeping sickness is believed to be a relatively new disease that has evolved only during the present century, presumably from *T. brucei* (Ormerod 1961). *T. gambiense* is probably evolutionarily older, having had time to become adjusted to its vertebrate host, man. It is advantageous to any parasitic organism to avoid destroying its host, and in the case of pathogens transmitted by blood-sucking insects, the longer they can maintain a population in a host individual, the greater is the likelihood of their transmission to another host. A sufferer from *gambiense* sleeping sickness may survive untreated for five years.

T. gambiense was discovered in 1901 at a time when a devastating epidemic of sleeping sickness was sweeping west Africa. Its main vector, *G. palpalis*, and also *G. tachinoides*, are riverine flies, and the epidemic spread rapidly up the Congo, eventually reaching the Nile and Lake Victoria. In the ten years between 1895 and 1905, it took toll of half a million lives. Morris (1965) writes: 'Miles of empty river banks bear silent witness to the tragedy even today . . . Recently I travelled long stretches of river devoid of population or with only an occasional fishing village or Government Station, which travellers, from 1860 to 1910, had described as densely populated by industrious tribes . . . A brief search along the overgrown riverbanks confirmed these stories by re-vealing the ruined villages and overgrown farms.' The epidemic of sleeping sickness coincided with, and probably resulted from, the increased passage of people along the waterways as the continent was opened up. The disease was eventually effectively controlled by spraying and clearing vegetation near to water, the haunt of the vectors, and also by administering drugs to large sections of the population. By 1960 total eradication was almost within sight.

Now, however, with the formation of independent African states, uniform vigilance against a resurgence of the disease is more difficult to achieve, and there are disquieting signs of its return. In Kenya, especially around Lake Victoria, the disease still claims hundreds of victims a year.

Africa is not the only continent to suffer from trypanosome pathogens. *T. cruzi* is an organism which affects man and other mammals in South America. Its life cycle is quite different from those of the pathogens of nagana and sleeping sickness. *T. cruzi* lives and multiplies intracellularly within the lymph nodes and other organs, and causes a condition known as Chagas' disease. Inside vertebrate cells it adopts a rounded body form without an undulating membrane, but some individuals may adopt a trypanosome form and pass into the blood stream from where they may be ingested by the insect vector, usually a blood-sucking bug (*Triatoma* or *Rhodnius*). *T. cruzi* multiplies within the gut of the insect and passes out in the faeces. The bugs generally defaecate at the site of a blood meal, and the trypanosomes gain entry to the vertebrate through the bug's feeding puncture. *Rhodnius prolixus* is a very important vector, whereas the allied *R. pallescens* is much less effective, probably because in the latter species up to two hours may elapse between feeding and defaecation.

Allied to *Trypanosoma* are species of *Leishmania*. Some are pathogenic in vertebrates. In the vertebrate they adopt a rounded, non-flagellated body form, but a flagellated stage is found in the insect vector. Vectors include sand-flies (*Phlebotomus*) and some other nematocerous Diptera. Human pathogenic *Leishmania* are found in all of the warmer regions of the world where they are responsible for kala-azar, oriental sore, and espundia.

Helminth Pathogens

The only platyhelminths which are regularly transmitted to man by insects are species of *Hymenolepis*, a very large genus of tapeworms. These species utilise fleas or flour beetles as their intermediate hosts, and for human infection to occur, the insects must be ingested.

Of much greater medical significance are filariid nematodes. As adults these worms live in the bodies of vertebrates, and the juvenile stages appear in the circulation from where they may be withdrawn by blood-sucking insects. Development continues in the vector and eventually infective larvae are produced. These accumulate about the proboscis of the vector, and when it next bites a vertebrate, they pass onto the definitive host and either enter the wound or bore their way through the skin to enter the blood stream. *Wuchereria bancrofti* inhabits the lymphatic system of man, where occlusion of the vessels may cause the painful and unsightly condition known as elephantiasis. Its vectors are various species of mosquitoes and the juvenile worms, microfilariae, appear in the peripheral blood circulation only at night when the vectors are biting. The microfilariae of another filariid, *Loa loa*, appear in the peripheral circulation during the day-time. The vectors for this nematode are day-biting Tabanidae. Hawking (1964), working with *W. bancrofti*, found that the periodic cycle of microfilarial activity in this species is probably controlled by the 24-hour temperature cycle of the host, a relatively low host body temperature at night bringing about a migration of microfilariae to the peripheral circulation. If the body temperature of the host is raised artificially during the night, the number of microfilariae in the peripheral circulation decreases.

A third filariid parasite of man to use an insect vector is *Onchocerca volvulus*. The worms live in skin nodules in man and are believed to sometimes attack the optic nerve and cause blindness. They are transmitted by simuliids, especially *Simulium damnosum*.

So far in this chapter the emphasis has been on the pathogens and their effects upon man, rather than upon the insect vectors. When the vector is merely a conveyor and inoculator of the pathogen, as it is in its association with viruses and bacteria, it suffers no direct damage. If, however, it conveys a rapidly fatal disease to its own host, then it is liable to destroy much of its food source. In Britain this has happened in the case of the rabbit flea, *Spilopsyllus cuniculi*, and the myxoma virus. The rabbit flea can probably breed only on the rabbit, and it therefore suffered a tremendous reduction in numbers when myxomatosis swept the country. *S. cuniculi* was almost certainly the most important vector of the virus during the epizootic. In the scarcity of its usual host, there is evidence (Rothschild and Ford 1965) that *S. cuniculi* may, in certain parts of England, be feeding regularly upon the hare, a host only sporadically parasitised before the myxomatosis epizootic. The hare is considerably less susceptible to the myxoma virus than is the rabbit.

In general, natural selection will tend to reduce mortality of all three organisms involved in a pathogen-vector-host relationship. When there is mortality, then it is likely that the relationship is of relatively recent origin; the epidemic typhus rickettsia-human louse-man association provides such an example, whilst, in contrast, the *Plasmodium vivax*-anopheline mosquito-man relationship involves quite a high degree of mutual tolerance. Sometimes, however, the pathogen depends for its transmission upon destruction of the vector, in a particular way, by man. This is true for the relapsing fever spirochaete, and also for infection by cestodes.

From the pathogen's angle it is most important to avoid rapid destruction of the definitive host in which the sexually reproductive stages are passed. It is here that the pathogen has invested its greatest reproductive potential. Early death of a proportion of the intermediate hosts need not be too serious. In the life cycle of *Plasmodium*, the mosquito is the definitive host and man the intermediate host, and it is man who is the more adversely affected of the two. *Wuchereria bancrofti*, on the other hand, passes its sexual stages in man and the mosquito carries only the larvae. Mosquitoes often die as a result of infection, whilst the adult worm has been known to survive in man for seventeen years.

This chapter is concluded with a table of insect vectors of human disease and the pathogens that they transmit. It is by no means an exhaustive list, and for further details a textbook of medical entomology should be consulted.

Insect vector	Pathogen	Disease
ANOPLURA		
Pediculus humanus	*Borrelia recurrentis* (Spirochaete)	Relapsing Fever
	Rickettsia prowazekii (Rickettsian)	Epidemic Typhus (Brills Disease)
	R. quintana	Trench Fever
	Pasteurella tularensis (Bacterium)	Tularaemia
HEMIPTERA		
Triatoma species	(Virus)	Encephalomyelitis
Panstrongylus (*Triatoma*) *megistus*, *Triatoma dimidiata*, *T. infestans*, *T. sanguisuga* and other species, *Rhodnius prolixus*, *R. pallescens*	*Trypanosoma cruzi* (Protozoan)	Chagas' Disease
DIPTERA		
Phlebotomus argentipes, *P. perniciosus*, *P. major* *P. orientalis*, *P. longipalpis* and other *Phlebotomus* (*Lutzomyia*) species	*Leishmania donovani* (Protozoan)	Kala-azar (Dumdum Fever)
Phlebotomus papatasi, *P. sergenti* and other species	*L. tropica*	Oriental Sore
Phlebotomus species	*L. braziliensis*	Espundia
Phlebotomus parfiliewi, *P. ariasi* and other species	*L. mexicana* and other species	Leishmaniasis
Phlebotomus verrucarum	*Bartonella bacilliformis* (Rickettsian?)	Verruga (Oroya Fever, Carrion's Disease)
Phlebotomus papatasi and other species	(Virus)	Pappataci Fever (Sandfly Fever)
Anopheles maculipennis, *A. quadrimaculatus*,		

Insect vector	Pathogen	Disease
A. gambiae, A. albimanus, A. punctulatus and many other species	Plasmodium vivax (Protozoan)	Benign Tertiary Malaria
	P. malariae (Protozoan)	Benign Quartan Malaria
	P. falciparum (Protozoan)	Malignant Tertiary Malaria
Anopheles subpictus (rossi), A. gambiae (costalis) and other species	Wuchereria bancrofti (Nematode)	Filariasis (Elephantiasis)
Aedes aegypti and other species	(Virus)	Yellow Fever
Aedes aegypti and other species	(Virus)	Dengue
Aedes caballus, A. tarsalis and other species	(Virus)	Rift Valley Fever
Aedes species	(Virus)	Encephalitis
Aedes polynesiensis, A. togoi and other species	Wuchereria bancrofti (Nematode)	Filariasis (Elephantiasis)
Eretmopodites species, Haemagogus species, Sabethes species	(Virus)	Jungle Yellow Fever
Culex fatigans	(Virus)	Dengue
Culex pipiens, C. fatigans, C. tarsalis and other species	(Virus)	Encephalitis
Culex univittatus, C. antennatus	(Virus)	West Nile Infection
Culex pipiens, C. fatigans and other species	Wuchereria bancrofti (Nematode)	Filariasis (Elephantiasis)
	Brugia malayi (Nematode)	Filariasis (Elephantiasis)
Mansonia titillans	(Virus)	Encephalitis
Mansonia species	Wuchereria bancrofti (Nematode)	Filariasis (Elephantiasis)
	Brugia malayi (Nematode)	Filariasis (Elephantiasis)
Culicoides austeni, C. inornatipennis, C. grahami and other species	Dipetalonema (Acanthocheilonema) perstans (Nematode)	Filariasis (usually non-pathogenic)

Insect vector	Pathogen	Disease
	Onchocerca species (Nematodes)	Onchocerciasis
Culicoides grahami	*Dipetalonema streptocerca* (Nematode)	Filariasis (usually non-pathogenic)
Culicoides furens	*Mansonella ozzardi* (Nematode)	Filariasis (usually non-pathogenic)
Forcipomyia townsendi, F. utae	*Leishmania* species (Protozoan)	Leishmaniasis
Simulium damnosum, S. neavei	*Onchocerca volvulus* (Nematode)	Onchocerciasis (River Blindness)
Tabanus striatus and other species	*Bacillus anthracis* (Bacterium)	Anthrax
Chrysops discalis	*Pasteurella tularensis* (Bacterium)	Tularaemia
Chrysops silacea, C. dimidiata	*Loa loa* (Nematode)	Loiasis (Calabar Swelling)
Stomoxys calcitrans?	(Virus)	Poliomyelitis
Stomoxys calcitrans?	*Bacillus anthracis* (Bacterium)	Anthrax
Stomoxys calcitrans	*Trypanosoma* species (Protozoan)	Trypanosomiasis
Lyperosia exigua	*Trypanosoma* species (Protozoan)	Trypanosomiasis
Glossina morsitans, G. swynnertoni, G. pallidipes	*Trypanosoma rhodesiense* (Protozoan)	Sleeping Sickness
Glossina palpalis, G. tachinoides	*Trypanosoma gambiense* (Protozoan)	Sleeping Sickness

SIPHONAPTERA

Insect vector	Pathogen	Disease
Xenopsylla cheopsis, X. astia	*Pasteurella pestis* (Bacterium)	Bubonic Plague (Black Death)
Xenopsylla species	*Rickettsia typhi* (Rickettsian)	Endemic (Murine) Typhus
Xenopsylla cheopis	*Hymenolepis nana* (Cestode) *H. diminuta* (Cestode)	

Insect vector	Pathogen	Disease
Nosopsyllus fasciatus	Pasteurella pestis (Bacterium)	Bubonic Plague (Black Death)
	Rickettsia typhi (Rickettsian)	Endemic (Murine) Typhus
	Hymenolepis diminuta (Cestode)	
Ctenocephalides canis	Dipylidium caninum (Cestode)	
	Hymenolepis nana (Cestode)	
Pulex irritans	Hymenolepis nana (Cestode)	

Section 2

Protelean Parasitic Insects

INSECTS IN WHICH only the immature stages are parasitic are termed protelean parasites. The larval period in an endopterygote insect's life cycle is predominantly a time of feeding and growth, and many have adopted the habit of protelean parasitism with free-living adults. Most often the host is another arthropod, usually an insect, but several Diptera utilise vertebrates as hosts for their larvae. These latter are endoparasites; it is difficult to imagine a dipterous maggot modified for a permanently ectoparasitic existence on a vertebrate, although as we shall see, a few Diptera have larvae which may be termed temporary ectoparasites in that they make periodic excursions from a resting place onto the body of their host for feeding, rather like adult mosquitoes and horse-flies. Likewise, those Diptera that have invertebrate hosts are also nearly always endoparasites, as are the majority of Hymenoptera. It is these two large orders that provide the great majority of examples of protelean parasitism. Their larvae are readily suited for endoparasitism, having become adapted in most cases to living in a mass of semi-liquid food. They have a more or less cylindrical body form without obstructing projections, a thin cuticle, few sense organs, and limited powers of locomotion. Some Hymenoptera are ectoparasites, but in such cases the host is nearly always living in a situation, like a plant gall or a leaf mine, in which it is completely enclosed by plant tissue.

Those protelean parasites that attack invertebrates nearly always destroy their hosts, and may be described as parasitoids to differentiate them from typical parasites which are much smaller than their hosts and adapted to inflict a minimum of damage. Approaches towards typical parasitic behaviour are, however, occasionally found among parasitic Hymenoptera. Some chalcid parasites of scale insects permit their hosts to lay eggs before killing them. More striking, some ichneumons (e.g. *Perilitus*) that parasitise adult beetles may complete their development and emerge from the host, leaving it in a condition from which it may recover to resume feeding and even egg-laying. In fact in the laboratory, some individual beetles have been able to support two generations of parasites (Timberlake 1916). In general, parasitoids behave in their early larval stages as typical parasites, and only when they grow larger does their feeding behaviour become predaceous.

Among exopterygote insects, protelean parasites with free-living adults are not known. When the immature stages are parasitic, as for instance in lice and some bugs, then so also are the adults.

8. Parasitic Hymenoptera

THE HYMENOPTERA ARE, biologically, the most varied of insect orders, and in number of described species they are second only to the Coleoptera. At a rough estimate there are probably about 200 000 species of Hymenoptera living today. In habits they range from the phytophagous sawflies with caterpillar-like larvae to the ants, bees, and wasps whose wonderfully organised societies are amongst the most remarkable products of invertebrate evolution. The majority of species, however, are protelean parasites of other insects.

Adult Hymenoptera stand apart morphologically from the other endopterygote insect orders. They are characterised by the possession of two pairs of membranous wings coupled together by a row of small hooks on the anterior edge of the hind wing, and by the fusion of the first abdominal segment (propodeum) to the thorax behind which (except in sawflies) is a narrow waist or petiole. The mouthparts are of the biting type, secondarily lengthened into a proboscis for drawing up nectar in the higher bees and masarid wasps. The females are equipped with an elongated ovipositor which in some groups Aculeata is used only as a sting. Metamorphosis is complete with a pupal stage that is either naked or enclosed in a cocoon. The larvae of sawflies are polypodous, usually quite active and sometimes brightly coloured. The larvae of the other groups are generally apodous, whitish, sedentary creatures living in the midst of food in an enclosed space. To prevent contamination of their environment, they do not pass faeces. The mid-gut is closed posteriorly and has no connection with the hind-gut. Waste matter is stored as urates, often visible in the larva as white spots in the fat body. The connection between mid- and hind-gut is established in the pupa, and the accumulated solid waste, the meconium, is expelled at the time of pupation.

Sex determination in Hymenoptera is unusual in that females are generally produced from fertilised eggs and are diploid, whilst males develop from unfertilised eggs (arrhenotokous parthenogenesis) and are haploid. This frequently results in a departure from an even primary sex ratio. Queen bees are able to control the fertilisation of their eggs and hence the sex of their brood, whilst among the parasitic Hymenoptera a number of factors operate to determine the extent of fertilisation and the primary sex ratio. These factors include the rate at which eggs are laid, which in turn may be a function of host density (Flanders 1946), the age of the female, and in some species at least, the structure of the female reproductive system (King 1961). In a number of sawflies and parasitic Hymenoptera of several families, males are exceedingly rare or absent altogether; in these species females develop parthenogenetically (thelytokous parthenogenesis).

There are three major taxonomic groups of Hymenoptera; the phytophagous sawflies (Symphyta), the Hymenoptera Parasitica, and the Hymenoptera Aculeata which includes ants and solitary and social bees and wasps. This classification gives a broad but not a precise indication of habits. In both Symphyta and Aculeata there are some parasitic groups, and phytophagous species (Cynipinae, some Chalcidoidea) occur in the Parasitica. Parasitic Hymenoptera are found in the following superfamilies:

        Suborder SYMPHYTA
          Superfamily Orussoidea
        Suborder APOCRITA
         Division Parasitica
          Superfamily Evanioidea
          Superfamily Trigonaloidea
          Superfamily Ichneumonoidea
          Superfamily Proctotrupoidea (= Serphoidea)
          Superfamily Chalcidoidea (many)
          Superfamily Cynipoidea (some)
         Division Aculeata
          Superfamily Bethyloidea
          Superfamily Vespoidea (some)

In other aculeate superfamilies there are social parasites, and these will be considered in a later chapter.

For additional reading, a very detailed account of parasitic Hymenoptera is included in C. P. Clausen's book *Entomophagous Insects* (1940), and Doutt (1959) and a paper by Hagen in De Bach (1964) give useful reviews.

The Ovipositor

A key component in the hymenopteran's equipment is the ovipositor. In parasitic species this has developed into an organ that is ideally constructed for placing eggs with precision into the host's body or into its concealed cell.

Paired abdominal appendages have but a transient existence in the embryos of most insects, with the exception of those at the posterior end of the abdomen. Here they are developed in the adult insect as gonopods in association with the reproductive apertures. A gonopod is envisaged as consisting originally of a basal coxa (coxopodite or gonocoxa) carrying a stylus externally and a gonapophysis internally (Figure 54a). In a female insect the gonopods

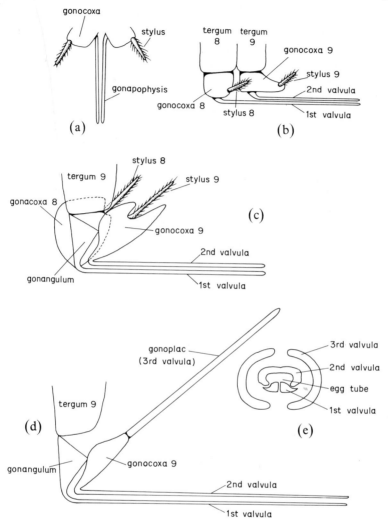

Figure 54. Diagrams to show the structure of the insect ovipositor. (a) Basic structure of a gonopod. (b) Female gonopods of *Petrobius* in lateral view. (c) Female gonopods of *Thermobia* in lateral view (adapted from Scudder 1957). (d) Hymenopteran ovipositor in lateral view (adapted from Scudder 1957). (e) Transverse section through a hymenopteran ovipositor.

are derived from abdominal segments 8 and 9.

The female external genitalia (ovipositor) of a pterygote insect consist of three pairs of valvulae, the first or anterior pair derived from the gonapophysis of segment 8, the second pair from the gonapophysis of segment 9, and the

third or posterior pair occupying approximately the position of the styli of
segment 9 but apparently representing new structures termed gonoplacs by
Scudder (1957). The styli of segment 8 have been lost.

The origin of the parts of the hymenopterous ovipositor is most readily
understood by considering first the ovipositor of the relatively unspecialised
Apterygota. Thus in *Petrobius* (Figure 54b) unmodified styli are present on
segments 8 and 9, and the ovipositor has only two valvulae. There is a similar
arrangement in the firebrat, *Thermobia* (Figure 54c), excepting that the first
valvulae do not articulate with the gonocoxae of segment 8 but with a sclerite,
the triangular plate or gonangulum, derived from the gonocoxa of segment 9
and itself articulating with tergum 9 (Scudder 1957).

From *Thermobia* to the condition in Hymenoptera (Figure 54d) is not a
great step. As in all winged insects possessing an ovipositor, there are three

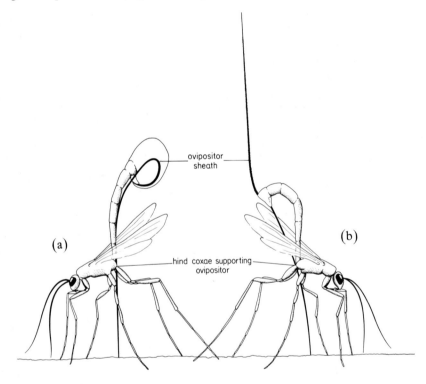

Figure 55. Ovipositing Ichneumonidae. In (a) both the hind
coxae and the ovipositor sheaths (third valvulae) support the
terebra during drilling, the sheaths gradually forming a loop
dorsally as the terebra penetrates the substrate (e.g. *Rhyssella*).
In (b) the hind coxae alone support the terebra, the sheaths
being held vertical after the initial stages of drilling (e.g.
Pseudorhyssa).

pairs of valvulae, the third pair (gonoplacs or ovipositor sheaths) replacing the styli of segment 9. There is no trace of the gonocoxae of segment 8 in Hymenoptera, although in some other orders the gonangulum and gonocoxa 8 fuse to form a composite sclerite (Scudder 1957). Gonangulum and gonocoxa 9 may be termed valvifers.

In Hymenoptera the second valvulae are fused, and together they form a sheath for the first valvulae. The second valvulae have longitudinal ridges which lock into complementary grooves on the first valvulae, an arrangement that permits the first valvulae to slide inside the second, the two pairs of valvulae behaving as a single functional unit. This unit is termed the terebra. The spatial relationship of the various parts can perhaps be better appreciated by reference to the diagram of a transverse section through a hymenopterous ovipositor (Figure 54e). The third valvulae, the ovipositor sheaths, are much broader than the terebra, less rigid, and covered on their outer faces with short hairs. Sense organs are situated at the apices of both the terebra and the ovipositor sheaths.

During oviposition, cutting ridges on the first and second valvulae enable the ovipositor to drill into the host's body or cell. The terebra may be supported by both the third valvulae and by the hind coxae, or by the latter alone (Figure 55). As well as depositing eggs, the ovipositor is frequently used to inject a paralysing toxin into the host. Also, it may secrete silk (in the chalcid *Eupelmus*) or be used to construct a feeding tube (page 137). The materials for these functions are provided by glands associated with the base of the ovipositor.

The ovipositor is long in most Hymenoptera. In Torymidae (Chalcidoidea) and some Ichneumonoidea it is a conspicuous structure which protrudes far beyond the apex of the abdomen. In many parasitic Hymenoptera however, a long ovipositor can be almost completely concealed inside the abdomen. Thus in Cynipoidea and many Chalcidoidea it is coiled inside the abdomen, which in consequence is deep and more or less keeled ventrally. An unusual device for accommodating the ovipositor inside the abdomen is found in *Inostemma* (Proctotrupoidea) (Figure 80a) which has a horn-like process curving from the front of the abdomen forwards over the thorax. The ovipositor can be retracted into this process.

Hyperparasitism

Hyperparasitism is the parasitism of a parasite. A hyperparasite may also be termed a secondary parasite. Hyperparasitism is quite frequent amongst the smaller parasitic Hymenoptera, and a few examples will be described. The American chalcid species *Perilampus hyalinus* (Perilampidae) lays its eggs upon foliage, and the first instar larva is of the active, planidium type (page 232 and Figure 109a). These planidia seek out and penetrate the bodies of caterpillars (e.g. *Hyphantria*). Inside the primary host they search for a larva of the primary parasite, *Ernestia* (Diptera, Tachinidae), which they in turn enter to complete their development. The relationships between these three species are thus:

Perilampus (hyperparasite or secondary parasite)
                  → *Ernestia* (primary parasite)
                       → *Hyphantria* (primary host)

Species of *Charips* (Cynipoidea) are also obligate hyperparasites, attacking larvae of the ichneumon *Aphidius* (Braconidae) feeding in aphids, but in *Charips* the adult female, and not the first instar larva, seeks out parasitised aphids. The female can apparently distinguish with her antennae whether or not an aphid contains *Aphidius*. The chalcid *Asaphes vulgaris* (Pteromalidae) is also an obligate hyperparasite of aphids through their primary aphidiine braconid parasites, but in this case the egg of the hyperparasite is laid on the surface of the braconid pupa after the braconid has completely consumed its aphid host. Further, *Asaphes* will feed upon *Aphidius* that are already parasitised by *Charips*, so that a rather complex web of potential relationships exists. This will be most readily understood by reference to Figure 56.

Asaphes, which attacks primary parasites that have already destroyed their primary host, is recognised as a hyperparasite. This being so, many of the chalcids which feed as larvae within galls of Cynipidae can be regarded as facultative hyperparasites. The food web centred upon the oak gall of *Cynips*

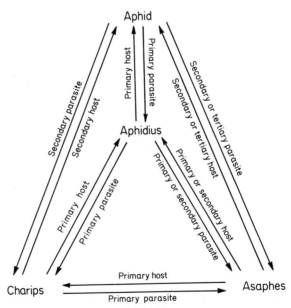

Figure 56. The interrelationships of some aphid parasites.

divisa (Figure 57) is of extraordinary complexity (Askew 1961b). With the exception of *Synergus, Syntomaspis,* and *Eudecatoma,* the rest of of the remaining species, all chalcids, are remarkably flexible in their dietary requirements and will apparently feed upon larvae of any species, including their own, in the gall. Thus there may be a series of occupants of a single gall, each feeding upon its predecessor, progressing long after the original gall wasp has been destroyed. The limit to such a series is imposed by the gradual diminution in size of the occupants, since at each stage in the series some food is

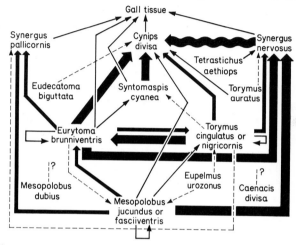

Figure 57. Food web of the inhabitants of the oak gall of *Cynips divisa*. The arrows point to the food source of each species and the thickness of the arrow indicates the frequency of the relationship. The sinuate arrow from *Synergus nervosus* indicates that it destroys but does not eat *C. divisa* (from Askew 1961b).

wasted and, of course, energy is continually being expended. Tertiary parasitism, however, is not infrequent, and parasitism of the fifth order has been recorded in this gall. It should be emphasised that such hyperparasitism is facultative; obligate hyperparasitism is virtually restricted to secondary parasitism, although Muesebeck and Dohanian (1927) mention a species of the chalcid genus *Pediobius* (= *Pleurotropis*) which may be an obligate tertiary parasite.

Multiparasitism

When two species of parasitic Hymenoptera lay their eggs on or in the same host individual, the almost invariable outcome is the destruction of one of the species by the other. If the victim were eaten by the survivor then it would be a case of hyperparasitism. However, when the two parasites oviposit about the same time, or are otherwise at similar stages of development, the survivor does not usually eat the other, and such simultaneous parasitism of an individual host by parasites of two or more species is termed multiparasitism or multiple parasitism (Smith 1916).

Ectophagous parasites may also be involved in a multiparasitic relationship since they do not, as a rule, discriminate between hosts which contain endo-parasites and those that do not. The endoparasite perishes through the destruction of its host, not as a result of direct attack by the ectoparasite. The relationship between *Asaphes* and *Charips*, referred to above, is of this type.

One of the most interesting examples of multiparasitism concerns the alder wood-wasp, *Xiphydria camelus*, and its two ichneumon parasites, *Rhyssella curvipes* and *Pseudorhyssa alpestris*. *Rhyssella* lays its eggs upon the wood-wasp grub and it may be followed in this by *Pseudorhyssa*. The first instar larva of *Pseudorhyssa* kills that of *Rhyssella*. This relationship is obligatory for *Pseudo-rhyssa*, not because its larva needs to feed upon a *Rhyssella* larva, but because it is unable to drill through wood with its own ovipositor to reach the wood-wasp larva (page 151), and it has to rely upon *Rhyssella* to bore a hole which it can later utilise. An essentially similar association exists between the chalcid *Eurytoma monemae*, the chrysidid *Chrysis shanghaiensis*, and the moth *Monema flavescens*, three Chinese species (Piel and Covillard 1933). *E. monemae* attacks the fully grown *Monema* caterpillar inside its very robust cocoon, and it is able to do so only by using the oviposition hole chewed in the cocoon by the *Chrysis*, a much larger insect.

A quite different example of what may be termed obligatory multiparasitism is provided by parasites of the fruit fly, *Dacus cucurbitae*. In this host the chalcid *Tetrastichus giffardianus* is able to avoid encapsulation only if the braconid *Opius fletcheri* has already parasitised it (page 176).

Obligatory multiparasitism is, however, exceptional. Further, multiparasitism tends to be avoided by many parasites since it involves possible destruction by another parasitic species. Fisher (1961, 1965) has made a detailed study of multiparasitism in the caterpillar of *Ephestia* by two endoparasitic Ichneumonidae, *Devorgilla* (= *Nemeritis*) *canescens* and *Diadegma chrysostictus*. Both parasites are solitary, only one individual being able to complete development in a host. Each is able to distinguish and lay fewer eggs in hosts previously parasitised by the other species. Fisher injected, by means of a micropipette, one egg of each ichneumonid, at the same time, into each of a series of *Ephestia* caterpillars. The eggs developed normally but, on hatching, the first instar larvae sooner or later (usually within a day) met and fought, one being eliminated. *Diadegma* was the victor in two-thirds of the encounters, and this was probably because of the rather longer incubation period of *Devorgilla*. In conflicts between first instar larvae of different ages, the older larva invariably won. On meeting, one larva sinks its mandibles, which are relatively very large in the first instar, into the other and holds on for up to thirty minutes. The wounded larva is not usually killed outright by this attack, but it is rendered susceptible to encapsulation by the host's haemocytes (page 174).

When there is a delay of three days or more between injections of the eggs of *Diadegma* and *Devorgilla*, the older parasite again survives, but it achieves sole occupation of the host by a different method. The younger parasite may fail to emerge from its egg, or if it does hatch no food is eaten and it gradually shrinks and dies, sometimes being encapsulated. There is no fighting. Instead, suppression of the younger parasite is physiological. By a series of experiments, Fisher was able to show that suppression could not be due to toxic secretions or specific inhibitors produced by the older larva, nor to food starvation. However, by increasing the oxygen to 50% in the atmosphere surrounding the parasitised host, he was able to keep the younger parasite alive. Not only this, the younger larva destroyed the older larva with its better developed mandibles. Therefore, by starving younger larvae of oxygen, older larvae are ensured of victory in such cases of multiparasitism.

Thus far only endoparasites have been considered. Ectoparasites engaged in multiparasitism can not destroy their competitors by oxygen starvation, both parasites being surrounded by atmospheric oxygen. The situation is in many ways analogous to Fisher's experiment in which the amount of available oxygen was increased, and very often the outcome is the same; survival of the younger parasite.

When there is no intrinsic superiority of one of the contestants, the result of competition through multiparasitism is usually decided at the time of oviposition. Among endoparasites the first one in the host survives; among ectoparasites the latest arrival generally completes its development.

Superparasitism

The parasitic Hymenoptera are commonly solitary, developing singly upon their hosts. Should more than one egg be laid upon the host, by one or more females of a solitary species, the supernumeraries perish. Or if, in a gregarious species, more larvae attempt to develop upon a host than can adequately be provided for, then some or all may die or the brood may produce undersized adults. This is superparasitism, and it is found among both endoparasites and ectoparasites.

The mechanics of suppression of supernumerary parasites are similar to those described under multiparasitism. Excess parasites in a solitary species are often destroyed by fighting amongst themselves. If two or more first instar *Aphidius* (braconid) larvae meet inside an aphid they attack each other with their relatively large mandibles and only one survives. If, however, an egg is laid in a host that already contains an advanced *Aphidius* larva, then the younger of the two parasites dies even though there is apparently no fighting. In this case death of the younger larva is probably due to oxygen starvation (Stary 1966).

A different method of suppression of supernumeraries is quoted in Clausen (1940). It concerns an *Elachertus* (Chalcidoidea) species attacking caterpillars of *Artona*. The parasite is ectophagous and solitary. 'When more than one egg is deposited on a host, the hatching of one causes immediate cessation of development of the remainder.' It is suggested that the young larva may produce a toxin which poisons its fellows; oxygen starvation is very unlikely to afflict an ectoparasite.

Some species of *Macrocentrus* (Braconidae) are polyembryonic (page 139), a single egg dividing to produce a number of cell masses that each develop into larvae which live gregariously within the host. The egg of *M. ancylivorus* divides to form a number of potential embryos, but only one completes its development; as soon as one becomes a first instar larva the development of the others stops (Daniel 1932).

One would expect superparasitism to be avoided, the only benefit conferred by the habit being a possible reduction in the likelihood of encapsulation by the host (page 176). There is, in fact, a great deal of evidence to show that parasitic Hymenoptera belonging to several families tend to avoid superparasitism. Much of the evidence amounts to the non-random distribution of parasite eggs in available hosts. Stoy (in Salt 1932) provides a formula for the random distribution of x parasites in N hosts:

$$y = N^x - Cp \left[\frac{1}{N} \right]^p \left[1 - \frac{1}{N} \right]^{x-p}$$

where y is the number of hosts containing p parasites. This formula is derived from a binomial distribution. A binomial distribution would be approached when many parasites are distributed at random over a limited number of hosts, but when the rate of parasitism is not too high a Poisson series, which is probably more easily calculated, may be used to obtain values for a random distribution. The probability of the occurrence of 0, 1, 2, 3, etc. parasites per host is then:

$$e^{-z}, ze^{-z}, \frac{z^2 e^{-z}}{2!}, \frac{z^3 e^{-z}}{3!} \quad etc.$$

where $e = 2.72$ (natural logarithm base), and

$$z = \frac{\text{total number of parasites}}{\text{total number of hosts}}$$

If the actual distribution of parasite eggs differs significantly from the calculated random distribution in the direction of more hosts than expected supporting only one parasite, and fewer than expected remaining unparasitised, then it can be said that the parasite exercises discrimination (Figure 58).

How do searching female parasites distinguish between parasitised and non-parasitised hosts? Salt (1934, 1937a) made a detailed investigation of *Trichogramma evanescens* as a solitary egg parasite of the moth *Sitotroga cerealella*, and found that the female chalcids could recognise when an egg had been merely walked on by another female. If such contaminated eggs are washed, *Trichogramma* will insert its ovipositor into them but it usually refrains from laying an egg. Parasitised hosts may be detected, therefore, in an initial examination of their surface by the antennae of *Trichogramma* or, if this fails to indicate parasitisation, then insertion of the ovipositor provides a

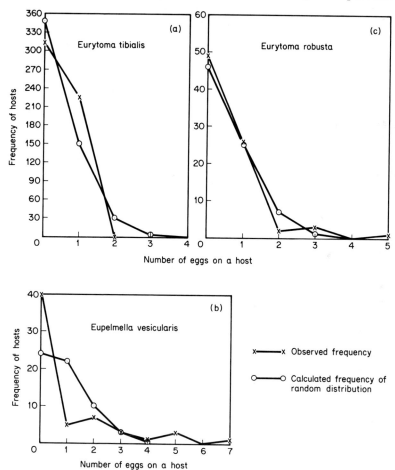

Figure 58. The distribution of eggs of three chalcid parasites of the knapweed gall fly (from Varley 1941). (a) *Eurytoma tibialis* (= *curta*) with non-random distribution and discrimination against already parasitised hosts. (b) *Eupelmella vesicularis* showing non-random, aggregated distribution. (c) *Eurytoma robusta* showing an almost random distribution.

second check. Salt (1937a) considers that ovipositing female *T. evanescens* may deposit an odoriferous secretion from their tarsal glands when they walk on a host. When hosts are scarce, the restraint of ovipositing females may break down and previously parasitised hosts be again attacked. Under these conditions it is the larger hosts that are most readily superparasitised.

Jackson (1966) demonstrated that the restraint exercised by *Caraphractus cinctus* (Chalcidoidea, Mymaridae) in avoiding superparasitism of water-beetle eggs was best developed in older females. Inexperienced females would oviposit in recently parasitised eggs. This species is a gregarious parasite, and is apparently able to estimate with its antennae the size of a host egg and to regulate accordingly the number of eggs laid. Only one egg is laid in the small egg of *Agabus*, two or three in the slightly larger egg of *Ilybius*, and several in the still larger egg of *Dytiscus*. Jackson discovered a gland and associated reservoir at the base of the ovipositor of *Caraphractus*, and postulated that this may be the source of a repellent secretion deposited at oviposition. *Caraphractus* is able to discriminate between parasitised and unparasitised hosts only after insertion of the ovipositor.

When a female *Trissolcus* (=*Asolcus, Microphanurus*) *basalis* (Proctotrupoidea, Scelionidae) attacks a batch of shield bug eggs, it marks each egg in which it has laid by drawing the tip of its abdomen across the cap of each egg in a figure-eight movement (Figure 59). This scratches the surface of the parasitised egg and serves as a warning that the egg has already been parasitised. This habit has been reported in several other Scelionidae and is probably widespread in the family (Safavi 1968).

In those instances where a host is detected as previously parasitised only

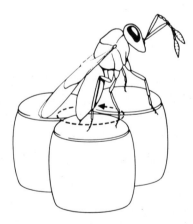

Figure 59. A female *Trissolcus basalis* (Proctotrupoidea, Scelionidae) marking an egg in which it has just oviposited. The tip of the ovipositor is being drawn across the cap of the egg.

after penetration, it is evidently sense organs on the ovipositor that are responsible. Such sense organs on the ovipositors of Chalcidoidea have been described by Fulton (1933) in *Habrocytus cerealellae* and by Varley (1941) in *Eurytoma tibialis* (=*curta*). Varley writes that 'the ovipositors of various other parasitic Hymenoptera have been examined, and all have sense organs of some kind near to the tip and also along the ovipositor itself.'

Some parasites, such as *Eurytoma tibialis*, which attack hosts buried deep within plant tissue, are unable to examine the surfaces of their hosts for traces left by previous parasites and they may, like *Caraphractus*, rely entirely on their ovipositors to discriminate between parasitised and healthy hosts. This is not always the case, however, for *Ibalia leucospoides*, a cynipoid parasite of larvae of the wood-wasp, *Sirex cyaneus*, can detect traces of previously ovipositing females on the surface of the wood (Salt 1937a).

Superparasitism, generally detrimental to a solitary parasite, is avoided by many species. Nevertheless, of the parasites of the knapweed gall fly (*Urophora jaceana*) investigated by Varley (1941), only the endoparasitic *Eurytoma tibialis* exercises discrimination against superparasitism (Figure 58a) whilst four ectoparasitic chalcid species either distribute their eggs randomly (Figure 58c) or even in an aggregated manner (Figure 58b). Varley points out that super-parasitism is only detrimental if the parasite's eggs so wasted might have been laid on unparasitised hosts, and it is really the ability to find hosts, rather than egg supply, which limits the increase in numbers of a parasite.

The ability to discriminate against previously parasitised hosts has not so far been recognised among parasitic Diptera.

Orussoidea

This is a relic group represented by a few species scattered over most regions of the world. Orussoids are classified with the sawflies very largely because the abdomen is broadly attached to the thorax, lacking a waist or petiole, but in other respects they differ markedly from sawflies. Rohwer and Cushman (1917), who are responsible for one of the few detailed studies of these insects, preferred to erect a new suborder Idiogastra for their reception, placing it between the suborders Symphyta and Apocrita. All sawflies except orussoids are phytophagous. Two American species of *Orussus* are known to be solitary endoparasites of larvae of beetles of the family Buprestidae. Morphologically also orussoids stand apart from other Symphyta and approach the Hymenoptera Parasitica in the reduction in number of their wing veins, the site of insertion of their antennae (resembling Stephanidae), and in the structure of the male genitalia. The ovipositor in the female is unique in form, being concealed within the body in an integumentary membrane that extends forwards into the thorax. Curiously, the female pupa has a visible ovipositor lying along the back and reaching to the head. The white, legless larva is very like the type of larva characteristic of the Apocrita, but this is most likely to be a result of convergent evolution. Nevertheless it has been suggested that at least some of the groups of Parasitica had an *Orussus*-like ancestor.

Evanioidea

The Evanioidea includes a few small and poorly known families. The Evaniidae live as larvae inside the egg cases or oothecae of cockroaches. An egg is laid within a host egg, but the larva soon becomes an ectoparasite and may consume other eggs in the ootheca. Only one parasite can develop in each ootheca. The adult evaniid is of unusual form with a minute, triangular abdomen attached to the thorax by a dorsally inserted petiole (Figure 60), an appearance that gives them their descriptive appellation 'ensign flies' in North America.

The Aulacidae are mostly parasites of wood-boring beetle larvae, although a few attack wood-wasps. *Aulacus striatus* (Figure 61a) lays its eggs in the egg of the alder wood-wasp, *Xiphydria camelus*, gaining access to the host by ovipositing through the hole in timber made by the egg-laying wood-wasp. Its larva emerges from the fully grown *Xiphydria* larva.

The Gasteruptiidae are parasites of solitary bees and wasps; species of *Gasteruption* (Figure 61b) have been reared as parasites of the bees *Osmia* and

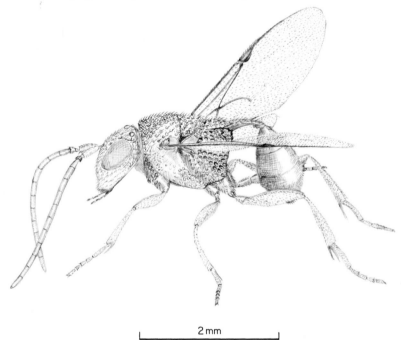

2 mm

Figure 60. Male *Brachygaster minuta* (Hymenoptera, Evanioidea), the only native British species of evaniid. It parasitises cockroaches (*Ectobius*) and is not uncommon on parts of the Dorset coast.

Prosopis, and of the wasps *Odynerus* and *Trypoxylon*. They are not usually host-specific. The egg is laid upon the host larva, or even upon its egg, and when this has been consumed the food store in the cell is eaten. Sometimes the parasite may break into an adjoining cell to eat a second host, behaviour more akin to a predator than a parasite. The Gasteruptiidae are especially well represented in Australia.

The few British species of Evanioidea have been monographed by Crosskey (1951).

The superfamily may be polyphyletic, although the Gasteruptiidae and Aulacidae appear to be quite closely related. The larvae of Evaniidae and Gasteruptiidae are very similar to those of aculeate Hymenoptera (Short 1952), and Crosskey (1962), listing other aculeate features of the Gasteruptiidae, concludes that there is little except their biology to link them with other Parasitica.

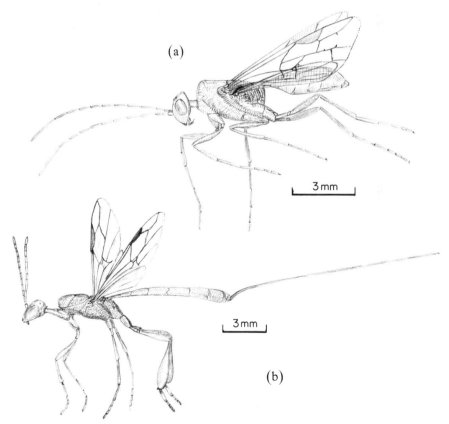

(a)

3 mm

3 mm

(b)

Figure 61. Hymenoptera, Evanioidea. (a) Male *Aulacus striatus*. (b) Female *Gasteruption jaculator*.

Trigonaloidea

Like the Orussoidea, this superfamily is represented by a single family, Trigonalidae, containing very few species and having a scattered but almost world-wide distribution. It also has in common with the Orussoidea the fact that its taxonomic affinities are obscure, although the majority view is that it is not too distantly removed from the Ichneumonoidea. However, affinities with the Vespoidea, an aculeate group, are suggested by the wing venation and structure of the reduced ovipositor.

Trigonalids are mostly either solitary endoparasites of the larvae of social wasps (Vespoidea), or they are hyperparasites of ichneumons and tachinid flies in caterpillars of Lepidoptera and sawflies. Each female trigonalid can lay several thousand eggs, and these are placed upon vegetation. The eggs, which remain viable for several months, do not hatch until they are eaten by a caterpillar, and the first instar larvae penetrate the intestinal epithelium and enter the haemocoel where they may encounter and enter a primary parasite. A similar route to the host is followed by some tachinid Diptera (page 205). Still unknown is the manner in which the first instar larvae of those trigonalids that parasitise wasps attain their hosts. One can only suppose that it is through wasp grubs being fed upon fragments of caterpillars containing trigonalid eggs.

Chalcidoidea

These very small insects, usually about two or three millimetres in length and coloured green or black with metallic reflections, abound in almost all terrestrial environments—'the green myriads in the peopled grass . . .' of Francis Walker. The number of described species is somewhere in the order of 25 000, and in the British Isles there are probably about 1500 species. Many await description, especially from tropical regions. The smallest of all insects, the Mymaridae (Figure 63) or fairy-flies, are included, albeit rather uneasily, with the chalcids. Mymarids are sometimes as small as 0.2 millimetres long, smaller than some Protozoa.

Though small, most chalcids have a robust form. The head is relatively large and broad, housing powerful mandibular muscles, and it is equipped with large compound eyes and a pair of 'elbowed' antennae which vibrate and test the air or substrate about the insect. They are active insects, and often jump several centimetres on being disturbed, frequently aided in this by their wings. The wing venation is very much reduced, but even so, most chalcids are capable aviators, being able to hover delicately about foliage provided that the air is reasonably still. A few species are apterous or brachypterous, sometimes in the female sex only. The body of a chalcid is furnished with a thick cuticle, generally reticulately sculptured and often with very beautiful structural colours.

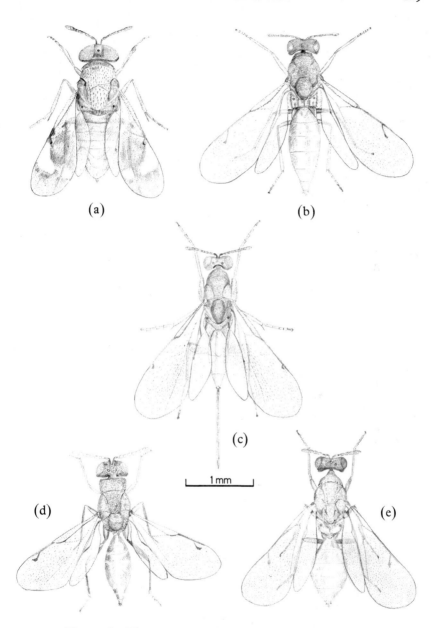

Figure 62. Hymenoptera, Chalcidoidea. (a) Female *Microterys tesselatus* (Encyrtidae). (b) Female *Mesopolobus jucundus* (Pteromalidae). (c) Female *Torymus auratus* (Torymidae). (d) Female *Eurytoma rosae* (Eurytomidae). (e) Female *Olynx arsames* (Eulophidae).

The female's ovipositor in most cases is concealed beneath a pear-shaped gaster, but in the Torymidae (Figure 62c) it is mostly exposed and in *Syntomaspis apicalis*, a common British species that parasitises larvae of *Biorhiza pallida* deep inside oak-apple galls, the length of the exposed part of the ovipositor is more than twice as long as the rest of the body. The ovipositor is used to penetrate tough layers of vegetable tissue or insect cuticle, but chalcids are handicapped by their small size so that deeply buried hosts generally remain inaccessible to them. However *Xiphydriophagus meyerinckii* (Pteromalidae), a parasite of alder wood-wasp grubs deep inside alder wood, reaches

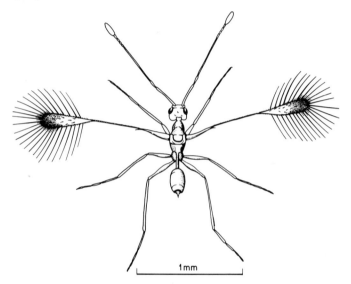

Figure 63. A fairy-fly, *Mymar pulchellus* female (Chalcidoidea, Mymaridae). This is one of the larger species in the family.

its host by virtue of its small size. The female *Xiphydriophagus* crawls deep into an alder trunk by way of old emergence tunnels of its host and then, by chewing its way through the intervening wood, it reaches a wood-wasp larva in an adjacent gallery (Thompson and Skinner 1961). *Rhyssella curvipes*, an ichneumonid parasite of the alder wood-wasp, must force its ovipositor through two centimetres or so of wood in order to attain the same objective.

There are sense organs on the tip of the ovipositor, presumably for assessing the suitability of the host, but the sense organs involved in host-finding are located on the antennae. The antennae of chalcids are sexually dimorphic. Those of the male are usually a little longer and more slender than those of the female, and in some Eulophidae and a few genera in other families they are equipped with long side branches.

The male genitalia of chalcids are fairly uniform in structure, which is disappointing for the taxonomist, but secondary sexual characters are occasionally developed. Notable amongst these are the decorations of the middle tibiae of some male *Mesopolobus* (Pteromalidae) species parasitic on cynipid galls on oak. These adornments consist of stripes of orange and sometimes brown, and a small, black apical flange or spot. *M. fasciiventris* has a flange but no stripes, *M. fuscipes* has the stripes but no flange, and *M. xanthocerus* and *M. tibialis* both have the stripes while the former has a flange and the latter a spot. During courtship these males pass their middle legs across the eyes of their partners, and the decorations evidently serve a sexual function. The male of *M. diffinis*, a parasite of gall-making Cecidomyiidae on herbaceous plants, has unadorned tibiae but unusual smoky spots on its hindwings. During courtship these hindwings are held downwards over the female's eyes, a position never adopted by other males in the genus, all of which have clear hindwings.

Not all chalcids are parasitic. The family Agaonidae includes the curious fig insects which inhabit the seeds of figs and are essential to the plant for its successful pollination. A large number of Eurytomidae and some Torymidae (most *Megastigmus*, some *Syntomaspis*, and the fig-wasp subfamily Sycophaginae) are also phytophagous as larvae, as are the Brachyscelidiphaginae (Pteromalidae) which induce the growth of quite elaborate plant galls. Some species of *Eurytoma* (Figure 62d) as larvae feed first on their insect hosts, but later they may turn to vegetable tissue. Larvae that subsist on plant tissue are generally furnished with robust mandibles bearing subapical teeth. The typical chalcid larval mandible is a simple, pointed structure lacking subsidiary teeth, but in the Eurytomidae it is bidentate and in *Megastigmus* there is a series of teeth. Since it is unlikely that these groups are primitively phytophagous, like sawflies, the teeth must be new acquisitions. Although the direct origin of chalcids from a phytophagous ancestor is only a very remote possibility, it is by no means certain that the group as a whole is monophyletic.

The host relationships of chalcids are extremely varied. The orders Lepidoptera, Diptera, Coleoptera, Hymenoptera, and Hemiptera provide the majority of host species, though few orders entirely escape chalcid attack, and a few species are parasites of arachnids (ticks, mites, spiders' egg cocoons). Chalcids attack all host developmental stages although adults are least often attacked and larvae most often. Each chalcid species, however, is usually restricted to attacking a particular host stage. Again, chalcids may be endoparasites or ectoparasites, solitary or gregarious, in their development. It is sometimes possible to clearly define the host range of a chalcid group. For instance, all Eucharitidae attack only ants and all Leucospidae seem to be parasites of solitary bees. Mymaridae and Trichogrammatidae are parasitic in the eggs of other insects. However most groups, at least above the generic level, defy precise statements regarding their host associations, and it is possible to make only generalisations which can usually be eroded by numerous exceptions. Of the three largest families of chalcids, the Encyrtidae (Figure 62a) are most often parasites of Hemiptera Homoptera, especially scale insects (but some attack Lepidoptera, Coleoptera, or ticks), the Eulophidae (Figure 62e) typically parasitise caterpillars of Lepidoptera (though many other insect host groups and spiders' eggs are used), and the Pteromalidae (Figure 62b) are very often parasites of the larvae of Diptera and Coleoptera (though again many insect groups from Homoptera to Siphonaptera figure amongst their hosts).*

At the species level it is rather rare to find strict host-specificity, although a strong preference for a particular host species is often apparent. Most chalcids have a host range which may be defined in terms of the systematic relationship of the hosts, although sometimes it is easier to define the host environment. For example, *Elachertus olivaceus* (Eulophidae) attacks only species of the moth genus *Coleophora* that feed on rushes, and *Olynx* (Eulophidae) species (Figure 62e) parasitise only Cynipidae that form galls on oak. Many similar examples could be cited. However, in contrast, *Sympiesis sericeicornis* (Eulophidae) is a parasite of leaf-miners on deciduous trees and attacks hosts belonging to several different lepidopterous families as well as some Hymenoptera. The host ranges of many of the chalcids that parasitise the inhabitants of cynipid oak-galls are governed by the gall's position (e.g. leaf, root, bud), form, and season rather than by its contents (Askew 1961b). In a few extreme cases it is difficult to find any affinity, either systematic or ecological, between the hosts of a species. *Eupelmus urozonus* (Eupelmidae), *Dibrachys cavus* (Pteromalidae), and *Cirrospilus vittatus* (Eulophidae) have apparently large and diverse host ranges embracing several orders of insects and including hosts of varied habits and varying stages of development. Such species should be closely investigated; what is thought to be a single species may instead be an aggregate of sibling species, each with a discrete host range, in which biological differences have not been accompanied by morphological distinctions. *Eurytoma rosae* (Eurytomidae) (Figure 62d) was at one time thought to inhabit as a larva cynipid galls on oak, rose, maple, and various Compositae. Adults from different plant sources are virtually indistinguishable in their external morphology, but when the eggs were examined it was found that strikingly clear-cut differences exist (Figure 64) (Claridge and Askew 1960).

Whilst most chalcids attack their hosts in the larval or pupal stages, some oviposit in eggs and a few even in adults. Development may proceed either inside the host (endoparasites, endophagous) or on the surface of the host (ectoparasites, ectophagous), the former being the more specialised condition since it involves adaptations to life in the very exacting internal environment of a host. Sometimes the same genus may contain both endophagous and ectophagous species, and for an illustration of biological diversity within a genus, one can do no better than consider *Eurytoma* (see also Figure 65). *Eurytoma serratulae* and *E. tibialis* are endoparasitic in larvae of trypetid Diptera

* An account, in tabular form, of the host associations of the families of chalcids and other parasitic Hymenoptera is included in the Royal Entomological Society of London *Handbooks for the Identification of British Insects*, Vol. VI, Part 1, by O. W. Richards

(*Urophora* species) in flower-heads (*E. tibialis*) and stems (*E. serratulae*) of Compositae whilst *E. robusta* is an ectoparasite of the same hosts (Claridge 1961). *E. rosae* has been described as a predator since it chews its way from cell to cell in the bedeguar rose gall of *Diplolepis rosae*, often consuming two or more larvae of the inquiline cynipid *Periclistus brandti* during the course of its development. Likewise, *E. oophagus* devours most of the eggs in cocoons of its spider host.

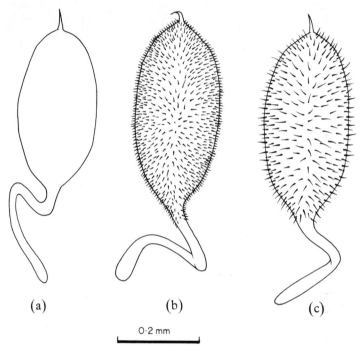

(a)   (b)   (c)

0·2 mm

Figure 64. Eggs of three species of *Eurytoma* (Chalcidoidea, Eurytomidae). The adult insects are scarcely, if at all, distinguishable on morphological characteristics. (a) *Eurytoma rosae.* (b) *E. brunniventris.* (c) *E. centaureae* (from Claridge and Askew 1960).

Another species which attacks eggs is *Tetrastichus mandanis* (Eulophidae). This starts its larval life as an endoparasite in the egg of the *Juncus* delphacid, *Conomelus anceps*, but emerges from the egg in its second instar, after having completely consumed the contents, and becomes a predator, sometimes burrowing through one centimetre of rush stem in its quest for more eggs of the delphacid or other Homoptera (Rothschild 1966).

An extreme case of diversity of habit is found in species of the family Aphelinidae, parasitic on scale insects and other Hemiptera Homoptera, where certain species show a sexual differentiation in their host relationships.

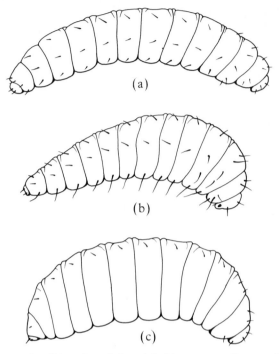

Figure 65. Diversity of larval habits among Eurytomidae (Chalcidoidea) has led to a variation in the morphology of the fully grown larvae. (a) *Tetramesa longicornis*, phytophagous in grass stems. (b) *Eurytoma rosae*, an ectoparasite in rose galls. (c) *Eudecatoma biguttata*, an endoparasite of oak gall-wasp larvae.

A simple example of this is provided by *Encarsia formosa*, a parasite of the greenhouse white-fly. Males of *E. formosa* are rare, but when they are produced they develop, not as primary parasites of the white fly like their sisters, but as internal hyperparasites of the larval females of their own species (autoparasitism). Among aphelinids which regularly produce arrhenotokous males, the divergent ontogenies of the two sexes may be more striking, and within the genus *Coccophagus*, parasites of scale insects, three distinct categories can be recognised (Flanders 1959, Zinna 1961):

(i) Male has same scale insect host species as female. The female is endophagous, the male ectophagous, and both sexes are primary parasites. e.g. *C. ochraceus, C. bivittatus, C. longifasciatus*

(ii) Male has same scale insect host species as female. Both sexes are endophagous, the female being a primary parasite and the male a hyperparasite through larval females of its own species (autoparasitism or obligate adelphoparasitism).

e.g. *C. scutellaris, C. capensis, C. nigritus*

(iii) Male develops as a hyperparasite, endophagous or ectophagous, of a primary parasite. Sometimes the male is hyperparasitic through its own female but usually it develops in a different host scale insect species (facultative adelphoparasitism).

e.g. *C. basalis, C. gurneyi, C. caridei, C. insidiator, C. cowperi, C. lycimnia*

In the genus *Prospaltella* an even greater disparity between male and female ontogenies has been discovered in some American species. The female has a scale insect host but the male is endoparasitic in lepidopterous eggs.

In *Encarsia formosa* there is no difference in the site of deposition of diploid and haploid eggs, the oviposition behaviour of the female remaining the same before and after mating. The same can be said for *Coccophagus basalis* and *C. gurneyi* in category (iii), both types of egg being laid in the sub-oesophageal ganglia of coccids. A haploid egg, however, does not hatch until a female larva has developed in the scale to provide the male larva with the food it requires. In *C. gurneyi* the male often develops on a female larva of its own species. The male of *C. basalis*, however, uses a female larva of a different species, since its own female does not completely consume the scale and the liquid environment is unsuitable for the male larva which respires through spiracles. The female larva breathes cutaneously.

In the remaining species listed in category (iii), and in those in category (ii), the egg-laying habits of the females are changed irreversibly with mating. Before mating, the female lays her haploid eggs either inside a primary parasite (category (ii) species) or on the surface of a fully grown primary parasite (*C. insidiator, C. caridei, C. cowperi, C. lycimnia*). The male larvae of category (ii) species lack open spiracles. After mating, the females of these species deposit their eggs as primary parasites and any haploid eggs among them fail to develop since they are in an unsuitable environment for male development. Likewise, the *Prospaltella* species, whose males develop as primary parasites of Lepidoptera eggs, change their oviposition habits after mating, when they attack the homopterous hosts of the females. Any male eggs laid in a homopteran host perish.

The category (i) species, such as *C. ochraceus*, use the same host species for male and female development. The male is not a hyperparasite through its female, but develops externally on the scale, whereas the female develops internally. Haploid and diploid eggs are laid in different places, but the oviposition site does not change with mating and is reversible, unlike the change in oviposition habit of the species in categories (ii) and (iii) (bar *C. basalis* and *C. gurneyi*). When the egg-laying female has a spermatheca full of sperm and the rate of egg-laying is relatively slow, all the eggs that are laid are diploid, and are placed inside the scale to produce females. Male eggs are laid when the egg-laying rate is rapid, as when hosts are plentiful, and these are placed externally. When the spermatheca is empty, all eggs laid are haploid, and they too are deposited on the surface of the scales to produce males. Thus the condition of the spermatheca affects the oviposition behaviour of the female.

Zinna (1962) regarded autoparasitism as a self-regulatory device for maintaining the population density at a suitable level. An autoparasitic aphelinid provides its own 'natural enemy' to prevent it from over-exploiting its scale insect host. The case of the *Prospaltella* species, whose males parasitise Lepidoptera, is seen by Zinna as a further attempt to reduce pressure upon the scale insect host population. Flanders (1967) considers that *C. ochraceus*, a species whose sexes are both primary parasites, evolved from an ancestral form that was autoparasitic. Perhaps in this case pressure upon the host could be increased. But these are speculations and the biological significance of autoparasitism remains obscure. The subject is one of great complexity and it has been possible here to present only a few details, these being largely confined to the one genus, *Coccophagus*. For a fuller account of disparate male and female development in Aphelinidae, the review of Flanders (1967), from which most of the data given here have been taken, is recommended.

Chalcids that attack larvae frequently paralyse their victim prior to oviposition. This is achieved by an injection through the ovipositor. Sometimes several thrusts of the ovipositor are required to effect complete paralysis. Host paralysis is the general rule for ectophagous chalcids, and its principal purpose would appear to be to eliminate movements of the host which might crush or dislodge the delicate chalcid egg or young larva. Although paralysed, the host continues to live for some time until killed by the feeding of the parasite, and its body does not therefore decay. Bee larvae stung by *Melittobia acasta* (Eulophidae) stay 'fresh' for up to nine months. Endophagous chalcids do not usually paralyse the host since their eggs and larvae are protected inside it. Some Eupelmidae, which are ectoparasites, do not paralyse their hosts either; instead they protect their eggs by spinning a silken web over them as they lie beside the host.

When the ovipositor is withdrawn after stinging or ovipositing, a drop of the host's body fluid exudes through the puncture, and chalcids of many families have been observed to feed upon this. In fact several species may sting a host apparently with the sole object of obtaining a meal, for oviposition does not always follow. *Metaphycus helvolus* feeds and oviposits on different individual scale insects because feeding renders the host unsuitable for oviposition. *Tetrastichus asparagi* (Eulophidae) and some other species which attack eggs or young nymphs or larvae may destroy more hosts by adult feeding than by parasitism.

An elaborate method of obtaining the host's body fluids has been developed by several species of Pteromalidae, Eurytomidae, Torymidae, Eupelmidae, and Eulophidae that are separated from the host they are attacking by a

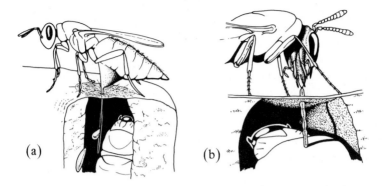

(a)   (b)

Figure 66. Female *Habrocytus cerealellae* (Chalcidoidea, Pteromalidae) ovipositing on (a) a grain moth caterpillar and, (b) feeding at the feeding tube constructed with its ovipositor (from Fulton 1933).

thickness of plant tissue or host cell. In these species a special feeding tube is formed, a process described in detail by Fulton (1933) from observations made on *Habrocytus cerealellae* (Pteromalidae). This species is a parasite of caterpillars of the grain moth *Sitotroga*. The female chalcid, after paralysing the caterpillar in its cell in a cereal grain, withdraws its ovipositor until only the extreme tip protrudes into the cell. A viscous fluid begins to ooze from the ovipositor tip, and this fluid is caused to form a thin layer around the ovipositor by twisting and vertical movements of the latter. The ovipositor is slowly worked back into the host cell until the paralysed host is reached, when the tip is inserted again into the original puncture. The fluid about the ovipositor partially hardens, and after a few minutes the ovipositor is slowly withdrawn. Body fluids from the host pass by capillarity up the feeding tube thus formed, and the parasite feeds upon them at the surface (Figure 66). It has been shown that this source of food may be essential for maturation of the eggs. If *Nasonia* (=*Mormoniella*) *vitripennis* (Pteromalidae) females are prevented from feeding at their hosts (blowfly pupae in puparia), they stop producing eggs and those already formed in the ovaries are resorbed. Host feeding by adult chalcids is an excellent example of insect economy in time and effort; a period of host-searching may yield food for both the adult and its progeny. *Nasonia* requires no other source of food than that provided by its host, and in this it is comparable with the Diptera Pupipara, and perhaps with those fleas which supply their larvae with blood via their faeces (page 37). *N. vitripennis* is a species very easily cultured in the laboratory, and consequently has been the subject of much research into its physiology, genetics, and behaviour (Whiting 1967).

Not all adult chalcids, however, feed upon their hosts, and some probably receive all necessary nutriments from feeding on nectar or honey-dew.

Flowers of Umbelliferae are especially attractive to adult chalcids. Exception-ally, more solid plant food may be taken; adults of a *Pediobius* (Eulophidae) species have been seen to consume strips of epidermis torn from leaves.

In ectophagous chalcids, the egg is laid either on the body of the host or else very near to its paralysed body. *Olynx euedoreschus* (Eulophidae) nearly always lays its egg on the ventral surface of the body of its gall wasp host. On the other hand, *Systasis dasyneurae* (Pteromalidae), a parasite of cecidomyiid larvae in buds of linseed, lays it eggs in the buds but often at some distance from the host, so that the first instar larvae have to move around in search of their hosts. This type of behaviour recalls that of the planidium larvae of Eucharitidae and Perilampidae. Species in these two families are exceptional in ovipositing upon vegetation and leaving it to their active first instar plani-dium larvae (Figure 109a) to locate their hosts. *Perilampus chrysopae*, a larval parasite of lacewings, lays its eggs on foliage infested with aphids. Aphids are the prey of its host, and by ovipositing in such situations *P. chrysopae* increases the likelihood of its larva finding a host. *P. chrysopae*, however, is atypical in that most species of Perilampidae are hyperparasites, usually of caterpillars through braconids, ichneumonids, or tachinids. If the planidium larva of such a species enters an unparasitised caterpillar, it fails to develop. Eucharitidae are parasites of ant pupae. The very numerous eggs are often laid on flower buds, and the planidia attach themselves to worker ants which transport the planidia to their nests. Inside an ants' nest, the planidia leave the worker ant and fasten themselves to ant larvae. When these larvae pupate, they are eaten by the eucharitids.

Endoparasitic chalcids usually lay their eggs in the haemocoel of the host, but a few are more precise, always selecting a particular site in the host's body. *Microterys clauseni* (Encyrtidae), for example, oviposits in the intestine and *Coccophagus basalis* in the suboesophageal ganglion of their scale insect hosts.

The mechanics of oviposition are interesting. The ovipositor is often tightly enclosed within the hole it has drilled through vegetable tissue or hard cuticle. It is therefore incapable of sufficient expansion to allow the egg, which is of much greater diameter than the egg passage of the ovipositor, to pass down. It is the egg which must be modified in shape. Generally the chalcid egg is ellipsoidal, and the narrower end passes down the ovipositor first, the whole egg being constricted and lengthened. When the egg apex emerges from the ovipositor tip, the contents are forced into it and this helps to pull the remain-der of the egg through the ovipositor. The eggs of Eupelmidae, Eurytomidae (Figure 64), *Megastigmus*, Aphelinidae, and some other groups are provided with long stalks which serve as reservoirs for the egg contents whilst the body of the egg is being squeezed down the ovipositor. These changes in egg shape are permitted by the elastic properties of the chorion.

Egg stalks (pedicels) sometimes serve a different function. In some Encyrti-dae, for example *Blastothrix*, the young metapneustic larva inside the host retains its posterior end inside the remains of the egg shell, the stalk of which protrudes through the host's body wall to maintain contact with the atmosphere (Figure 67). Air reaches the young chalcid larva through the lumen of the egg stalk. In other encyrtids, for instance *Ooencyrtus* and *Microterys*, the egg stalk is plugged distally, and the larvae rely upon air diffusing along an aeroscopic plate or rib which runs the length of the egg stalk and well onto the body of the egg.

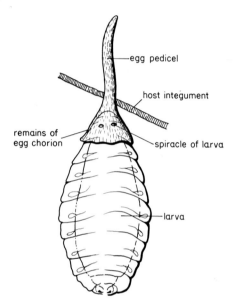

egg pedicel

host integument

remains of
egg chorion

spiracle of larva

larva

Figure 67. First instar larva of *Blastothix* (Chalcidoidea, Encyrtidae) with posterior part of body, which bears the spiracles, embedded in remains of its egg, the pedicel of the egg maintaining contact with the atmosphere through the host's integument (after Imms 1931).

The number of parasites developing on a host depends very much upon the size of the host. Small leaf-mining microlepidopterous caterpillars usually support only a single chalcid, whilst chalcids attacking larger noctuid caterpillars are generally gregarious, often twenty or more developing on each host. However, the record for large numbers of individuals developing together on a single host is indisputably held by some encyrtids which practice polyembryony. These are species of *Litomastix*, *Copidosoma*, and related genera, all parasites of Lepidoptera laying their eggs in those of the host and emerging as adults from the fully grown host larva. Polyembryony is the development of several individuals from a single egg. The egg divides into a large number of cells aggregated as morulae and enclosed in a membrane, the whole structure being elongated and termed an 'embryo chain'. Eventually, in the haemocoel of the caterpillar, the chain fragments and the morulae, which number about five hundred, each develop into a young chalcid larva. Often more than one egg is initially laid in a host so that as many as three thousand chalcids may emerge from a parasitised caterpillar (Figure 68). The individuals derived from a single egg are all of one sex.

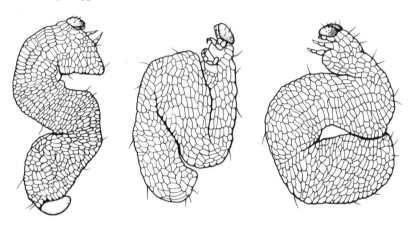

Figure 68. Caterpillars of the silver-Y moth parasitised by *Litomastix truncatellus* (Chalcidoidea, Encyrtidae). The chalcid egg reproduces polyembryonically (from Silvestri 1906).

It is important that all soft parts of the host be consumed before the parasites pupate, since decaying host remains might easily infect adjacent parasite pupae. Chalcid larvae, in common with those of other parasitic Hymenoptera, do not contaminate their usually confined environment with faeces until their growth is complete. The accumulated waste (meconium) is evacuated just prior to pupation. An interesting adaptation of some eulophid larvae which parasitise leaf-miners is the construction of a circle of little faecal pillars about themselves. The larvae pupate within this circle and the pillars harden and serve as 'pit props', preventing the collapse of the host mine as the plant tissues dry out (Figure 69).

Cocoon formation is rare in chalcids, but is known in *Euplectrus* (Eulophidae) which pupate on leaf surfaces, in some *Coccophagus* (Aphelinidae) which, contrary to the general rule, do not consume all of the soft parts of their scale insect hosts, in *Systasis* (Pteromalidae), and also in a few species of Encyrtidae. In *Euplectrus* the silk is a product of the malpighian tubules and emerges from the anus; in Ichneumonoidea the cocoon silk issues from the mouth, being a product of the labial glands.

Ichneumonoidea

Ichneumonoids, commonly referred to as ichneumons, are generally larger insects than chalcids but of more slender proportions. They have relatively

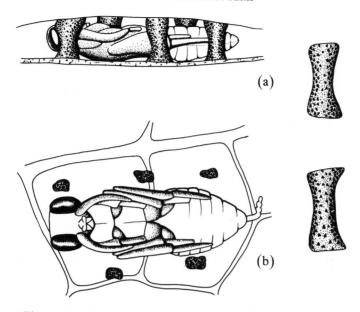

Figure 69. Supporting props of faecal matter made by the larva of *Chrysocharis gemma* (Chalcidoidea, Eulophidae) in a leaf-mine. (a) Section through the mine. (b) View of the ventral surface of the pupa of *C. gemma* lying in the mine and surrounded by five props. (All from Viggiani 1964.)

long legs, and mobile heads and abdomens. The abdomen is provided with a narrow petiole between it and the thorax, permitting much freedom of movement, and very often the female has a conspicuous ovipositor. The ovipositor of a species whose hosts feed deep in plant tissue is often longer than the rest of the body, and Bischoff (1927) figures an ichneumonid (Figure 70) and a braconid whose ovipositors are respectively about eight and six times as long as their bodies; how these insects manipulate such apparently unwieldy weapons baffles the imagination. Other species have an ovipositor that is scarcely visible externally. The integument of ichneumons is less heavily sculptured than that of chalcids and not usually metallically coloured, though some species, especially of Ichneumonidae, are banded or spotted with white, yellow, or red which relieves the predominant black or brownish colouration of the group. The wings are usually well-developed, rather elongated, and with many more veins than the chalcid wing. Some ichneumons, however, are apterous or brachypterous, and they then may bear a superficial resemblance to ants. The antennae of ichneumons are longer than those of chalcids, consisting of many more segments, and they are not sharply 'elbowed' between the scape and flagellum. This is correlated with a difference in the way in which the antennae are used. The chalcid searching for a host beats a rapid

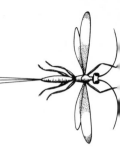

10 mm

Figure 70. An unidentified species of Ichneumonidae from Peru which has an exceedingly long ovipositor (drawn from a photograph in Bischoff 1927).

tattoo on the substrate with the tips of its antennae, exploring the surface just in front of the head. The ichneumon, on the other hand, spends much time with its antennae held out horizontally, quivering slightly like a diviner's rod, or probing with them into nooks and crannies that might conceal a host. The tips of the antennae explore the surface or test the air usually well in front of the head, and their movement increases to a maximum near to a host. When a host concealed within a mine or burrow has been located, the ichneumon walks or leans forwards, raises its thorax as far from the substrate as its legs will permit, and brings the apex of its ovipositor to bear on the exact spot indicated by the antennae.

Ichneumons are abundant in rank vegetation, especially at the edges of woodland or in hedgerows. They may also be seen searching for hosts in woodland canopy or over the litter of forest floors, but they are rather less frequent than chalcids in dry habitats such as chalk downland or sand dunes, and they tend to avoid activity in very hot, summer weather. A few genera of Ichneumonidae and Braconidae are nocturnal, and these are usually yellowish in colour. *Netelia* (=*Paniscus*), a large ichneumonid parasite of moth caterpillars, is readily attracted by lights and frequently caught in mercury vapour moth-traps. The abdomen of *Netelia* is very flexible and armed with a short, stout ovipositor which is capable of piercing human skin.

There is rather less morphological diversification among ichneumons than chalcids, reflected in the fact that the superfamily Ichneumonoidea is frequently represented as including only two very large families, the Ichneumonidae and the Braconidae. These two families are not readily distinguished in the field. Perhaps the best diagnostic character is the difference in wing venation (Figure 71). They are also biologically similar. Many insect orders are represented among their hosts as well as arachnids (spiders, pseudoscorpions), but lepidopteran caterpillars are the most frequent hosts of both ichneumonids and braconids. Being generally larger than chalcids they tend to attack larger species of Lepidoptera, and hyperparasitism is practised by only a few genera. Also, they are more frequently solitary parasites, especially

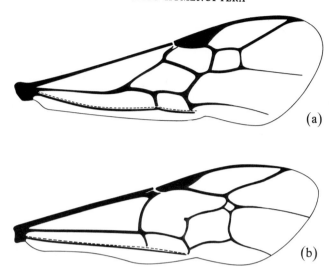

Figure 71. Ichneumonoid forewing venation. (a) Braconidae.
(b) Ichneumonidae.

the Ichneumonidae. There are no known phytophagous Ichneumonoidea.

Many species are known to be polyphagous, and Cushman (1926a) has drawn attention to the fact that the limits of the host range of a species are often less related to host taxonomy than to the habitat or food plant of the host.

However, although ichneumons may only search for hosts upon a particular food plant or in a particular environment, they may move into other habitats for other purposes. In England, the ichneumonid *Itoplectis maculator* has one generation a year (Cole 1967). It lays its eggs singly in pupae of small or medium-sized Lepidoptera, searching for them on foliage of deciduous trees, especially oak, and in undergrowth. Adults emerge in July and mate, and the males then die. The females feed upon honey-dew of aphids, which is particularly abundant at this time on leaves of sycamore and oak, and they then aestivate. The site of aestivation is probably in dark, moist places near to the ground. In September the females reappear to again feed upon honey-dew, but as autumn advances they seek out hibernation sites in which to pass the winter. The most usual hibernation site appears to be amongst the needles of coniferous trees. In spring the females that have successfully overwintered return to deciduous trees to seek out pupae of the spring flush of Lepidoptera.

An example of the importance in host relations of the host food plant is provided by Taylor (1932) who recorded that the caterpillar of the cosmopolitan *Heliothis armigera* (the American cotton boll worm), which feeds on a wide variety of plants, is attacked in South Africa by *Microbracon brevicornis* (Braconidae) only when it feeds upon *Antirrhinum*.

The effect of the habitat of the host's food plant is emphasised by Stary (1966) in describing the host relations of braconid parasites of *Aphis fabae*, the bean aphis, in Czechoslovakia. On its primary host plant, the spindle tree (*Euonymus europaeus*), this aphid is attacked by *Ephedrus plagiator*, *Trioxys angelicae*, and *Praon abjectum*, whereas on secondary host plants such as common beet (*Beta vulgaris*) it is parasitised by *Aphidius* (= *Lysiphlebus*) *fabarum*. The primary host plant is a member of the forest community; the secondary host plants are members of the steppe community. Should aphids have primary and secondary host plants that live in the same community, they are often attacked by the same parasite species on both food plants. In other words, the habitat has a very strong influence in determining the qualitative nature of parasitism. In general, when host aphids migrate to another habitat at the end of spring, their parasites either pass the time when they are without fresh hosts in a state of diapause, or else they may attack alternative hosts that remain in or enter the habitat. They do not follow their original host.

Odour is of importance in releasing the stimulus to oviposit. The odour may emanate from the host itself, from its cocoon, or from frass that it has deposited. The braconid *Cardiochiles nigriceps* is stimulated by mandibular gland secretion from caterpillars of *Heliothis* (Vinson 1968); *Devorgilla* (= *Nemeritis*) *canescens* (Ichneumonidae) will only attack larvae of the wax moth, *Meliphora*, after they have been contaminated by the odour of its usual host, the flour moth *Ephestia* (Thorpe and Jones 1937).

It is also often necessary for the host to be in its natural situation to elicit an oviposition reaction. Potentially suitable host pupae removed from their cocoons may be ignored by ichneumons. Chalcids similarly usually ignore hosts *ex situ*. Arthur (1966) demonstrated that *Itoplectis conquisitor* (Ichneumonidae) is capable of associative learning. In the laboratory this species learns to associate the colour of crêpe paper 'cocoons' with the presence of hosts. Later (1967) the same author showed that the ichneumon can recognise other physical characteristics of the artificial shelter, such as its orientation, length, and diameter, and can be conditioned to them.

The flight period is another factor that delimits the host range of an

Mirid host	% *distribution of* Leiophron *species*		
	L. heterocordyli	L. orthotyli	L. apicalis
Heterocordylus tibialis	100	0	0
Asciodema obsoletum	3.4	95.8	0.8
Orthotylus virescens	0	85.9	13.1
O. adenocarpi	2.7	18.9	78.4
O. concolor	0	5.6	94.4

Table 8. The parasitisation of broom Miridae by species of *Leiophron* (Braconidae) (from Waloff 1967).

ichneumon. Three species of *Leiophron* (Braconidae) attack plant bugs (Miridae) on broom (Waloff 1967). The female *Leiophron* generally oviposits in young nymphs, lifting them off the substrate with her legs in the process. Older nymphs are either too fast or too strong to be seized in this way. The three *Leiophron* species have slightly different flight periods, and their mirid hosts likewise are at a vulnerable stage at rather different times. This, to a large extent, accounts for the different parasite populations supported by the broom mirids in England (Table 8).

Although many ichneumons are polyphagous, it is possible to generalise about the typical host relationships of the various subfamilies, bearing in mind that there are always exceptions. Considering only the larger subfamilies, the prevailing associations can be set out as follows:

BRACONIDAE
Aphidiinae (sometimes considered a distinct family) e.g. *Aphidius, Praon, Trioxys, Ephedrus*
    Solitary endoparasites of Hemiptera Homoptera (aphids).
Euphorinae e.g. *Perilitus, Euphorus, Leiophron, Microctonus, Syntretus*
    Solitary or gregarious endoparasites, mostly of adult Coleoptera, but also of Psocoptera, Hemiptera, and Hymenoptera.
Macrocentrinae e.g. *Macrocentrus, Meteorus*
    Solitary or gregarious endoparasites of Lepidoptera larvae.
Blacinae e.g. *Blacus, Eubadizion, Triaspis*
    Endoparasites of larval Lepidoptera and Coleoptera.
Microgasterinae e.g. *Apanteles* (Figure 72), *Microgaster, Microplitis*
    Endoparasites, often gregarious, of unconcealed Lepidoptera larvae.
Cheloninae e.g. *Ascogaster, Chelonus*
    Solitary endoparasites of Lepidoptera larvae, ovipositing in eggs.
Alysiinae (= Dacnusinae) e.g. *Alysia, Dacnusa, Aspilota*
    Solitary endoparasites of Diptera larvae, emerging from puparia.
Opiinae e.g. *Opius*
    Endoparasites of larval Diptera, especially Agromyzidae and Trypetidae, emerging as adults from puparia.
Braconinae (= Vipioninae) e.g. *Vipio, Bracon, Habrobracon, Microbracon*
    Ectoparasites, often gregarious, of concealed Lepidoptera larvae.
Rogadinae e.g. *Rogas*
    Solitary endoparasites of Lepidoptera larvae.
ICHNEUMONIDAE (after Townes 1969)
Ephialtinae (= Pimplinae in part) e.g. *Scambus, Pimpla, Polysphincta, Ephialtes, Itoplectis, Rhyssa*
    Ectoparasites (except *Ephialtes, Itoplectis*) of concealed larvae and pupae in plant tissue or cocoons, especially Lepidoptera, and of spiders (*Polysphincta*).

Tryphoninae e.g. *Phytodietus*, *Netelia* (= *Paniscus*), *Polyblastus* (Figure 73), *Tryphon*, *Cteniscus*, *Exenterus*
Ectoparasites of exposed larvae of Lepidoptera and sawflies.

Gelinae (= Cryptinae) e.g. *Gelis* (= *Pezomachus*), *Hemiteles*, *Phygadeuon*, *Itamoplex* (= *Cryptus*)
Ectoparasites of insects (mostly Lepidoptera) in cocoons, many hyper-parasitic through Braconidae or other Ichneumonidae. Some species of *Hemiteles* and *Gelis* attack spider cocoons.

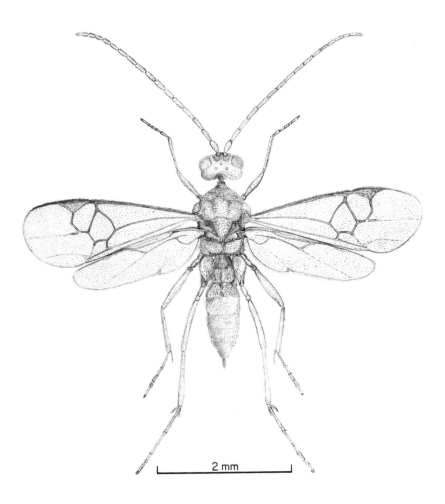

2 mm

Figure 72. Female *Apanteles cajae* (Ichneumonoidea, Braconidae).

Banchinae e.g. *Glypta, Lissonota, Exetastes, Banchus*
   Endoparasites of Lepidoptera larvae.
Scolobatinae e.g. *Ctenopelma, Perilissus, Mesoleius*
   Endoparasites of sawfly larvae.
Porizontinae e.g. *Campoplex, Charops, Devorgilla* (= *Nemeritis*), *Diadegma*
   (= *Horogenes*)
   Endoparasites, usually of Lepidoptera larvae.
Ophioninae e.g. *Ophion*
   Endoparasites of Lepidoptera larvae. Adults mostly nocturnal.
Diplazontinae e.g. *Diplazon, Syrphoctonus*
   Endoparasites of larvae of Syrphidae (Diptera), sometimes ovipositing in
   eggs, emerging from puparia.
Ichneumoninae e.g. *Phaeogenes, Alomya, Platylabus, Diadromus, Amblyteles,*
   Ichneumon, Cratichneumon, Coelichneumon, Barichneumon
   Endoparasites of Lepidoptera pupae, sometimes ovipositing in larvae.

From this summary it is apparent that ichneumons are much more frequently endoparasites than ectoparasites. This is related to their regular
parasitisation of lepidopterous larvae and other hosts in exposed situations.

The ichneumon egg is most commonly laid in or on the larval stage of the
host, and generally the host is not allowed to progress beyond this stage, being
destroyed by its parasite when fully grown, often after spinning a cocoon.
However, a surprising number of ichneumons oviposit in the eggs of their
hosts. *Oocenteter* (Ichneumonidae) and *Chelonus* (Braconidae) may be cited
as examples, and both complete their development in the host larvae. *Diplazon*
may lay its own egg in that of its hover-fly host and the adult ichneumon
emerges from the fly puparium, an unusually extended period inside the host.
Adult insects are seldom attacked by ichneumons, but *Perilitus* may oviposit
in adult beetles and *Syntretus* attacks adult Ichneumonidae (Cole 1959b) and
bumblebees (Alford 1968). Both of these genera are braconids of the tribe
Euphorini.

Cushman (1926b) has categorised the oviposition habits of ectoparasitic
Ichneumonidae, recognising four major types. The simplest involves the
deposition of an egg on or near a host that is usually paralysed and which is
concealed inside a cocoon or feeding gallery. *Rhyssa*, a parasite of wood-wasp
larvae, is an example. Next there is the firm attachment of the egg to the host's
body by a mucus secretion, the young larva later anchoring itself to the ad-
herent egg shell by means of spines at its posterior end. With this type of
ectoparasitism, the host is at most only temporarily paralysed. An example is
provided by *Polysphincta*, a spider parasite. The third type involves a more
elaborate fixing of the egg to the host, and is commonly found in the subfamily Tryphoninae. Here the egg is provided with a pedicel which is pushed
through a puncture in the host's integument, the body of the egg remaining on
the surface of the host. In *Tryphon* the end of the pedicel may carry a trans-

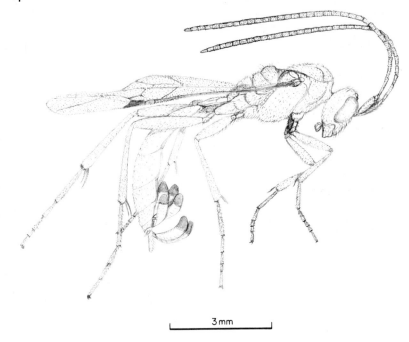

Figure 73. Female *Polyblastus cothurnatus* (Ichneumonoidea, Ichneumonidae). Fully incubated eggs are carried on the ovipositor until a host is found.

verse bar (Figure 74) to prevent it from slipping out of the host. *Exenterus* is another tryphonine genus. Three species attack larvae of the sawfly, *Neodiprion sertifer*, and each adopts a different site for the placement of its eggs (Figure 75) (Pschorn-Walcher 1967). *E. abruptorius*, the most abundant species, lays small eggs fairly evenly spread over the host's body. These eggs are deeply embedded in the integument and so are inaccessible to the mandibles of the sawfly larva. *E. amictorius* lays large, protruding eggs, and *E. adspersus* lays eggs that lack the anchoring mechanism found in the other two species. Eggs of both of these latter two species could be dislodged by the host and it is important, therefore, that they are placed where the host is unable to reach them. Both species, in fact, oviposit mainly on the thorax, *E. amictorius* dorsally and *E. adspersus* ventrally. Cushman's fourth category includes *Grotea* and a few allied genera whose species oviposit in the cells of bees. The young ichneumon larva eats the host's egg or young larva and then feeds upon the provisions stored within the cell. This is an example of inquilinism which is the subject of a later chapter.

The host may or may not be paralysed by the ovipositing ichneumon; if it is paralysed it may recover within a few hours or it may die. Each species of parasite is constant in the treatment of its hosts. In general, ectoparasites

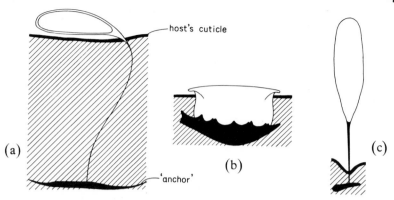

Figure 74. Eggs of tryphonine Ichneumonidae *in situ*. (a) *Tryphon semirufus*. (b) *Anisoctenion alacer*. (c) *Tryphon incestus*. (All after Clausen 1940.)

usually induce at least a partial paralysis whilst endoparasites do not, as a rule, permanently paralyse their hosts. Paralysis of the host serves to prevent it from dislodging an ectoparasite, at the same time preserving it in a fresh condition, and it may also prevent the host from moulting and shedding its parasites with its old integument.

The egg of many endoparasitic species of Braconidae is remarkable for its increase in volume within the host. Ogloblin (1913) records a one thousand-fold increase in volume of the egg of *Perilitus coccinellae* before eclosion (hatching) of the first instar larva. Such an increase could be tolerated only by an exceedingly elastic chorion, and this property also permits the major deformations of the eggs of many Ichneumonidae during oviposition. In those

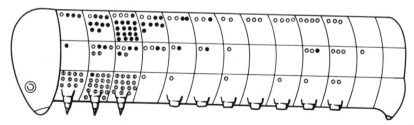

o　Exenterus abruptorius

●　E. amictorius

⊙　E. adspersus

Figure 75. Diagrammatic representation of the distribution of eggs of three species of *Exenterus* (Ichneumonidae) on the larva of the sawfly *Neodiprion sertifer* (adapted from Pschorn-Walcher 1967).

species whose hosts lie in tunnels in wood several millimetres below the surface, the egg must be squeezed down an ovipositor whose valves are prevented by the wood from gaping to ease its passage. The egg tube between the inner valves has a diameter distinctly smaller than that of the ovarian egg. To overcome these difficulties the egg is provided with a long pedicel, and during oviposition much of the fluid contents of the egg pass temporarily into this pedicel whilst the body of the egg becomes much attenuated. Whether the pedicel or the body of the egg emerges first from the ovipositor depends upon the species.

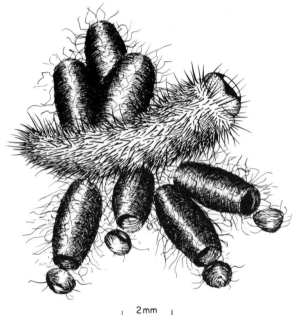

2mm

Figure 76. Empty cocoons of *Apanteles cajae* (Braconidae) grouped about the remains of a tiger-moth caterpillar.

Rhyssa persuasoria is the largest British ichneumonid with a body in the female up to four or five centimetres in length terminating in an equally long ovipositor. Its host is the larva of the wood-wasp *Sirex* (= *Urocerus*) *gigas* which burrows in coniferous wood, and the full length of the ovipositor is generally required to reach it. The ovipositor is furnished with sharp cutting ridges at its apex, and is forced down through the wood by pressure and twisting movements of the abdomen. The whole procedure of egg-laying may take about thirty minutes during which time the female ichneumon is in an exposed and vulnerable situation, so that it is of the utmost importance that the aim of the ovipositor be well-directed. In fact, hosts are contacted with great accuracy, a testimony to the efficiency of the antennae.

At the time of egg-laying, *Sirex* injects spores of a symbiotic fungus into the wood and the fungal hyphae eventually permeate the burrow of the wood-wasp larva and the surrounding wood. It is a substance present in this fungus that guides *Rhyssa* to its host (Spradbery, in press). Frass from siricid galleries is very attractive to *Rhyssa*, especially the fresh frass produced by a fully grown wood-wasp grub. The attractive substance derived from the fungus is present in high concentration in the frass, and Spradbery (1968) has devised a laboratory technique for culturing wood-wasp parasites that depends upon their response to wood-wasp frass. Honey-bee larvae are placed on siricid frass in paper-covered cells and *Rhyssa* readily oviposits through the paper upon these unnatural hosts, misled by the presence of the frass into acting as if the larvae were wood-wasp grubs. *R. persuasoria* will develop to maturity on honey-bee larvae, although a single larva is too small and additional larvae have to be supplied.

Rhyssella curvipes is a parasite of the alder wood-wasp, *Xiphydria camelus*, and its problems are similar to those of *Rhyssa persuasoria*. Another parasite of *X. camelus* is *Pseudorhyssa alpestris*, and the biology of these two Ichneumonidae has been investigated by Thompson and Skinner (1961). The female *Rhyssella* detects with the tips of its antennae a wood-wasp gallery deep in alder wood. It drills down to the gallery with its ovipositor, which is supported during drilling by both the ovipositor sheaths or outer valves and by the hind coxae. The ovipositor sheaths do not enter the egg shaft and they gradually form loops on either side of the abdomen as the inner and second valves of the ovipositor penetrate the wood (Figure 55). *Pseudorhyssa* is allied to *Rhyssella* but its ovipositor is more slender and equipped with only rudimentary cutting ridges. It locates the oviposition holes of *Rhyssella* and inserts its own ovipositor down them. Since oviposition for *Pseudorhyssa* does not involve drilling, its ovipositor sheaths are not required for support, and they are held away from the body in the resting position once the inner valves have entered the oviposition tube of *Rhyssella* (Figure 55). The egg of *Pseudorhyssa* is laid beside that of *Rhyssella* on the paralysed wood-wasp larva, and the two first instar Ichneumonids hatch about the same time. The larva of *Pseudorhyssa* has a larger head than that of *Rhyssella* and much larger mandibles. When the two larvae meet, *Pseudorhyssa* seizes *Rhyssella* and destroys it so that it can develop without competition upon the host.

Dotocryptus bellicosus (= *macrocercus*), a cryptine ichneumonid, attacks larvae of the solitary wasp *Odynerus* lying in earthen cells. It penetrates the very hard wall of the cell with the aid of a fluid secretion from its ovipositor (Janvier 1933).

Most species of Braconidae and Ichneumonidae attack hosts that are either unshielded or protected merely by a thin-walled cocoon. Aphidiinae attack exposed aphids, Alysiinae attack agromyzids beneath a thin leaf epidermis, and Euphorinae attack adult beetles or nymphal bugs, often lifting them from the substrate in their legs to insert the ovipositor into the softer ventral surfaces of the hosts. Exposed larvae of Lepidoptera and sawflies are attacked by several groups. Even though these hosts are easily reached, they are not as defenceless as might be thought. Caterpillars usually wriggle violently when touched by an ichneumon, and they may drop from the food plant on a silken thread. This manoeuvre usually eludes the parasite, although instances have been observed of parasites running down the silken threads in pursuit.

Aphids attacked by Aphidiinae (Figure 77) kick, and this may repel the parasite or cause it to oviposit in a leg, in which case the parasite fails to develop (Starý 1966). *Trioxys* has developed spines upon the last sternite which help to hold the aphid during oviposition. Another aphid defence mechanism is the production of a waxy secretion from the siphunculi which may repel the parasite. Many aphids are tended by ants which drive off predators, but Starý noted that ants do not molest ovipositing Aphidiinae. Indeed some Aphidiinae (e.g. *Paralipsis enervis*) mimic ants and are fed by them on regurgitated food.

The ichneumonid *Tryphon* and allied genera insert the caudal pedicels of their eggs into the cuticles of their host larvae, usually sawfly grubs, with single rapid thrusts of their ovipositors. The chorion of the egg of *Tryphon* is inflexible, and only the pedicel passes down the ovipositor egg-tube, the body of the egg moving outside the valves. The contrast between *Tryphon* and *Rhyssa* is striking.

Polysphincta is an ichneumonid that attacks spiders, placing its egg on the host's opisthosoma. This is an act fraught with danger and is accomplished with great speed. However in addition to laying its egg, the female *Polysphincta* paralyses the spider with a sting from its ovipositor and it is then able to feed at leisure on fluids exuding from the host (Cushman 1926b). Several other ichneumons are known to feed from their hosts, but only some braconids have been recorded as constructing special feeding tubes in the manner of some chalcids. The ichneumonid *Exeristes comstockii* is unable to mature its eggs until it has fed on the host (Bracken 1965), and it is probable that a great many species are unable to realise their full reproductive potential until they have had a meal on the host's body fluid. Leius (1961) showed that the fecundity of *Itoplectis conquisitor* is improved when pollen as well as host tissues are provided as food. Many ichneumonids take carbohydrates in the form of honeydew or nectar.

In the great majority of endoparasitic species, the egg is laid in the host's haemocoel. Exceptions are *Heteropelma calcator* and *Amblyteles subfuscus* which oviposit in the gut and salivary glands respectively, and some braconids which place their eggs in nerve ganglia. *Monoctonus paludum*, an aphidiine parasite of aphids, places its egg with precision in the host's compound thoracic ganglion. Most endoparasites pierce the host's integument through an arthrodial membrane.

The habit of retaining eggs for a period of incubation, common among

Diptera, is very rare among female Hymenoptera. It is, however, practised by Tryphoninae, which carry the eggs for a while at the apex of the ovipositor. The eggs are held in position by their pedicels which are clamped between the ovipositor valves. Sometimes, for example in *Polyblastus* (Figure 73), well-developed larvae can be seen inside the eggs. The functional significance of this habit is not certain, but Kerrich (1936) suggests that *Polyblastus*, by reducing the period of time spent on the host as an early developmental stage, lessens the risk of being cast off the host on a moulted skin.

The association between mother and brood is extended further by *Cedria paradoxa*. This species is a gregarious ectoparasite of certain caterpillars in parts of Asia, and is unique in the parasitic section of the Hymenoptera in practising maternal care. The female remains beside her offspring until they have attained adulthood, and is thought to protect them from hyperparasites.

Young larvae of several euphorine Braconidae retain the trophamnion about themselves for a short while after eclosion from the egg. On disintegration of the membrane, the trophamniotic cells float in the host's haemocoelic fluid absorbing food materials, particularly fats, and they increase considerably in size. Those associated with *Leiophron* may attain a diameter of 0.1 millimetres (Waloff 1967). Eventually these food-laden cells (teratocytes) accumulate at some point in the host's body, generally in the posterior regions, and the by then well-grown braconid larva may browse upon them (Jackson 1935). An embryonic membrane also encloses the first instar larva of the ichneumonid *Therion morio*, but this is thought to serve a protective function by preventing the host's phagocytes from reaching the parasite (Tothill 1922).

The fully grown larvae of many species of Ichneumonoidea pupate beside the remains of their host inside silken cocoons. The mass of yellowish cocoons of the braconid *Apanteles glomeratus* beneath the shrivelled remains of a large white butterfly caterpillar are a familiar sight (see also Figure 76). The cocoons can be very tough, and the silk threads, a product of the labial glands, can be teased apart only with difficulty. A seasonal dimorphism in cocoon structure has been noted in some Ichneumonidae (e.g. some *Ephialtes*, *Eulimneria*) (Clausen 1940), analogous to the dimorphism of *Eulophus* (Chalcidoidea) pupae (page 172). A tough, thick-walled cocoon is formed to house over-wintering pupae, whilst the summer cocoon is flimsy and less darkly coloured.

The larva of an aphidiine kills its aphid host in the final nymphal instar, but before doing so it bites a hole in it ventrally and glues it to the leaf surface by a secretion from its salivary glands. The dead aphid (Figure 77c) has a bloated appearance and is brownish or whitish in colour, quite distinct from unparasitised aphids in the colony. Inside such an aphid 'mummy' species of *Aphidius* spin a cocoon in which to pupate. Larvae of *Praon*, however, leave the mummy to spin a tent-like cocoon beneath it (Figure 77b). In Aphidiinae also there is sometimes a seasonal dimorphism in cocoon structure.

A few Braconidae (e.g. *Meteorus*) and Ichneumonidae (*Charops*) have

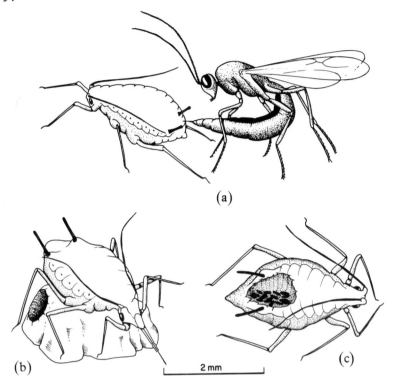

(a)

(b)   (c)

2 mm

Figure 77. Braconidae, Aphidiinae. (a) *Aphidius* ovipositing
in an aphid (from Clausen 1940 after Webster). (b) Cocoon of
Praon beneath the shell of the host aphid. (c) Aphid 'mummy'
from which an *Aphidius* has emerged.

cocoons suspended from vegetation by long, silken cords. This may serve to
keep them out of reach of crawling predators. Braconid cocoons seem generally
to be ignored by birds. A few Lepidoptera such as the arctiid *Deilemera
antinorii* exploit this by spinning imitation braconid cocoons on the surface of
their own cocoon (Poulton 1931).

Cocoons of some ichneumonids (*Bathyplectes*, *Spudastica*, *Phobocampe*, and
others) jump. A jump is caused by the sudden straightening of the larva within
a cocoon, and it may clear a height of two or three centimetres, several times
the cocoon's own length. Since it is the larva which causes the cocoon to jump,
the faculty is lost after a few days when the larva pupates. A species of
Bathyplectes is a common parasite of the weevil *Phytonomus plantaginis* in
southern Britain, and its cocoon, which is brownish with a white median
girdle, may be found inside the very flimsy cocoon of its host, attached to the
stems of grasses and herbaceous plants. In the laboratory this cocoon jumps
when a bright light is shone upon it, and in the field sunlight probably has the

same effect. What is the biological significance of the habit? Wherever I have found this species of *Bathyplectes*, quite a considerable proportion of its population has been attacked by the gregarious hyperparasite *Sceptrothelys grandiclava*, a chalcid of the family Pteromalidae. This chalcid searches for its host on the stems of plants, and it seems very probable that if the *Bathyplectes* cocoon can jump out of the weevil cocoon and fall to the ground, it will escape from the hyperparasite.

Proctotrupoidea

This superfamily, for which Serphoidea is a synonym, is neither so large nor so biologically diverse as the two preceding. In size of individuals they tend to lie, very approximately, between chalcids and ichneumons. They are usually black or brown in colour, the females have concealed ovipositors, and their wing venation varies very much from family to family. The great majority of proctotrupoids are endoparasites, only a few (*Lygocerus* and allied genera) are hyperparasites, and they attack mostly insects but also spiders' egg cocoons and centipedes. The more important families are the following:

Platygasteridae e.g. *Platygaster*, *Inostemma* (Figure 80a)
   Endoparasites, solitary or gregarious, of Diptera (Cecidomyiidae) larvae, or sometimes of nymphs of Hemiptera Homoptera, ovipositing in host eggs.
Scelionidae e.g. *Telenomus* (Figure 80d), *Trissolcus* (= *Asolcus*), *Hoplogryon*, *Scelio* (Figure 80e)
   Endoparasites, usually solitary, of insect eggs.
Ceraphronidae (= Calliceratidae) e.g. *Lygocerus*, *Conostigmus*, *Aphanogmus* (Figure 80c)
   Many are hyperparasites of Hemiptera Homoptera (aphids, scale insects) (*Lygocerus*); others are endoparasites of Diptera larvae (*Conostigmus*).
Diapriidae e.g. *Platymischus* (Figure 80b), *Loxotropa*, and *Galesus* (Diapriinae), *Belyta* and *Aclista* (Belytinae)
   Mainly gregarious endoparasites of Diptera pupae, sometimes larvae.
Proctotrupidae (= Serphidae) e.g. *Phaenoserphus* (Figure 79), *Codrus* (= *Exallonyx*)
   Mainly endoparasites of beetle larvae, solitary or gregarious.
Heloridae e.g. *Helorus*
   Endoparasites of lacewing (Neuroptera) larvae.
Pelecinidae e.g. *Pelecinus* (Figure 78)
   Endoparasites of beetle larvae.

Outstanding morphological features in the superfamily are the excessively attenuated abdomen of *Pelecinus* (Figure 78), and the forwardly curving horn on the abdomen of *Inostemma* (Figure 80a) that accommodates the long ovipositor when it is not in use. Both of these features are devices to ensure that eggs can be laid in hosts buried deep in plant tissue.

Proctotrupoids have some very interesting biological adaptations. Platygasterids, for instance, usually oviposit in the egg of the hosts but do not complete their development until the host has become a fully grown larva, a habit shared by *Diplazon* (Ichneumonidae), *Entedon ergias* (Chalcidoidea), and a few other parasitic Hymenoptera. Within the host egg the platygasterid egg is often located in a particular region of the embryo, a position which varies between species but which is frequently in the central nervous system (Figure 81). Platygasterids therefore attack eggs in an advanced stage of development.

Several species of *Platygaster* that are parasitic on gall midges (Cecidomyiidae) are able to reproduce polyembryonically. Since the hosts of these species are small they are unable to support more than about twenty individuals, a figure that does not compare with that of some polyembryonic Encyrtidae (page 139). A number of Platygasteridae show evidence of incipient polyembryony but the potential is not realised, a situation previously described for the braconid *Macrocentrus ancylivorus*. Such cases may be

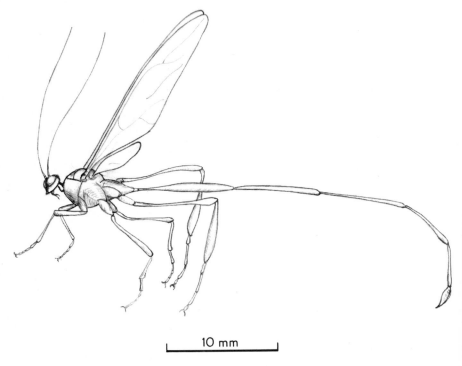

10 mm

Figure 78. Proctotrupoidea, Pelecinidae. A female *Pelecinus* (from Borror and DeLong 1966).

explained by the past existence of an ancestral host capable of supporting more than a single parasitic individual. In the species that practice poly-embryony, the trophamnion forms a sheath about the egg, and it grows as the egg divides. Eventually each embryo becomes enveloped by a piece of tro-phamnion. This is essential to the growth of the embryo, providing it with food materials absorbed from the host. The mechanics of this operation are not clearly understood. The first instar larva, when it escapes from the trophamniotic sheath, is often of extraordinary appearance suggestive of the crustacean *Cyclops*, with an inflated 'cephalothorax' carrying relatively enor-mous, curved mandibles, and a very narrow abdomen which terminates, in some species, in a bifurcated 'tail' (Figure 81).

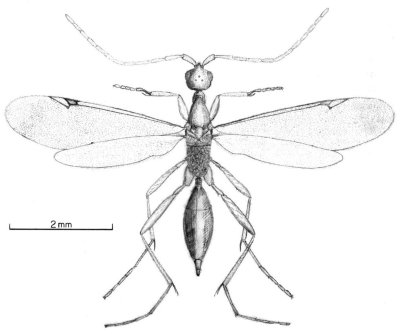

2 mm

Figure 79. Proctotrupoidea, Proctotrupidae. A female *Phaenoserphus calcar*.

The Scelionidae also oviposit in insect eggs, but unlike the Platygasteridae, they complete their development in this host stage. Insects of many orders serve as hosts, although scelionids often require to oviposit in freshly laid eggs for the subsequent development of their brood. The habit of phoresy has been developed by some species to enhance their opportunities for this (page 179).

Species in the other families of Proctotrupoidea do not oviposit in their host eggs. Diapriids mostly lay their eggs in the puparia of Diptera, the Heloridae in the larvae of lacewings. Proctotrupids usually oviposit in young beetle

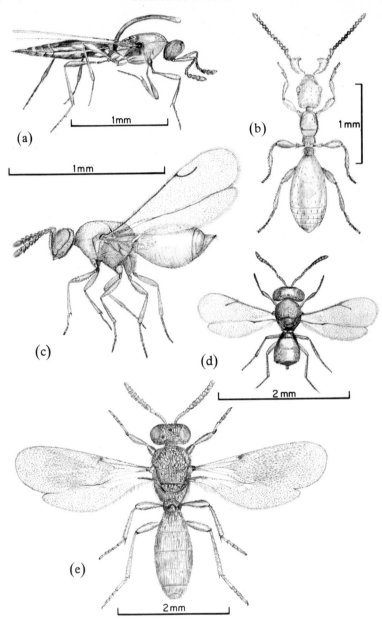

Figure 80. Hymenoptera, Proctotrupoidea. (a) *Inostemma boscii* (Platygasteridae) female. (b) *Platymischus dilatatus* (Diapriidae) male. (c) *Aphanogmus* sp. (Ceraphronidae) female. (d) *Telenomus punctatissimus* (Scelionidae) male. (e) *Scelio walkeri* (Scelionidae) female.

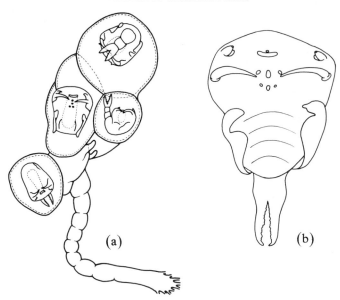

Figure 81. *Inostemma piricola* (Proctotrupoidea, Platygasteridae). (a) Four first-instar larvae in the brain of a gall-midge
larva (*Contarinia pirivora*). (b) Enlarged drawing of first instar
larva in ventral view (both after Marchal 1906).

larvae, generally those inhabiting leaf litter such as Carabidae, Staphylinidae,
and Elateridae. *Phaenoserphus viator* has been studied in detail by Eastham
(1929). The eggs are laid in first or second instar larvae of the ground beetle,
Pterostichus, and development is very slow until the host larva approaches
maximum size. The parasite larvae, which may number about thirty, then
feed rapidly and destroy the host. Each parasite makes its own hole in the
integument of the host's ventral surface, and crawls out for about three-
quarters of its length, leaving only its posterior end anchored. The larvae
then pupate without forming a cocoon, and all of the pupae are orientated the
same way with their heads directed towards the anterior end of the remains of
the host larva (Figure 82).

Cynipoidea

Cynips is a genus of gall wasps, insects whose larvae induce the formation of
galls on plants in which they feed on vegetable tissue. Gall wasps belong to the
subfamily Cynipinae. They attack a variety of plants, especially oak trees, and
their galls are often striking and elaborate structures. However, gall wasps are
phytophagous and fall outside the scope of this book, with the exception of
those species that behave as inquilines (page 244). Belonging to other groups
of Cynipoidea, however, are species that behave as parasites. With few

exceptions, these have been much neglected by entomologists. Species of Cynipoidea are generally small, mostly black or brown in colour, often with a deep, ploughshare-shaped abdomen and large, triangular radial cells on the forewings.

The Ibaliidae are solitary endoparasites of wood-wasp larvae. A European species, *Ibalia leucospoides*, has been introduced to New Zealand for the control of *Sirex cyaneus*. It oviposits in the host egg using the oviposition hole of the host, exactly like *Aulacus striatus* (Evanioidea). It locates the oviposition

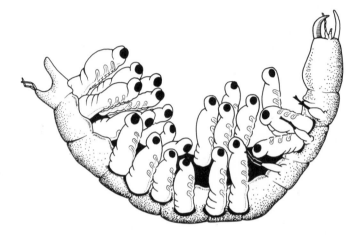

Figure 82. The corpse of a ground-beetle larva from which are protruding pupae of *Phaenoserphus viator* (Proctotrupoidea, Proctotrupidae) (after Eastham 1928).

site of *Sirex* by detecting the presence of the symbiotic fungus that *Sirex* deposits at oviposition. Wood-wasp drills about two weeks old are most attractive, and no discrimination is shown between empty drills and those containing eggs provided that the fungus is present. Fungus grown on a perspex plate induces an oviposition response in *Ibalia* (Madden, in press). *I. leucospoides* waits until the host has become a fully grown larva before destroying it.

Species of Figitidae and Eucoilidae are solitary endoparasites, nearly always of Diptera, ovipositing in young larvae and emerging as adults from the host puparia. *Idiomorpha* (=*Cothonaspis*) *rapae*, a eucoilid parasite of the cabbage fly, *Hylemyia brassicae*, will oviposit only in darkness and is attracted to the whereabouts of its host by the odour from putrifying, infested cabbages (James 1928).

Other genera in these two families include *Eucoila* (=*Psilodora*) and *Kleidotoma* (Eucoilidae) and *Figites, Melanips, Anacharis,* and *Aegilips* (Figitidae). Species of *Anacharis* and *Aegilips* mostly parasitise lacewings

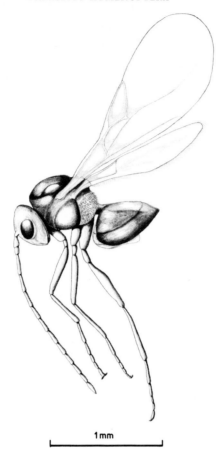

Figure 83. Cynipoidea, Cynipidae. Female *Charips victrix*, a
hyperparasite of aphids (drawing by Ferris Neave).

(Neuroptera) (e.g. Kerrich 1940, Selhime and Kanavel 1968), and these
genera are placed in the subfamily Anacharitinae of the Figitidae.

The family Cynipidae includes two large subfamilies, the Cynipinae or gall
wasps and the Charipinae (=Allotriinae). The latter, of which *Charips*
(Figure 83) is an example, are hyperparasites of aphids through their primary
braconid parasites.

Bethyloidea

Bethyloids fall within the division Aculeata of the suborder Apocrita. Aculeates
include ants, and solitary and social bees and wasps. Bethyloids are in many
ways the least specialised of the aculeate groups, and included in their number
are wasps that exhibit a transitional range of habits between typical parasitism

and indications of social life. The superfamily, which tends to be a receptacle for a rather miscellaneous collection of groups of obscure affinities, contains the families Dryinidae (Figure 84), Bethylidae, Chrysididae, and Cleptidae.

All dryinids are parasites of Hemiptera Homoptera, almost exclusively of Delphacidae and Cicadellidae (leaf-hoppers). Females of most species are remarkable for their chelate front legs, one tarsal claw being much enlarged and apposable against the modified fifth tarsal segment (Figure 85). This apparatus, worked by a femoral muscle, is used to grasp the host which may be either a nymph or an adult. An egg is laid between two overlapping sclerites, frequently on the abdomen, and the host may be paralysed for a few minutes after oviposition. The young parasite larva is at first invisible from the outside, but eventually its position is marked by the appearance of a blackish sac (Figure 86). The sac is composed of cast larval skins and is added to at each ecdysis, of which there are usually five. The host remains active whilst its parasite is developing, but it does not moult. When the dryinid larva is fully

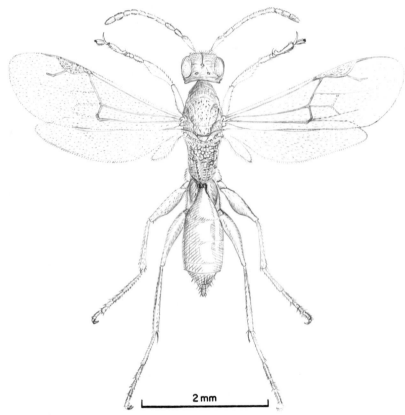

2 mm

Figure 84. Bethyloidea, Dryinidae. Female *Anteon flavicorne*.

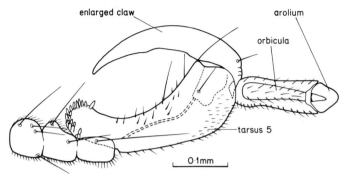

Figure 85. The distal part of the chelate front leg of a female *Anteon flavicorne* (Bethyloidea, Dryinidae) (after Richards 1939).

grown, the host is killed, and usually it dies firmly fixed to its food plant. However, the parasite larva emerges from its sac and crawls away to pupate in a silken cocoon, either on the plant or amongst leaf litter on the ground.

The life history of *Aphelopus* is rather different from that of other dryinids. The females do not have chelate front legs and they usually attack only very young nymphs. The egg is laid internally and is surrounded by the hypertrophied trophamnion. The trophamnion persists until the parasite larva is in its final instar, and all food consumed by the larva must pass through it. The larva breaks out of the trophamnion in its second instar to project from the host, but inside the host the layer continues to isolate the parasite from the host's haemocoel. As in the other genera, the part of the larva protruding from the host is enclosed in a sac made of larval exuviae, but in *Aphelopus* this sac is soft and yellowish in colour. As usual, however, there is an exception. *A. theliae*

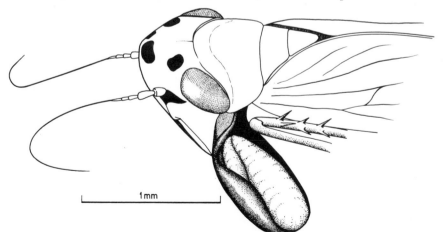

Figure 86. Dryinid sac attached behind the head of a leafhopper.

is entirely endoparasitic and forms no external sac. *A. theliae* is also remarkable for the fact that it is the only dryinid known to attack Membracidae; in addition the species is polyembryonic, one egg dividing to produce up to seventy young larvae (Kornhauser 1919).

Richards (1939), in a monograph of the British Dryinidae and Bethylidae, recognises three different evolutionary trends in females:

(i) reduction of wings and associated thoracic segments,
(ii) reduction of mouthparts, and
(iii) development of larger and more mobile chelae on the front legs.

The loss of wings gives many dryinids an ant-like appearance. It has been suggested (Donisthorpe 1927) that this is an example of mimicry, but Richards points out that the gait of a dryinid is quite distinct from that of an ant.

The Bethylidae are 'in habits intermediate between true parasites and fossorial Hymenoptera' (Richards 1939). The female paralyses the host (or prey), sometimes drags it to a sheltered position, and oviposits upon it. Often more than one egg is laid and the female may stand guard over her brood, sometimes (some *Sclerodermus* species) until the larvae are fully grown. The larvae feed externally, grow very rapidly, and spin cocoons in five or six days. The female bethylid quietens the host by biting as well as by stinging, and she feeds upon the body fluids that accumulate about the wound. The dragging of the host to a sheltered position and the guarding of it are habits typical of nest-making Hymenoptera. Most solitary wasps move their prey to a cell, and the guarding of the brood foreshadows the maternal care of social wasps, ants, and bees. Sometimes more than one individual bethylid lay their eggs upon the same host, and the ensuing co-operative guarding is strongly reminiscent of communal nesting in some polygynous wasp societies.

All bethylids attack larvae of Coleoptera and Lepidoptera, frequently those living in grain, fruits, leaf-mines and rolls, or other moderately concealed situations. Species of *Bethylus* parasitise Lepidoptera, whilst *Cephalonomia* are parasites of *Cis*, *Tribolium*, *Calandra*, and other beetles of similar habits.

Like dryinids, many female bethylids are apterous or brachypterous, but males are generally fully winged. Soon after emergence from their own cocoons the males tear open those of their sisters, and as a result there is extensive inbreeding in many species. This has the effect of restricting the exchange of genetic material within the population, and consequently of creating circumstances in which the fragmentation of an original population into an aggregate of sibling species would be facilitated. Sib-mating is widespread amongst parasitic Hymenoptera and is probably of considerable evolutionary significance (Askew 1968b).

The Cleptidae and Chrysididae are closely allied but quite unlike bethylids.

The Cleptidae are solitary ectoparasites of fully grown sawfly larvae in cocoons, and the Chrysididae are solitary ectoparasites of fully grown larvae of vespoid and sphecoid wasps and of bees, and occasionally of Lepidoptera larvae in cocoons. A small group of chrysidids behave as endoparasites in stick insect eggs. Chrysidids are beautifully coloured insects, mostly metallic green, blue, and red, and they have a very thick, strongly sculptured cuticle. This latter is no doubt a protection against the stings of the wasps whose nests they attack, as also is their ability, when disturbed, to curl up into a ball which shields their more vulnerable ventral surfaces. Most chrysidids are true parasites, feeding only on the larval wasp and not consuming the provisions left in the cell by the female wasp. They often delay feeding on their host larva until it has grown to a good size. There are, however, some inquiline chrysidids which lay their eggs in newly formed wasp host cells and subsist entirely on the stored provisions (spiders).

Vespoidea

Among Vespoidea, as among Bethyloidea, biologies are diverse. True parasitism merges into the behaviour of solitary nest-making wasps, which eventually leads on to the highly developed social life of a vespid nest.

The larvae of dung beetles and chafers (Scarabaeoidea) living in soil are hosts to a number of Vespoidea. Wasps of the families Tiphiidae and Thynnidae burrow in search of them, paralyse them with a sting, and oviposit on them without attempting to move them. They therefore behave as typical parasites (or parasitoids). Species in another family, the Scoliidae, also attack scarabaeoid larvae in their burrows, but after oviposition they dig deeper into the soil, pulling the paralysed larva after them, and eventually build a cell for its reception. A distinction sometimes used between a parasitic and a non-parasitic hunting wasp is whether or not it makes a cell for the host (or prey). On this basis scoliids are not parasites but tiphiids and thynnids are. This distinction is convenient because of its precision, but the only reason behind it is that a host not enclosed in a cell is able, upon recovery from paralysis, to resume a relatively active life like the host of a more typical parasite. Even so, the paralysis inflicted upon such a host may be permanent and, on the other hand, prey enclosed in a cell may recover sufficiently to move around in what space is available. In the family Pompilidae, spider-hunting wasps, there are some species (e.g. *Homonotus*) which enter the lair of the host to paralyse it whilst some others drive the spider from its nest in order to attack it, but later drag the spider back into its retreat. The majority of pompilids first go hunting for a spider and only after they have paralysed one do they build a cell to contain it. In most other hunting wasps the cell is constructed first, its location 'remembered', and prey is brought back to it. Spiders attacked by *Homonotus* partially recover from their paralysis, but the majority of pompilids inflict a permanent immobility.

Other vespoids that do not construct a cell for their prey are the Mutillidae, most of which attack larvae or pupae of bees and wasps in their own cells although a few parasitise pupae of Diptera inside puparia. Since the host is already contained in a suitable cell, there is no necessity for the mutillid to construct one for it. The same applies to the Sapygidae, wasps that lay their eggs in bees' cells and often directly on the bees' eggs. The young sapygid larva is very active like that of the Parasitica and unlike that of other Aculeata with the exception of the Chrysididae. It sucks out the contents of its host's egg and then lives for the remainder of its larval life upon the stored provisions of pollen and honey, an exceptional diet for a wasp grub.

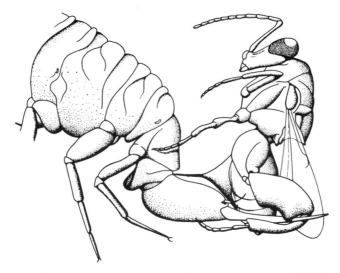

Figure 87. A chalcid, *Lasiochalcidia igiliensis* (Chalcididae), ovipositing in an ant-lion larva. The host is a predatory insect and the parasite provokes it into emerging from its burrow and seizing its legs, whereupon the parasite is able to insert its ovipositor into the membrane between head and thorax (from Steffan 1961).

The Methocidae, whose larvae feed upon the larvae of tiger beetles (Cicindelidae), may also be regarded as marginally parasitic. The female *Methoca ichneumonides* searches for a burrow of a tiger-beetle larva, allows herself to be seized by the predatory larva, and then endeavours to sting it in the thorax before being damaged herself. She then waits at the entrance to the burrow until the paralysing effect of her sting has acted, whereupon an egg is laid on the beetle larva. Afterwards the *Methoca* fills in the burrow with soil and gravel and finally camouflages its position with pieces of twigs and grass. The same seemingly hazardous method of gaining access to the host is employed by the chalcid *Lasiochalcidia igiliensis* (Chalcididae), a parasite of ant lions (larvae of the neuropteran *Myrmeleon* (Figure 87) (Steffan 1961)).

In the other large superfamily of wasps, the Sphecoidea, parasitism is a much less common habit than it is among vespoids, but it is found in the genus *Larra* (Larridae). This wasp lays its eggs on mole crickets after driving them from their subterranean burrows and paralysing them. The mole cricket soon recovers from its paralysis and resumes an active life. Other genera of Larridae, attacking grasshoppers and crickets, make a cell to which they drag the host. The large size of a mole cricket has rendered this inexpedient for *Larra*.

The difference in behaviour between an ichneumon and a fossorial wasp is not great, and several examples cited in the section on Bethyloidea and in the present section indicate ways in which the transition could have been made. Fossorial wasps, having developed the technique for transporting prey back to a nesting site, largely through the ability to 'memorise' where they have made their nest, are in a position to make repeated hunting expeditions for the purpose of provisioning a single nest. This enables them to stock their nests with many individuals of a small prey species, instead of with one large individual. Insects that are captured in numbers by nest-making wasps include a variety of smaller Diptera and Coleoptera, adults of species whose immature stages are frequently parasitised by the parasitic Hymenoptera.

Aquatic Hymenoptera

Many insects pass their immature stages in water, but there are relatively few aquatic adult insects. Those bugs and beetles that spend a large part of their adult lives in water are very obviously adapted to this environment. They are stream-lined in shape with their membranous hindwings, if present, concealed beneath robust wing covers; their legs are laterally expanded and fringed with hairs for propelling them through the water, and they have specialised means of obtaining oxygen. None of the aquatic Hymenoptera enjoy these modifications.

Jackson (1964) has investigated the biology of the chalcid *Mestocharis bimacularis*, a species which is aquatic in a marginal sense only. It never voluntarily enters water, and drowns quickly if immersed. Nevertheless, it is a parasite of the eggs of dytiscid water-beetles which are partially embedded in the foliage of water plants. A fluctuating water level is essential to it. As ponds dry out, some water-beetle eggs may be left above the water surface, and *Mestocharis* is then able to reach them. The chalcid can walk upon the surface film. An egg of *Dytiscus* may support up to twelve *Mestocharis*, and if the host remains out of the water, adult *Mestocharis* emerge from it within two months or so of oviposition. If the water level rises to submerge the host egg after oviposition, the *Mestocharis* develop only as far as fully grown larvae and they remain at this stage, sometimes for over a year, to complete their development only after re-exposure to the atmosphere. Alternate flooding and drought

favour *Mestocharis*; prolonged drought could lead to the exhaustion of the
supply of host eggs. The possible protraction of the larval stage, plus the fact
that adult females can live for ten months or more, serves to carry the species
over periods of unfavourable conditions.

Some ichneumons that attack aquatic hosts may make forays into water,
walking slowly over stones or vegetation, but never remaining submerged for
very long. The ichneumon *Agriotypus armatus* (Figure 88a), sole representative of the family Agriotypidae, is such a species. It is an ectoparasite of

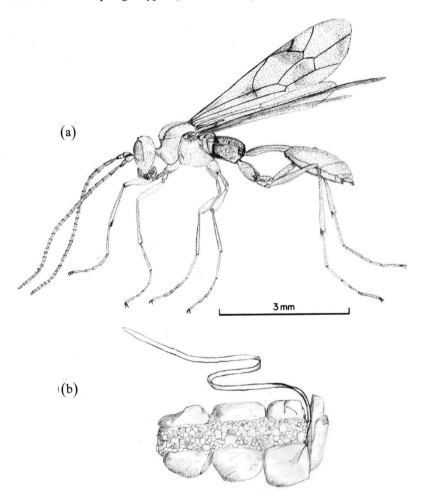

(a)

3 mm

(b)

Figure 88. *Agriotypus armatus* (Ichneumonoidea, Agriotypidae). (a) Adult. (b) Parasitised pupal case of the caddis
Silo showing the ribbon by means of which *Agriotypus* is able
to respire underwater.

prepupae and pupae of the case-making caddis *Silo* and *Goera*. The female *Agriotypus* walks into the water and is enclosed in a film of air trapped by the body pubescence. This air enables the insect to stay submerged for about thirty minutes. Its antennae are held horizontally along its back and probably play little part in host location. The *Agriotypus* larva spins a cocoon inside the caddis case, and to the cocoon is attached a ribbon of silk about three centimetres long and one millimetre broad. Internally the ribbon consists of loosely arranged threads and it contains air. This ribbon, which extends freely into the water from the caddis case (Figure 88b), is the respiratory apparatus of the pupa and of the adult before it emerges. *A. armatus* passes the winter as an adult inside its cocoon, and it dies if its ribbon is removed. The volume of air contained in the ribbon is much too small to supply the insect's needs, but the ribbon acts as a plastron. That is, it resists compression so that the internal volume remains constant, and oxygen can diffuse into it from the surrounding water to replace that which is used by the parasite. If the ribbon were not incompressible, nitrogen would gradually leak out until eventually it would contain no gas at all.

Another chalcid parasite of dytiscid eggs studied by Jackson is *Caraphractus cinctus* (Jackson 1958). This is one of the minute fairy flies (Mymaridae) and is a completely aquatic insect, passing nearly all of its life submerged. Movement under water is effected by swimming with its wings, a fact first discovered by Sir John Lubbock (1862): 'As the motion in *Polynema natans* [*Caraphractus cinctus*] is caused by the wings, it might also be called a flight; owing, however, to the density of the medium and partly to the direction in which the wings act, the movement, though not inelegant, is slow, and rather a succession of jerks than a continuous progression.' The swimming of *Caraphractus* has a rhythm not unlike that of an ephyra larva of a jellyfish. Both mating and oviposition are performed in water. Unlike *Mestocharis*, *Caraphractus* ignores host eggs above the water level. The species is able to skim rapidly over the surface of water, but females seldom fly; in fact many have rudimentary or shortened wings. Jackson records that female *Caraphractus* can survive immersion for several days. The respiratory system is typical of that of a terrestrial insect with gas-filled tracheae, but the very small size ($1\frac{1}{2}$ millimetres long) and thin cuticle of *Caraphractus* permit adequate respiratory exchange between the water and the general body surface.

A third chalcid egg parasite of Dytiscidae, and also of Hemiptera and Odonata, is the trichogrammatid *Prestwichia aquatica*. This species was found by Lubbock at the same time that *Caraphractus* was observed, in a pond at Chislehurst in Kent. Lubbock (1862) writes: 'when under water, [*Prestwichia*] holds its wings motionless, and uses its legs as oars. Though they are neither flattened nor provided with any well-developed fringe of setae, still they seem to serve their purpose pretty well, and the motion of this species is more rapid than that of [*Caraphractus*].' Thus *Caraphractus* swims with its wings and

Prestwichia with its legs. *Prestwichia* is even smaller than *Caraphractus*, measuring less than a millimetre in length, and as many as seventy individuals have been reared from a single *Dytiscus* egg. Mating may be accomplished inside the host egg.

Other minute, aquatic Hymenoptera include species of *Limnodytes* (Proctotrupoidea, Scelionidae) which parasitise pond-skater (*Gerris*) eggs, and species of a New World genus of Trichogrammatidae, *Hydrophylita*. Insects in both of these genera are said to swim with their wings in the manner of *Caraphractus cinctus*.

Dimorphism

Dimorphism is the existence within a species of two distinct forms of the same developmental stage. The most familiar example is the dimorphism imposed by bisexuality, sexual dimorphism, but it is with other forms of dimorphism that we are concerned here.

One of the most striking examples amongst parasitic Hymenoptera was discovered by Salt (1937b) when he demonstrated the host-determined dimorphism of male *Trichogramma semblidis* (Chalcidoidea). Male *T. semblidis* reared from alder fly (*Sialis*) eggs are apterous, whereas those reared from lepidopterous eggs are macropterous. There are also differences between the two forms in antennal and leg structure (Figure 89a). Host size may influence parasite size (Figure 89b), and the latter may be correlated with relative lengths of wing veins or antennal segments, the strength and extent of sculpturation, or, as Oldroyd and Ribbands (1936) noted, the number of wing hairs. However these characters may show a continuous variation between two extremes, in which case they do not, of course, constitute a dimorphism.

Nutritional disparity plays a large part in dimorphism in *Melittobia chalybii*. This chalcid is a parasite of bee larvae and exists in two forms (Schmieder 1933), the first with well-developed wings and eyes, the second with reduced wings and, in the female, a thinner cuticle. The first form develops upon the more solid tissues of the host, and its life cycle extends over about three months. In contrast, the second form subsists mainly on body fluid, and its life cycle is of only about two weeks duration. The first brood of parasites developing on a host consists of both forms. Individuals of the second form develop quickly and produce another brood on the original host, this time made up entirely of individuals of the first form, so that later there is an emergence composed of adults of the first form derived in part from the second generation and in part from the initial parasitisation. In this manner *M. chalybii* is able to make maximum use of a host in terms of offspring supported by it.

Seasonal dimorphism is not very common, although it is known in *Olynx gallarum*, a chalcid whose spring and summer generations have differently shaped abdomens (Askew 1961a), and in another chalcid, *Torymus auratus*, there is a striking dimorphism in ovipositor length (Figure 90). In the autumn generation of *T. auratus* the ovipositors are short, but in the spring generation some females have short ovipositors and others long ovipositors. This is correlated with the situations of the hosts attacked by these two generations, the autumn generation ovipositing inside small spangle galls, and the spring generation attacking galls ranging in size from very small to a centimetre or more in diameter (Askew 1965).

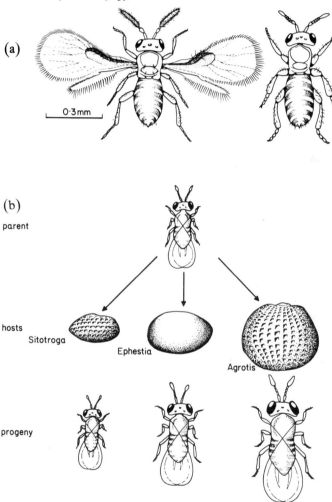

Figure 89. The effect of the host upon *Trichogramma* (Chalcidoidea, Trichogrammatidae). (a) Male *T. semblidis* reared from an egg of a moth (left) and an alder-fly (right). Note differences in wing development, antennae, and legs. (b) The relative sizes of *T. evanescens* reared from moth eggs of three different genera. (Both from Salt 1941.)

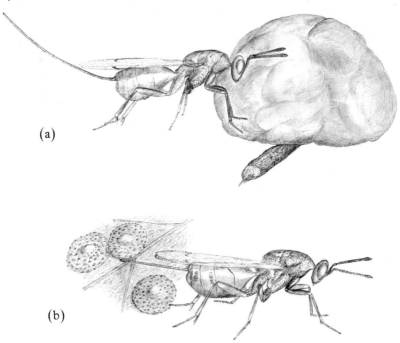

(a)

(b)

Figure 90. Dimorphism in *Torymus auratus* (Chalcidoidea, Torymidae). A high proportion of first generation females (a) have long ovipositors enabling them to parasitise hosts in large galls such as the oak apple, whereas second generation females (b) which attack only small spangle galls all have short ovipositors.

Seasonal variation in colouration (seasonal dichroism) is a more frequent phenomenon, especially amongst chalcids. It often happens that adults resulting from overwintering immature stages are more extensively darkened than those appearing in summer or autumn from preceding stages of the same year. Chalcids that exhibit this tendency include *Eulophus larvarum* (Gradwell 1958), *Anagyrus schoenherri* and *Aphycus albicornis* (Murakami 1960), *Olynx gallarum* (Askew 1961a), and some species of *Torymus* (Askew 1965). Flanders (1931) showed that the amount of dark pigmentation in *Trichogramma minutum* is determined by the duration of exposure to low temperatures during the early pupal period.

In *Eulophus larvarum* the pupae as well as the adults exhibit dichroism, the overwintering diapausing pupae being black, those producing adults a few weeks after their formation being pale in colour (Gradwell 1958). In some Ichneumonidae, the overwintering cocoons are of a more robust build than those constructed in summer and from which adults emerge the same year.

Host Defences

There is very little information about the avoidance of parasitism through hosts taking measures that reduce the chances of their detection by parasites. The habits of feeding inside mines and galleries in vegetable tissue or pupating in concealed situations fall in part into this category, but more subtle devices might be expected to exist. The jumping cocoon of *Bathyplectes* (page 154) may be mentioned again in this connection.

Once a host has been located, it may still possess mechanisms which hinder its penetration by the parasite's ovipositor. Violent struggling is one such mechanism, and it is often effective when the parasite is a good deal smaller than its host. Caterpillars wriggle vigorously when attacked, and aphids kick. Their struggles may repel the parasites or cause them to fall from foliage to safety. In addition to these defences, a host may have passive defences such as a hard cuticle or, in aphids, a covering of wax.

Lepidopterous pupae, held in the confines of a cocoon or silken girdle, have no means of getting out of range of a parasite attack. Cole (1959a) describes some of the defences of such pupae against attack by Ichneumonidae. When *Apechthis* contacts pupae of the small tortoiseshell butterfly (*Aglais urticae*), the abdomen of the pupa wriggles violently 'so that the Ichneumonid was often thrown into the air as soon as it alighted.' *Apechthis* stands on its host's pupa in order to oviposit, and even if it succeeds in maintaining a foothold on the shining surface of the butterfly pupa, it has great difficulty in pressing with its ovipositor on any part for more than a second or so. In practice, attacks upon newly formed pupae with a still-soft cuticle are most likely to be successful. The pupa of the speckled wood butterfly (*Pararge aegeria*), unlike that of the small tortoiseshell, is capable of only weak movements. It relies for its defence upon the smoothness and toughness of its cuticle. Ichneumonids that attack pupae enclosed in cocoons (i.e. moth pupae) do not usually have any difficulty in maintaining a foothold on the rough, immobile surface of the cocoon, but they have an extra barrier, which in some species is very hard, to pierce with their ovipositors. Even if the cocoon is not hard, as when it is made of folded leaves, an ichneumonid may still have difficulty in parasitising the contents by virtue of the agility of the pupa and the hardness of its cuticle. Again, these properties are developed only after about twenty-four hours, and newly formed pupae readily fall victim to the attack. Cole (1959a) points out that the most vulnerable parts of a pupa are the articulating membranes between the abdominal segments. Unfortunately for the pupa, its mobility is dependent upon these membranes, so that there is a conflict between mobility on the one hand and a reduction of vulnerable area on the other. *Apechthis* is equipped with an ovipositor that has a down-curved tip, an adaptation enabling it to be inserted beneath the hard sclerites of the pupal abdomen and through the articulating membrane.

Another pupal protective device depending upon mobility of the abdomen is the 'gin trap', found in the pupae of several beetles and a few moths (Hinton 1955). The gin trap in its most usual form consists of a slight deepening dorsally of an abdominal intersegmental groove coupled with the development of a sharp, sclerotised margin (Figure 91) which constitutes the jaws of the trap. At rest the jaws are open, but on upwards flexure of the abdomen the trap is closed sharply, nipping anything stationed between the jaws. Gin traps, it is suggested, may deter parasitic Hymenoptera from walking on the surface of the pupa.

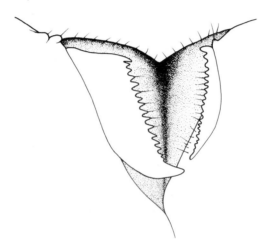

Figure 91. A gin-trap over an intersegmental division in the abdomen of a pupa of a tenebrionid beetle (*Alphitobius diaperinus*) (after Hinton 1955).

Even after a parasite has succeeded in laying its egg inside a host, the host may yet employ measures, this time of a physiological nature, which prevent the parasitisation from being successfully completed. Physiological defence reactions in parasitised insect hosts have been known for a long time, but our present understanding of the subject is due largely to the several contributions of Salt, brought together in a review published in 1963. The most commonly observed reaction is an aggregation of the host's haemocytes (blood cells) about the parasite, a response that is termed an encapsulation. The haemocytes move on to the parasite in large numbers, flattening themselves against its surface and later against the envelope of haemocytes already surrounding the parasite. The capsule is at first irregular in shape but later develops a smoother outline. A fully formed capsule often appears lamellate in cross section, especially towards the inside, owing to the flattening of the cells of which it is composed (Figure 92). Later the capsule may become differentiated into an outer, semi-opaque layer in which the cells retain their individuality, and an inner, translucent layer where the cellular structure has more or less degenerated. The inner layer surrounding the now much-distorted remains of the parasite eventually becomes transformed into a thin layer of non-living connective tissue, formed either from the cytoplasm of the haemocytes or from a secretion of the haemocytes, and at the same time cells may move from the outer layer back into the host's haemolymph. Very often in this form of encapsulation, melanin is deposited about the parasite as the inner cells of the capsule disintegrate. Phagocytosis of the parasite gradually takes place, at least during the earlier stages of capsule formation.

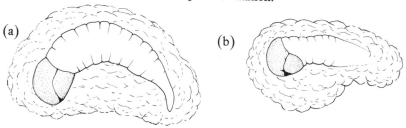

Figure 92. Encapsulation of a first instar larva of *Devorgilla canescens* by a caterpillar of *Diataraxia oleracea*. (a) 24 hours after the parasite egg was injected into the host, the parasite is surrounded by a capsule of blood cells. (b) 48 hours after injection the capsule has shrunk, the parasite larva is partly destroyed, and a melanin deposit has formed about the parasite's egg shell and in the mouth of the larva (from Salt 1963).

A rather different form of capsule, called a 'sheath capsule', has been described by Schneider (1950) and others. A sheath capsule forms very quickly, in about twenty-four hours, and it has the form of a thin but tough envelope which is brownish in colour owing to the presence of melanin. An accumulation of haemocytes precedes the formation of the sheath capsule as well as the cellular capsule.

Encapsulation leads to the eventual death of a hymenopterous parasite unless the capsule is incomplete; in this case vigorous wriggling might free the parasite from its investiture. Active movement by a parasite may also prevent formation of the capsule in the first place.

It seems most probable that death of encapsulated parasites results from suffocation. Branches of tracheae have frequently been observed to have grown towards capsules, indicating that there is an oxygen shortage in their vicinity. The susceptibility of young hymenopterous larvae to oxygen deficiency has already been discussed in connection with superparasitism (page 121).

An encapsulatory defence reaction has been recorded in fourteen orders of insects, and in all developmental stages except the egg. It is known to be provoked by parasitic insects belonging to the following families: Diptera—

Tachinidae; Hymenoptera—Ichneumonidae (except Ichneumoninae, Ephi-altinae), Braconidae, Encyrtidae, Eulophidae, Aphelinidae, Pteromalidae, and Eucoilidae. The form of capsule produced by a host species against a given parasite is usually constant. Different parasite species in the same host gener-ally also evoke a similar response. In contrast, the same parasite in different host species may elicit defence reactions which are visibly different. Bess (1939), observing encapsulation of *Coccophagus gurneyi*, an aphelinid chalcid parasite in species of *Pseudococcus* (Homoptera), noted that in *P. gahani* the capsule is reddish and it melanises slowly, in *P. citri* it is white and quickly darkens, and in *P. maritima* it takes the form of a thin sheath. Similarly, *Devorgilla canescens* (Ichneumonidae) produces different and characteristic responses in various species of microlepidoptera (Salt 1955). The egg of *Monoctonus paludum*, a braconid parasite of aphids, is enclosed in a sheath capsule by *Aulacorthrum circumflexum*, but in three other aphid species it is destroyed by other means, possibly by a humoral response of the host. In *Nasonovia ribisnigri* it develops successfully. *M. paludum* attacks all five aphid species equally readily (Griffiths 1961).

There are records of a parasite population being heavily encapsulated in one species of host but not at all in an allied host. Even different genetic strains of *Drosophila melanogaster* achieve different percentages of encapsula-tion against the cynipoid *Pseudeucoila bochei* (Walker 1959). Walker also demonstrated that the capacity to encapsulate this parasite varied with the age of the larva of *D. melanogaster*, maggots in the middle of their second instar being most successful in this respect. Other workers have found that, in general, the success of host larvae in encapsulating parasites tends to increase with age, presumably at least partly because they contain more haemocytes later in larval life. Sickly hosts tend to be less effective in disposing of their parasites by encapsulation and, conversely, moribund or inactive parasites may be more readily encapsulated.

The likelihood of a parasite being encapsulated may be inversely related to the degree of superparasitism (Puttler 1967, Askew 1968a) which suggests that the host's powers of encapsulation are limited and liable to exhaustion if too great a strain is placed upon them. A curious example of variability in expression of the defensive reaction is described by Pemberton and Willard (1918). *Tetrastichus giffardianus* (Chalcidoidea) is normally encapsulated by larvae of the melon fly, *Dacus cucurbitae*, but this reaction is suppressed when the host is also parasitised by the braconid *Opius fletcheri*. The braconid may weaken the host to the extent of preventing it from making its defensive reaction against the chalcid (Salt 1968).

The subject of encapsulation is complex, involving as it does so many variable considerations. Certain generalisations have, however, been made: 'On the basis of present knowledge it can be said that in general hymenopterous parasites are not encapsulated in their normal hosts, and that they do elicit defence reactions in unusual hosts' (Salt 1963). This generalisation cannot be extended to dipterous parasites of the family Tachinidae, for the larvae of many of these flies, living in their usual hosts, are enclosed by a respiratory sheath which is believed to be a modification of the host's defensive reaction (page 209); 'The great majority of tachinid larvae successfully meet the defence reactions of their hosts by diverting them' (Salt 1968). Neither does the generalisation seem applicable to the eulophid *Chrysocharis melaenis* which is encapsulated in *Agromyza demeijerei* but not in *Phytomyza cytisi*, even though the former host is preferred by the female chalcid and is, excepting for the defence reaction, more suitable for its development (Askew 1968a). However, accepting Salt's generalisation for most parasitic Hymenoptera, it seems that host species possess a defensive reaction which does not require conditioning to a particular parasite, and that the parasite species overcomes this reaction in a limited number of hosts—the 'usual' hosts. There must be strong, opposing, selective pressures operating; pressure to overcome the parasite on the part of the host and pressure to evade the defensive mechanism on the part of the parasite. Endoparasitic Hymenoptera generally have a narrower host range than ectoparasitic species, doubtless because their environment is intimately controlled by the host and their adaptations need to be more rigidly defined. One such adaptation must be resistance to the defensive reaction.

Our rather limited knowledge of how parasites resist encapsulatory defence reactions by their hosts has recently been summarised by Salt (1968). *Devorgilla canescens* eggs and first instar larvae are not encapsulated by *Ephestia*, the usual host. However, an egg whose surface has been physically or chemically altered promptly evokes a haemocytic reaction. Eggs removed from the ovarioles of *Devorgilla* are also encapsulated when injected into *Ephestia*, but eggs taken from the calyx and oviduct are immune. The calyx is a distal swelling of the oviduct and the ovarioles open into it. Rotheram (1967) showed that *Devorgilla* eggs gain a relatively thick, translucent coating to which adhere numerous particles about 1300 Å in diameter, in their passage through the calyx. It is believed to be this coating which confers on the egg immunity from encapsulation in *Ehestia*, although not in other hosts. The first instar larva of *Devorgilla* likewise owes its ability to hold the host's defensive reaction at bay to a property of its surface. By removing larvae prematurely from their eggs, Salt demonstrated that their protective surface was acquired between sixty-four and sixty-six hours of being laid. The first instar larva gradually loses some of its immunity, probably because growth causes extension of its cuticle and hence 'dilution' of its resistant covering. Such extension is maximal about the middle of the trunk, and girdles of encapsulation around this region observed by Salt (1964) may be thus accounted for.

Another hypothesis concerning parasite resistance proposed by Salt (1968) involves teratocytes. These are the cells, derived from the trophamnion in euphorine braconids and some other parasites, which float freely in the host's haemocoel, absorb food, and increase about ten times in diameter. The parasite larva, when it is approaching full size, feeds upon the teratocytes, and they have generally been considered to serve only a trophic function. Salt suggests instead that they may weaken the host by depriving it of food materials, and thereby reduce the likelihood of it being able to make an effective defence reaction. The situation may thus be analogous to the reduction in the defensive response in superparasitised hosts.

It is only when parasites are in the haemocoel of the host that they are at risk from encapsulation. Some endoparasites avoid direct contact with the haemocoel. The dryinid *Aphelopus* uses its trophamnion as a defensive barrier, and the platygasterid *Inostemma* is ensheathed in a membrane of connective tissue derived from the host's nerve cord. The membrane surrounding the embryo chain in polyembryonic Encyrtidae and *Platygaster* may be another device for warding off the host's haemocytes. Some Encyrtidae lay their eggs in host organs rather than in the haemocoel, and the larvae emerge only when they are vigorous enough to challenge the host reactions successfully, or possibly after they have gained immunity (Salt 1968). Tachinids (Diptera) whose eggs are eaten by caterpillar hosts, rapidly make their way, as first instar larvae, from the alimentary tract to the safety of a nerve ganglion or salivary gland. Here they remain for a time, apparently developing an immunity which they did not previously possess (Strickland 1923).

Another site in which endoparasites are not at risk from encapsulation is in host eggs. It is suggested by Salt that egg-larval parasites, species that oviposit in eggs but whose larvae complete their development in the host larva, may acquire immunity whilst in the egg. Such parasites might be able to enjoy a wider host range than allied larval parasites. Some support for this view is provided by the egg-larval eulophid parasite *Entedon ergias*. Of the nineteen palaearctic species of *Entedon* for which hosts are known, *E. ergias* is the only one recorded from hosts belonging to more than three genera and two families. *E. ergias* is known to parasitise beetles from six genera and three families (Bouček and Askew 1968), and it is the only species in the genus known to oviposit in eggs.

It is not yet known how a period spent in a host egg, or for that matter in a host organ, may prevent encapsulation at a later stage in the host's haemocoel.

Phoresy

The practice of phoresy, the purposive use of one species by another for transport, has already been described in connection with lice. Parasitic Hymenoptera provide further examples of the habit. Either first instar larvae or adults may be the phoretic stages.

Chalcids of the family Eucharitidae, and those of the allied Perilampidae, are said to be hypermetamorphic because there are two very different types of larva in the life cycle; an active first instar planidium larva which moults into a sedentary, typical, hymenopterous grub once parasitism has been effected. The function of the planidium is to gain access to the host which, in the Eucharitidae, is always an ant larva or pupa. Many eggs are laid by each female, and they are inserted with the ovipositor into incisions made in leaves, flower buds, or fruit, well away from the eventual host. Plants selected for oviposition are usually infested with aphids, and worker ants crawl over the foliage in large numbers, attracted by the honeydew. The planidia, on hatching, attach themselves to these ants and are carried to the nest, where they are able to move on to the ant larvae and complete their development.

Transport of adult parasites by adults of the host species sometimes occurs when the parasite needs to attack newly laid eggs and when the adult host is large enough not to be unduly hampered by its unwanted passenger. Several species of Scelionidae (Proctotrupoidea) are known to be phoretic. All scelionids are egg parasites, and phoresy has been noted in species that attack Orthoptera, Hemiptera, and Lepidoptera. Adult females cling to the female host and remain there until eggs are laid, whereupon they quickly detach themselves to insert their own eggs into those of the host. *Phanurus beneficiens* has been found in numbers on the moth *Schoenobius* (van Vuuren 1935). The wasps are located only on female moths, attached mostly to the wings. In this case the relationship is not obligatory.

A much more intimate association exists between *Rielia manticida* and its praying mantid host (Rabaud 1922). Female *Rielia* are at first winged, but on attaching themselves to the mantid they discard their wings and cling with their mandibles to the abdomen or wing bases. Most, though not all, infest female mantids. As well as using the mantid as a means of transport to its host, a mantid egg, the female *Rielia* feeds on the body fluids of the mantid, and should the latter die, the *Rielia* also quickly perishes. The mantid egg masses are enveloped in a frothy secretion which rapidly hardens into a firm covering. *Rielia* is able to oviposit in egg masses only whilst the covering is still fluid, and hence it needs to be on the spot when the eggs are laid. The fact that *Rielia* feeds upon its carrier transforms the relationship of the adult insect into one of parasitism rather than phoresy, and *Rielia* is one of the very few adult insects that may be said to parasitise an invertebrate host.

Another mantid egg parasite, *Podagrion*, a chalcid of the family Torymidae, is also sometimes phoretic but only facultatively so, and it does not feed upon its carrier. Species of the chalcid family Trichogrammatidae are, like the Scelionidae, exclusively egg parasites, and another example of phoresy is provided by *Oligosita xiphidii*. Ferrière (1926) records finding females of this trichogrammatid attached to the wings of a long-horned grasshopper, *Xiphidion*.

Pteromalus puparum, a familiar chalcid parasite of the large white butterfly (*Pieris brassicae*), exhibits phoresy of a rather different type. The chalcid is unable to oviposit in butterfly pupae that have hardened, and occasionally female *Pteromalus* may be seen riding on the backs of fully grown caterpillars, awaiting their pupation.

Diapause and the Synchronisation of Parasite and Host Life Cycles

It is often the case that a parasite with a relatively short reproductive period attacks a host that is available for only a brief period in the year. In such a situation it is obviously vital for the parasite to be synchronised with its host. Since a parasite is at least slightly smaller than its host, it has an inherent capacity to complete its development more quickly. Often the parasite must delay its own development so that its imaginal state coincides with a vulnerable host stage. The most direct method of synchronisation is for the development of the parasite to be controlled by the hormones produced by the host for the regulation of its own metamorphosis.

Chelonus annulipes is a polyphagous braconid parasite of moth larvae, laying its eggs in the eggs of its host. The parasite eggs soon hatch, but development is held at the first instar larval stage until the host larva is fully grown and ready to pupate. Physiological changes occurring in the host just prior to pupation stimulate the parasite to resume its development. In this way life cycles of host and parasite are of approximately the same duration, and adult parasites emerge at the time when moths are laying eggs. If the host is a species which passes through a number of generations in the year, then *C. annulipes* is multivoltine also; if the host has a single annual flight period, then *C. annulipes* is univoltine (Bradley and Arbuthnot 1938). Several other species of parasitic Hymenoptera (Salt 1941) develop as far as first instar larvae and then await a stimulus from the host before proceeding further.

In addition to ensuring synchrony of life cycles, the delay in development of the parasite beyond its first instar permits the host to continue growth and thus ultimately provide the parasite with sufficient food.

The chalcid *Trichogramma cacoeciae* as an egg parasite of the tortricid moth *Cacoecia rosana* provides another example (Marchal 1936). *C. rosana* is univoltine. Adults lay eggs in July and these remain in diapause (i.e. arrested development) until the following March. Development then progresses rapidly, and adult moths are on the wing by July. *T. cacoeciae* completes two generations a year. It lays its own eggs in freshly laid moth eggs in July. The larvae soon hatch, and feed rapidly to become fully grown after a few weeks, but they remain in this condition until the following March when they pupate. Adult *Trichogramma* emerge in April to attack unparasitised eggs in the same host egg batch from which they themselves emerged. The progeny of this second generation are in hosts which are not in diapause, and they develop without delay to emerge as adults in July. The difference between eating a diapausing host egg and one which is developing determines whether or not the parasite will go into diapause. This is clearly shown by the fact that *T. cacoeciae* may pass through a succession of five or six generations a year in the eggs of the cabbage moth, *Mamestra brassicae*, which do not diapause.

Schoonhoven (1962) found that development of a tachinid (Diptera), *Eucarcelia rutilla*, is similarly controlled by the hormones of its host, the moth *Bupalus piniarius*.

An ichneumon parasite of hover-flies, *Diplazon fissorius*, has been investigated by Schneider (1950, 1951). *D. fissorius* does not develop beyond the first instar until its host has pupated. By transplanting first instar larvae into hosts with either a different type of life cycle or at a different stage of development, Schneider was able to retard or advance the development of the parasite, thus showing that the hosts exert a direct effect upon their parasites. The cuticle of the first instar larva is thin and permeable to host hormones, and it is very probable that these hormones regulate the metamorphosis of the parasite.

Diapause is a state of arrested development; it is not an immediate response to an unfavourable environment like the quiescence caused directly by chilling an insect, but a hormonal effect which is induced at a receptive or sensitive stage of the insect's life cycle. In northern temperate regions many insects enter diapause in response to the influence of low autumnal temperatures or, more commonly, shortening periods of daylight. They remain in a condition of diapause throughout the winter until another environmental stimulus supervenes to 'break' diapause and allow development to recommence. The breaking of diapause is itself regarded as a physiological development which takes place only under specific conditions. The winter diapause of many insects is broken by a period of low temperature. Diapause provides the means by which the life cycle of an insect is kept in step with seasonal conditions in its environment.

Bradley and Arbuthnot (1938), followed by Andrewartha (1952) and Lees (1955), do not consider the delays in development of parasites described at the beginning of this section to be examples of diapause. This is because the quiescent state is induced by a direct influence of the environment, the host's hormones, not intermediated by the parasite's own hormonal system. Nevertheless, as Danilevskii (1965) points out, it is analogous in physiological mechanism to a diapause condition.

Diapause indisputably occurs in the development of *Apanteles glomeratus* (Braconidae) (Maslennikova 1959). Larvae of *A. glomeratus*, inside caterpillars of the cabbage white butterfly, *Pieris brassicae*, are sensitive to the length of daylight (photoperiod), and when, as autumn advances, this falls below a critical level, they are induced to enter a prepupal diapause. That this is a response to the external environment and not effected through the host's hormonal system was shown by Maslennikova in a series of refined experiments, the results from which can be seen in Figure 93. *P. brassicae*, like *Apanteles*, is sent into diapause by short photoperiods, and there is a geographical variation in the critical photoperiod which stimulates diapause. The

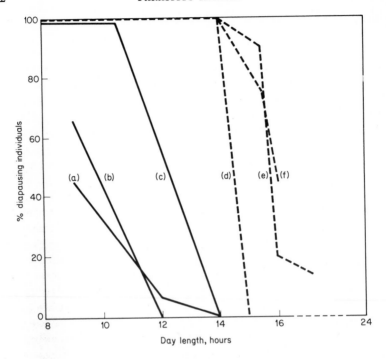

Figure 93. The effect of day length on the initiation of diapause in *Apanteles glomeratus* and its host, *Pieris brassicae*, from different regions of the U.S.S.R., reared at a constant temperature of 18°C. (a) Sukhumi *Apanteles* in Sukhumi *Pieris*. (b) Sukhumi *Apanteles* in Leningrad *Pieris*. (c) Unparasitised Sukhumi *Pieris*. (d) Unparasitised Leningrad *Pieris*. (e) Leningrad *Apanteles* in Leningrad *Pieris*. (f) Leningrad *Apanteles* in Sukhumi *Pieris* (from Danilevskii 1965 after Maslennikova).

hosts used by Maslennikova were from populations from Sukhumi in Georgian U.S.S.R. (43°N) and from Leningrad (60°N). At 18°C the Sukhumi *P. brassicae* enter diapause with a photoperiod less than twelve hours and the Leningrad *P. brassicae* with a photoperiod under fifteen hours. This is correlated with the longer periods of daylight at higher latitudes in summer; at 60°N a decrease in photoperiod below fifteen hours heralds the time of year when temperatures become too low for steady development. The *A. glomeratus* populations from Sukhumi and Leningrad also respond to different photoperiods, and in their case the critical periods are nine hours and sixteen hours respectively. The independence of *Apanteles* from the condition of its host was clearly demonstrated when Maslennikova reared, at 18°C, Sukhumi *Apanteles* in Lenigrad *Pieris* and *vice versa*. The parasites responded to the critical photoperiods of their regions of origin, irrespective of the host strain they were in.

In addition to being able to diapause as prepupae, larvae of *A. glomeratus* may diapause in their first instar. *Aporia crataegi*, the black-veined white butterfly, unlike *P. brassicae*, is a univoltine host of *A. glomeratus* which over-winters as a diapausing larva. *Apanteles* in an *Aporia* larva passes the winter as a first instar larva and remains at this stage until the host resumes develop-ment in spring. No matter how external conditions of photoperiod and tem-perature are adjusted, the first instar *Apanteles* larva will not develop so long as its host remains in diapause. As a parasite of *Aporia*, *A. glomeratus* is there-fore univoltine and synchronised with its host.

A. glomeratus is a parasitic species adapted to the voltinism of its hosts through the response of its first instar larva to the host's hormones. In addition, its independent response to photoperiod as a fully grown larva ensures that it does not get out of step with seasonal conditions. This dual regulation of development, which might be true for a wide range of parasites, enhances ability to utilise a variety of host species (Danilevskii 1965).

Caraphractus cinctus, a mymarid endoparasite of eggs of water-beetles, diapauses as a prepupa in eggs of *Agabus*. It has been shown by Jackson (1963) that diapause is induced by relatively short photoperiods of less than fifteen hours a day. Direct development occurs when the parasite is reared under a regime of between $16\frac{1}{2}$ and 24 hours light a day. The chorion of the *Agabus* egg is quite transparent and light of a very low intensity (0.03 to 0.5 foot candles) is adequate to avert diapause. It is interesting to compare *C. cinctus* with *Trichogramma cacoeciae*, the other egg parasite described in this section. *Trichogramma* relies upon its host egg to determine the course of its develop-ment. *Caraphractus* is dependent, instead, upon the external environment, and it passes through a true diapause state. Eggs of *Agabus*, the food of *Cara-phractus*, do not diapause. Diapause of *C. cinctus* is broken by low tempera-tures.

An example of temperature-induced diapause is provided by the work of Simmonds (1948) on *Spalangia drosophilae*, a chalcid ectoparasite of pupae of *Drosophila melanogaster*. Diapause occurs in the last larval instar of *Spalangia*, but its incidence is very variable. The lower the temperature in which the larvae develop, the higher the incidence of diapause. Simmonds also found that the temperature at which the female *Spalangia* is reared influences the number of her progeny which diapause. If she is reared at a low temperature, then few of her offspring diapause. In other words, diapause is most likely in a larva that has been reared at a low temperature but whose mother was reared at a high temperature (Figure 94). This is an example of the delay that there might be between the stimulus to diapause and its manifestation.

Another pteromalid chalcid, *Nasonia vitripennis*, may diapause as a fully grown larva. The number of diapausing larvae is increased by chilling the

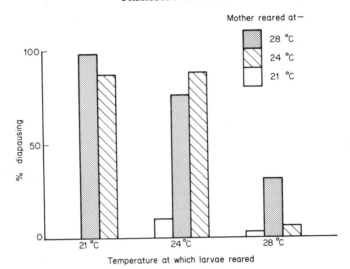

Figure 94. Diapause in final instar larvae of *Spalangia drosophilae* (Chalcidoidea). The lower the temperature at which the larvae are reared, the higher is the incidence of diapause, unless their mother was also reared at a low temperature (data from Simmonds 1948).

mother during oogenesis, or by providing her with few hosts (Schneiderman and Horwitz 1958); also, as females get older, they generally produce broods with an increasing proportion of diapausing progeny (Saunders 1962).

So far, examples have been described of how parasites are synchronised with their hosts by allowing their development to be controlled by the host. There are a few instances of the reverse situation; parasites affecting the development of their hosts. Larvae of the dipterans *Lipara lucens* and *Urophora jaceae* overwinter as diapausing larvae. However, if they are parasitised, diapause is terminated in the autumn and they pass the winter as pupae (Varley and Butler 1933). Mechanical wounding of the larvae during parasite oviposition may be instrumental in causing the premature pupation. The parasites perhaps benefit by virtue of the protection afforded them during winter by the hard host puparium.

Another example of a parasite influencing its host is provided by the endoparasitic braconid *Aphidius* which may cause an aphid nymph to develop into an apterous instead of a winged adult. This is believed to be the result of juvenile hormone escaping from the *Aphidius* into the host's haemocoelic fluid (Johnson 1959).

9. Protelean Parasitic Diptera

THE DIPTERA ARE second only to the Hymenoptera as practitioners of protelean parasitism. The larva of Brachycera and Cyclorrhapha, typically cylindrical in shape and lacking legs or other protuberances that might hinder movement inside the body of another animal, and with mouth hooks capable of lacerating living tissue, is a most versatile insect form that has several times become adapted to a parasitic life. It has even been argued (Keilin 1915) that parasitism was the ancestral larval habit in the Cyclorrhapha. Among Nematocera parasitism is of much less frequent occurrence.

An organ of fundamental importance in the armoury of the Hymenoptera is the six-valved ovipositor. The Diptera do not possess such a weapon and in consequence their hosts are usually species that are not protected in galleries in wood, leaf-mines, or galls. In addition, the absence of biting mouthparts in adult flies would make escape from hosts in such places very difficult. A number of families of Diptera have developed what may be described as a substitute ovipositor. In females in these families, the terminal abdominal segments are drawn out to form an elongated, slender tube which carries the female genital opening near to its apex. Accessory structures may also be developed on the abdomen, for example in some Tachinidae (Figure 103a) and Conopidae, to aid in the act of larviposition or oviposition.

The dipterous antenna is vastly different from the hymenopterous antenna. The latter is long and flexible, ideally suited to probing and testing the substrate, and for conveying to the parasite information about the immediate surroundings of a potential host or about the host itself. The very much shorter antennae of Diptera are unable to do this. Diptera do have sense organs located on their tarsi for the reception of stimuli emanating from the substrate, but these are hardly likely to be so effective as the questing antennae of parasitic Hymenoptera in host location and examination.

It may be a result of these differences in ovipositor, antennae, and mouthparts that proportionately more parasitic Diptera than parasitic Hymenoptera practise hypermetamorphosis and transfer the onus of precise host location from the adult to the first instar larva.

Several families of Diptera are exclusively parasitic, and these tend to be closely restricted to taxonomically clearly-definable host ranges. An exception is the family Tachinidae in which is found something of the diversity of host relationships more usually associated with parasitic Hymenoptera. Also exceptional are the Calliphoridae, a family that shows great adaptability in exploiting a range of feeding habits which includes saprophagy, parasitism, and predation. This versatility is sometimes expressed even at the species level.

Although the parasitic Diptera are not quite so biologically diverse as the parasitic Hymenoptera, their hosts come from more animal groups. Other insects are the most usual hosts, but slugs, earthworms, snails, centipedes, and spiders are also attacked by occasional species or genera. In addition, a large number of Diptera, which are the subject of another chapter, are facultative or obligatory larval parasites of vertebrates. Unlike the parasitic Hymenoptera, very few parasitic Diptera parasitise members of their own order.

Chironomidae (Nematocera)

Nearly all chironomids have aquatic larvae, and several of these are regularly associated with other animals. In the subfamily Chironominae the association is with sponges, bryozoans, and molluscs, whilst in the subfamily Ortho-cladiinae several species live on the larvae of other insects (Thienemann 1954). Sometimes the association is simply a commensal one. *Epoicocladius*, for instance, merely eats the remnants of its mayfly host's food. Often, however, the relationship is an ectoparasitic one. *Symbiocladius* also lives on larval mayflies, but it feeds upon the host's body fluids. In many cases the exact nature of the association has not been investigated, but among the Chiro-nominae *Chironomus limnaei* and species of *Parachironomus* are parasites of water snails, and species of *Demeijerea* and *Xenochironomus* are endoparasites of sponges (Steffan 1965).

Cecidomyiidae (Nematocera)

The larvae of most Cecidomyiidae live in plant tissue where they induce the formation of galls, but a few behave as typical protelean parasites of Hemiptera Homoptera. A species of *Endopsylla*, investigated by Lal (1934) in Scotland, lays its eggs on the wings of *Psylla*. The eggs are attached by a stalk which is embedded in the wing membrane. At first the young larvae are ectoparasitic, but after about four days they penetrate an intersegmental membrane to consume the host from inside. When fully grown the larvae emerge and pupate in the soil. *Endaphis perfidus*, as a parasite of the sycamore aphis, *Drepanosiphum platanoides*, has a similar life history (Barnes 1929, 1930).

The genus *Miastor* exhibits a very curious reproductive phenomenon called paedogenesis. Since the vertebrate foetus is sometimes likened to a parasite of its mother, I somewhat hesitantly mention *Miastor* here. The ovaries of *Miastor* develop precociously whilst the midge is still a larva. They contain a few large eggs which hatch whilst still inside the parent larva, and the secondary larvae devour their 'hosts' like endoparasites. Several larval generations may be passed in this manner before an adult midge is produced.

Acroceridae (Brachycera)

This family is also known as the Cyrtidae. It contains only about 250 described species, and all of these, so far as is known, are internal parasites of spiders. Early records of predation in spiders' egg cocoons are almost certainly erroneous. The species tend to be rather polyphagous, and they mostly attack

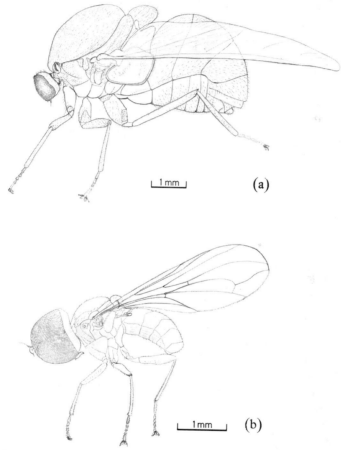

Figure 95. Adults of two Diptera whose larvae are parasitic. (a) *Ogcodes gibbosus* (Brachycera, Acroceridae). (b) *Pipunculus campestris* (Cyclorrhapha, Pipunculidae). *Ogcodes* has a planidium larva and lays its eggs apart from the host whereas *Pipunculus* oviposits directly in the host. The relatively enormous eyes of *Pipunculus* (contrast *Ogcodes*) are probably used to locate a host.

spiders that do not spin large silken webs. The families Lycosidae, Clubionidae, and Salticidae are especially prone to attack.

The adult fly is characteristically hunch-backed in appearance, having a
large, rounded thorax, very small head, and globular abdomen. The eggs are
laid in small clusters or rows, usually in exposed situations such as on fence
posts or dead tree stumps, and several females may congregate at these oviposition sites. Whilst standing in areas of limestone grassland in southern
England, I have had, on two occasions, about twenty *Ogcodes gibbosus* (Figure
95a) hovering about me. They appeared to be especially attracted by a white
sweep-net. In southern Wales I have seen equally large numbers of *Acrocera
globulus* hovering about the lower branches of isolated oak trees. These aggre-
gations were all observed in mid-summer in warm sunshine following showers
of rain.

Acroceridae lay large numbers of eggs. *Pterodontia flavipes* has been seen to
lay 2 300 eggs in less than an hour. The sites selected for oviposition are not
obviously associated with the proximity of hosts, and adult flies pay no attention to spiders. This is related to the small head lacking large sense organs, and
the high reproductive potential. The first instar larva is a planidium. It moves
by 'looping' or jumping, propelled by its caudal cerci. The planidium of one
species has been observed to creep along the silken strand of a spider's web.
Whilst waiting for a host the planidium stands erect on the tip of its abdomen.
Contact with the host is fortuitous. The larva is able to survive for about a
week without food, but inevitably a high mortality is suffered during this stage
of the life cycle. Once it has made contact with a spider, the planidium enters
it through an articulating membrane on the host's leg and then makes its way
to the opisthosoma. Here development proceeds at a rather slow rate until the
parasite larva is fully grown. It then emerges from the body of the doomed
spider, via the respiratory opening, and quickly pupates, sometimes amongst
silk produced by the dying host. Occasionally two acrocerids have been
observed to emerge from a single spider (Eason *et al.* 1967).

Bombyliidae (Brachycera)

The Bombyliidae is the largest of the three families of Brachycera that include
parasitic representatives. There are about 3 000 species of bee-flies distributed
over most parts of the world. They are sun-loving, medium-sized flies, usually
densely hairy with darkened areas on their wings, and often with a bee-like
appearance. The Systropinae, however, appear bare and are elongated flies
which mimic sphecoid wasps. Many have a very elongated proboscis directed
straight forwards and used for extracting nectar from flowers.

The commonest of the dozen species found in Britain, *Bombylius major*
(Figure 96), is an insect of the early spring and can often be seen hovering in
front of primrose flowers from which it takes both nectar and pollen. It is able
to fly when the temperature climbs to about 16.6°C, a lower flight threshold

temperature than that of the smaller, less hairy muscoid flies about at the same time of year (Knight 1968). As the ambient temperature approaches the flight threshold value, *B. major* engages in rapid wing-whirring, like a hawk moth, to boost its internal temperature. The first instar larva is an active planidium which, like that of the nemestrinids discussed below, is capable of burrowing in soil. Knight observed that *B. major* most frequently drops its eggs whilst hovering over bare patches of ground colonised by solitary bees (*Andrena*,

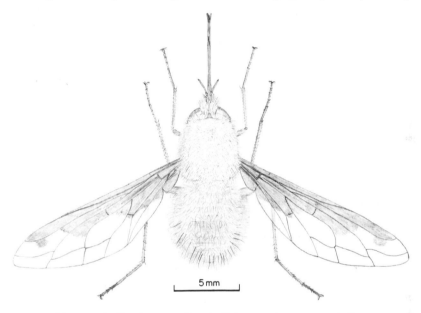

5 mm

Figure 96. Brachycera, Bombyliidae. A female *Bombylius major*.

Halictus) which are its hosts. It may also oviposit, however, whilst it is hovering in front of flowers or walking on the ground.

Most larval bombyliids are either ectoparasites of immature aculeate Hymenoptera, sometimes feeding on the food stored within the cell as well as on the host itself, or else they are predators, feeding upon the eggs of locusts. Species of *Systoechus* provide examples of the latter habit. A few species are ectoparasites upon beetle larvae living in soil, especially Scarabaeidae. The tiger beetle *Cicindela* is the host of *Anthrax analis* (Batra 1965). Some hosts belonging to other orders are also attacked but all have in common the fact that they are situated in light soil or sand. *Thyridanthrax* includes species which are endoparasites of tsetse fly pupae. Some Bombyliidae are polyphagous. *Villa alternata*, for instance, has been reared as an endoparasite of pupae of both noctuid moths and tenebrionid beetles.

Species that feed on noctuid moths enter the caterpillars as planidium larvae and emerge as adult flies from the pupae. The fly pupa may be armed with spines to enable it to partially force its way out of the moth pupa.

As in all groups having a first instar planidium larva, this body form is lost at the first larval moult, to be replaced by a sedentary maggot adapted for feeding on the host.

Nemestrinidae (Brachycera)

Nemestrinids are most abundant in the tropics. They are allied to the Bombyliidae and, like the bee-flies, the adults have a long proboscis. About 250 species have been described and the majority of these are probably endoparasites of beetle (especially chafer) larvae or pupae. A number of species, however, develop inside locusts. As in the other two brachyceran families of parasitic flies, it is the first instar planidium larva which must make contact with the host. The numerous eggs are laid in cracks and crevices in wood, and frequently at the entrances to the burrows of wood-boring beetles and bees. Wind is a direct stimulus to oviposition in *Trichopsidea clausa*. A wind speed of a little under 3 m.p.h. is sufficient to start egg-laying, and it is perceived by sense organs on the substitute ovipositor (Spencer 1958). The young larvae drop to the ground, although often they are dispersed by wind. The first instar larvae of *T. clausa* and *Neorhynchocephalus sackeni* have been observed (Spencer 1958) to jump into the air when a breeze strikes them.

Trichopsidea ($=$ *Symmictus*) is a genus whose species parasitise adult locusts of the genera *Locusta* and *Schistocerca*. In this case the host is so large in comparison to its parasite that it is often able to survive parasitism for a considerable period, even to the extent of laying eggs. The larva of *Trichopsidea* is associated with a respiratory tube (Figure 97) inside the host which recalls the tachinid respiratory funnel (page 209) but whose origin is uncertain. The tube is brown, probably cuticular, and it connects the parasite larva with the body wall of its host. A small, brown scar on an intersegmental membrane of the locust marks the point of entry of the planidium, and from this point the tube extends internally, gradually increasing in diameter until it ends in the locust's haemocoel, enclosing the posterior end of the *Trichopsidea* larva. The respiratory tube is spirally thickened and it widens and lengthens as the parasite larva grows. A respiratory tube of similar structure is associated with the other genus of grasshopper parasites, *Neorhyncocephalus* (Spencer 1958). When fully grown, the *Trichopsidea* larva emerges from the host and burrows into the ground. It may enter diapause and defer pupation until the ground becomes damp. This plays a part in synchronising the life cycle with host availability, because invasion of an area by locusts, and their subsequent breeding, is closely associated with increased rainfall (Greathead 1958). Hynes (1947) found that a bombyliid, *Systoechus somali*, is similarly induced to pupate by rainfall; *S. somali* attacks the egg pods of *Schistocerca gregaria* and can survive as a larva under dry conditions for at least three years.

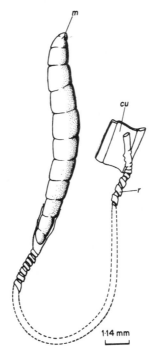

Figure 97. A third instar larva of *Trichopsidea costatus* (Brachycera, Nemestrinidae) with respiratory tube attached to the host's integument. *m*= mouth, *r*= respiratory tube, *cu*= host cuticle (from Greathead 1958).

Phoridae (Cyclorrhapha)

Placed here in the suborder that includes the 'higher' Diptera, the Phoridae nevertheless have certain features in common with the Brachycera, and their systematic position is the subject of debate. They are small flies and in life readily recognised by their very rapid, scuttling gait and characteristic, hump-backed thorax. The family is a large one with about 250 species known to occur in Britain. Phorids have a rather similar range of larval feeding habits to the Calliphoridae, most species being associated with decaying organic material, particularly of animal origin. Several of the species that habitually feed in excrement or on dead animals have also been reared from still-living but mori-bund animals, and the frequency with which this is recorded suggests that many phorids are facultative, polyphagous parasites. It is thought that most of the species that have been reared from invertebrate hosts initially enter the body through small wounds. Termites, bees, crickets, caterpillars and pupae of moths, and fly larvae have all had Phoridae reared from them. A few British species that are known to be parasitic will be mentioned.

Megaselia elongata (= *cuspidata*) appears to be a regular parasite of the large millipede *Schizophyllum sabulosum,* laying its eggs in minute wounds in the host's integument (Picard 1930). Adult coccinellid beetles are attacked by *Phalacrotophora* species, and their pupae by *Megaselia fasciata.* The latter species oviposits on the surface of the pupa and the first instar larvae penetrate it to develop gregariously internally before emerging to pupate. The female *M. fasciata* has a substitute ovipositor with which it stabs the pupa to obtain fluid on which it feeds. *Borophaga* (= *Hypocera*) *incrassata* is a solitary parasite, feeding inside the larva of the St. Mark's fly, *Bibio marci* (Morris 1922), one of the few examples known of a fly parasitising a fly.

Large numbers of Phoridae are associated with ants, many being nest scavengers, but there are also some extremely interesting commensal and parasitic associations. Donisthorpe (1927) describes in detail the behaviour of the very small British species *Pseudacteon formicarum* which parasitises worker ants of the genus *Lasius.* 'The little fly hovers over the ants, flying very steadily, and getting nearer and nearer to an ant, which it strikes at. If one keeps quite still it will strike at ants on one's hand; and it is very amusing to watch an ant, which has become aware of the presence of the fly, making a dash for safety pursued by the fly. The female *P. formicarum* possesses a very sharp pointed ovipositor, which is somewhat bent forward, and can be exserted and retracted. With this instrument she lays her eggs between the free segments of the ant's gaster . . . She is attracted first by the sense of smell, as a fly will strike at one's hands (when no ants are there) after they have become scented by the ant's acid; and subsequently by sight, when near enough to see the ants.'

The species of *Aenigmatias* (= *Platyphora*) are also parasites of worker ants, and the females are curious, wingless flies well-adapted to crawling about in ants' nests. *Metopina pachycondylae* lives in nests of ponerine ants in Central America. Its larva lives entwined about the neck of an ant larva. Whenever a worker ant feeds the ant larva, the *Metopina* interposes itself and steals a portion. The association is a commensal one and the ant and fly pupate side by side, the fly pupa enclosed in the silken cocoon spun by its host. The ant is the first to emerge from pupation and its cocoon is then carried from the nest by other worker ants and discarded. The phorid, when it emerges, has therefore little to contend with.

Another species of Phoridae, found in the West Indies, pounces on worker leaf-cutter ants (*Atta*) and quickly lays an egg on the neck, at the same time somehow inducing a short-lived paralysis in the ant. The phorid larva feeds in the head of the stricken ant and consumes its brain. An interesting stratagem employed by *Atta* workers as a defence against this phorid, which is attributed to the genus *Apocephalus*, has recently been described (Eibl-Eibesfeldt and Eibl-Eibesfeldt 1968). Workers collect pieces of leaves and carry them back to their nest in their mandibles. At such a time they are in no position to defend themselves against the sudden strike of *Apocephalus*. However they are pro-

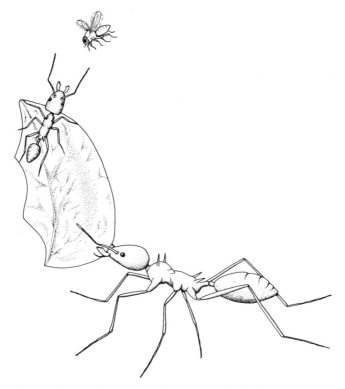

Figure 98. Small worker leaf-cutter ant defending a larger worker against the attack of *Apocephalus* (Cyclorrhapha, Phoridae) (redrawn from Ovenden in Eibl-Eibesfeldt and Eibl-Eibesfeldt 1968).

tected by a very small type of worker (*minima* worker) which rides on the piece of leaf, stationing itself on the top edge, and confronting any attacking *Apocephalus* with gaping jaws (Figure 98).

Pipunculidae (Cyclorrhapha)

After the miscellany of phorid biology it is a contrast to turn to the undeviating habits of the Pipunculidae, or Dorilaidae as they used to be called. Pipunculids (Figure 95b) are small, dull-coloured flies that are rendered outstanding by their enormous heads. The head is globular and composed mostly of the compound eyes. The neck is slender, giving mobility to the head, but at the same time making pipunculids susceptible to decapitation, at least as dried specimens in a collection. Considerable prowess in the air is another hallmark of the family: 'the most exquisite fliers that exist in Diptera' (Verrall 1901). They are accomplished hoverers and will sometimes fly backwards for short distances. *Eudorylas zonatus* has been observed to copulate in mid-air (Bristowe

1950). In their flight Pipunculidae demonstrate their affinity with the hover flies (Syrphidae), but unlike the latter they do not spend much time feeding at flowers.

All known species of Pipunculidae, numbering about 400, are solitary endoparasites of Hemiptera Homoptera, especially of Cicadellidae (= Jassidae) (leaf-hoppers), Cercopidae (frog-hoppers), and Delphacidae. Coe (1966) summarises the known host relations of the eight genera contained in the family. Almost invariably (*Eudorylas* is the exception) each genus confines itself to a single host family and many species are host-specific, although some range over several genera.

Genus	Host family
Chalarus, Alloneura,	
Pipunculus, Eudorylas	Cicadellidae
Eudorylas	Flatidae
Verrallia	Cercopidae
Cephalops	Delphacidae
Undescribed species?	Membracidae
Dorylomorpha, Nephrocerus	unknown

The pipunculid female pounces on a nymph (or adult in the case of *Verrallia* which attacks hoppers whose nymphal stages are protected by 'cuckoo-spit') of the host, and flies with it into the air. Vision probably plays a large part in host location. Jenkinson (1903) describes hunting by *Verrallia aucta:* 'As soon as a *Verrallia* saw a frog-hopper it poised itself in the air (like a kestrel hovering, but with a certain intensity perceptible in its motionlessness) and if the position of its victim was favourable it pounced upon it immediately.' The female pipunculid has a substitute ovipositor which is inserted through an intersegmental membrane on the ventral surface of the host's abdomen, and the egg is laid in the haemocoel. After oviposition the parasitised hopper is dropped, and it makes its way back to its food plant.

In Mauritius the sugar-cane hopper (*Perkinsiella saccharicida*), a delphacid, is parasitised by *Cephalops mauritianus* which has been studied by Williams (1957). The host is generally attacked in its late nymphal stages and only exceptionally as an adult. The pipunculid has only two larval instars. When fully grown, the larva occupies nearly all of the host's abdomen, which in consequence appears much distended. Eventually the parasite splits the dorsal integument of the host to crawl out and soon afterwards pupate in the soil. Some other pipunculids pupate on vegetation. The host quickly dies. Parasitised ('pipunculized') hoppers often have aberrant genitalia and malformed gonads, similar to those of stylopised hosts (page 241). Male and female hoppers are both attacked, apparently impartially. Pipunculized females never mature eggs. Williams never reared more than one fly from a hopper, and he suspected that females might be able to detect parasitised hoppers and avoid superparasitism. The biology of *C. mauritianus* is probably fairly typical of that of the family in general. However, very few species have been studied in any detail, which is a little surprising in view of the economic importance of many of their hosts.

Conopidae (Cyclorrhapha)

This is another family that is exclusively parasitic, the species usually attacking adult bees or wasps. Oviposition very often occurs in flight, the flying host being pursued and pounced upon. It may be held for a few seconds by a pincer-like organ at the apex of the female fly's abdomen. The egg is inserted into the host's abdomen by means of a substitute ovipositor. Raw (1968) witnessed *Leopoldius coronatus*, a southern European species, attacking workers of *Vespula* and *Polistes*. When the host was sighted, it was approached by the conopid in a zig-zag course which resembled the flight of the wasp, quite distinct from the straight flight adopted at other times. *L. coronatus* flew on to the back of its victim and remained there ovipositing for five to ten seconds.

Conopids (Figure 99), like pipunculids, rely a great deal upon vision in host location, and their compound eyes are very large. Since the eyes contribute to the major part of the head, this too is relatively very large in both of these families. In contrast the Bombyliidae and Acroceridae, which as adults do not contact their hosts, have strikingly small heads (Figure 95).

Many of the Conopidae have a wasp or bee-like appearance with an abdomen narrowed basally and often striped in black and yellow. The antennae are longer and more conspicuous than in most Diptera. Whether this appearance aids them in parasitising their hosts is very doubtful; rarely is there any close agreement in form or pattern between the conopid and the particular hymenopteran that it parasitises.

Oviposition is apparently internal, at least in most species, although observations on this point are limited. The eggs are often provided with stalks terminating in hooks, presumably for anchorage in the host's cuticle as in the ichneumonid subfamily Tryphoninae. In Conopidae, however, the egg pedicel arises from the cephalic end, whereas in Tryphoninae it is caudal. The parasite larva feeds in the host's abdomen, its posterior spiracles in contact with the host's tracheal air sacs (Cumber 1949). An afflicted bee or wasp remains active for a long time, since its locomotory muscles in the thorax are undamaged until the terminal period of parasite feeding. However, the anterior end of the larva is much attenuated and when the abdominal contents of the host have been exhausted, the larva inserts its anterior end through the narrow petiole of the bee or wasp and into the thorax. At this stage the host dies. Soon afterwards the parasite pupates inside its abdominal shell, having first withdrawn its anterior end from the thorax (Smith 1966).

Bohart (1941) drew attention to the fact that *Myopa* attacking *Andrena* may

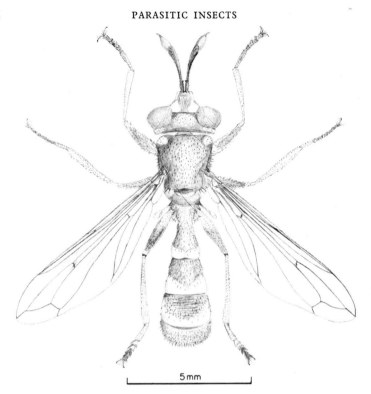

Figure 99. Cyclorrhapha, Conopidae. Female *Conops flavipes*.

be considerably larger than their hosts. On the face of it, this is an impossible situation, and Bohart suggests that the conopid larva feeds on ingested nectar to supplement the inadequate amount of food provided by the host's tissues, and that it expands considerably on emergence.

Smith (1966) tabulates the known host-parasite relationships of the family. Hosts are recorded for 43 of the 500 or so known species of Conopidae. The genera *Physoconops*, *Conops*, *Physocephala*, *Leopoldius*, *Sicus*, *Zodion*, *Myopa*, *Thecophora*, and *Dalmannia* have all been associated with bees or wasps. The most frequent hosts are species of *Bombus*, *Andrena*, *Halictus*, and *Vespula*, although several other genera have also been recorded. In addition, *Myopa picta* and *Physocephala vittata* are suspected of being predators on the eggs of locusts (Greathead 1963). If this is confirmed, the Conopidae will show a remarkable similarity in their host range to the Bombyliidae and coleopterous family Meloidae which also parasitise Hymenoptera and Orthoptera. In the case of the Bombyliidae and Meloidae, the fact that the larvae search for hosts on the ground or low vegetation presumably accounts for their association with both host orders. But this cannot apply to the Conopidae, many of which oviposit in flying Hymenoptera.

The genus *Stylogaster* (Figure 100a) was, for a long time, thought to follow the usual family habits in parasitising aculeate Hymenoptera. It is a tropical genus and has frequently been observed to follow columns of army ants (Dorylinae) and to dive into the vegetation, apparently to oviposit. Carpenter (1915) was the first to suggest that it is not the ants that *Stylogaster* attacks. He watched the conopids pouncing upon cockroaches flushed by the ants. Carpenter's observations have recently been confirmed and, in addition, other insects flushed by the ants have been seen to be attacked. Rettenmeyer (1961)

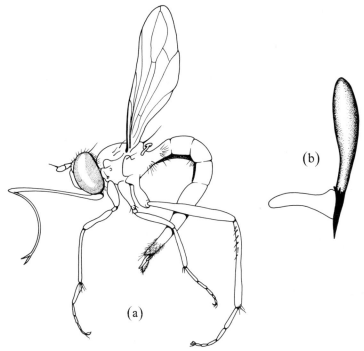

Figure 100. *Stylogaster malgachensis* (Cyclorrhapha, Conopidae). (a) adult female. (b) egg showing the pointed tip for penetration of the host, and the barbs and eversible sac which hold the egg in position after penetration (after Smith 1967).

showed that tachinids are parasitised; Stuckenberg (1963) and Smith (1967, 1969) added dung-breeding Muscidae and Calliphoridae to the host list. Thus *Stylogaster* departs radically from the usual host relationships of the Conopidae. Morphologically also species of *Stylogaster* are exceptional in having much attenuated substitute ovipositors. The host is stabbed with this organ and the eggs (Figure 100b) are anchored to its cuticle by a barbed point at one end and further secured by an eversible sac which inflates under osmotic pressure. It has been argued that *Stylogaster* should be placed in a separate family of its own.

Pyrgotidae (*Cyclorrhapha*)

Pyrgotids are quite large flies and they often have attractively patterned wings. They are very poorly known. Inhabitants of the warmer regions of the world, their hosts are adult chafer beetles (Scarabaeidae). The female fly oviposits through the rather thin cuticle of the dorsal surface of the beetle's abdomen. This region is exposed when the beetle is flying and has its elytra raised; it is at this time that the pyrgotid oviposits. Many chafers are nocturnal or crepuscular, and the pyrgotids which attack them have matching periods of activity. The host dies about two weeks after initial parasitisation, by which time the parasite has formed a puparium inside it.

Sciomyzidae (*Cyclorrhapha*)

Until quite recently Sciomyzidae or marsh flies were believed by most entomologists to feed as larvae rather generally upon decaying plant and occasionally animal material. However, intensive research by a team led by Professor C. O. Berg of Cornell University has now revealed a widespread association with snails and, much less commonly, slugs. There are about 440 species in the family and so far 140 of these, belonging to 32 genera (Knutson 1966) have been reared solely upon pulmonate molluscs. In spite of this relatively restricted diet, sciomyzid larvae have a range of habits which clearly emphasises the difficulty of defining what is meant by a parasitic insect. Berg, Foote, and Neff (1959) contrast predatory species, which quickly destroy the snail, with parasitoid species, which are intimately associated with the host and do not kill it for several days. There are, of course, species that fall between these two extremes.

The predatory species are found mostly in the subfamily Tetanocerinae. Their larvae are aquatic, hatching from eggs laid on vegetation, and seeking snails which they swiftly attack and kill. As a rule they are not particular about the species of snail eaten, and will attack several different species during their larval life, feeding on each victim for only an hour or so. One larva may destroy more than thirty snails. They will not eat decaying tissue. Several predatory larvae may feed together on a single snail. When fully grown, the larvae form puparia which are adapted for floating.

In contrast is the biology of a specialised parasitoid, *Pteromicra inermis*. The female fly lays eggs on the shells of snails of the genus *Lymnaea*. No other type is suitable and the snails are found in leaf-litter at the edges of ponds. Although several eggs may be laid on each snail shell, only one larva develops successfully in each host. The snail survives for five or six days with the sciomyzid feeding upon it, and when the host finally succumbs, the larva finishes its period of feeding by devouring a little of the dead tissue. It then pupates, never having left the body of its host, within a puparium which is neatly coiled to fit inside the spiral of the host's shell.

Most of the parasitoid species are terrestrial and include probably all of the subfamily Sciomyzinae. About half attack land snails and the remainder attack aquatic snails whilst·they are out of water. Not all are so specialised as *P. inermis*. In some species the fully grown larvae leave the host's shell to pupate in the earth, whilst in most the eggs are laid on vegetation and the young larva has to seek out a host. Fully grown larvae of several *Pherbellia* species, which attack regularly-exposed aquatic snails, produce a calcareous secretion from their malpighian tubules and this is used to construct a hard septum sealing the parasite in the host's shell (Knutson *et al.* 1967). The degree of host-specificity varies from species to species.

There seems to be an almost complete spectrum of habits between typical predator and specialised parasitoid. Berg, Foote, and Neff (1959) suggest that the two extremes in larval habits are the product of two diverging evolutionary lines stemming from a common, unspecialised ancestral type. Such a type could have had the characteristics of *Atrichomelina pubera*. This species lays eggs both on vegetation and a variety of snail shells, the larvae either living solitarily as parasitoids on a single snail or gregariously as predators, moving to fresh hosts as soon as the food supply is exhausted (Foote, Neff, and Berg 1960).

The snail-killing activity of Sciomyzidae may be exploited in the control of snails serving as intermediate hosts to helminth parasites of man and livestock.

Cryptochaetidae (*Cyclorrhapha*)

Originally included in the family Agromyzidae whose larvae are phytophagous, the curious genus *Cryptochaetum* is now placed in its own family. All known species are parasitic upon scale insects of the family Margarodidae. They are very small flies, and bear some superficial resemblance to certain Encyrtidae (Hymenoptera) which also attack scale insects. *C. iceryae* is an Australian species that was imported to the Californian citrus plantations to assist in the

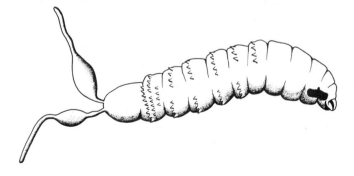

Figure 101. Second instar larva of *Cryptochaetum iceryae* (Cyclorrhapha, Cryptochaetidae) (after Thorpe 1931).

control of the cottony cushion scale, *Icerya purchasi* (page 214). The fly has an ovipositor with which it lays an egg in a young scale. The *Cryptochaetum* larva (Figure 101) is remarkable for its two very long caudal filaments. These function as tracheal gills (Thorpe 1941). The larva lacks mouthparts and obtains food by diffusion through its integument like a strepsipteran larva (page 238). The fully grown larva pupates inside the remains of its host.

Calliphoridae (*Cyclorrhapha*)

Typical representatives of this large and widely distributed family are blue-bottles and flesh-flies, whose larvae feed upon decaying organic matter. The majority of species are, in fact, saprophagous, although parasitism in the form of myiasis by representatives of the family will be discussed in a later chapter. Other species behave as typical protelean parasites of invertebrates.

The systematics and composition of this family have been variously interpreted. In the present work the family is considered to include the subfamilies Rhinophorinae, Polleniinae, Calliphorinae and Sarcophaginae.

The Rhinophorinae (= Melanophorinae) are smallish, rather slender flies that show considerable affinity with the Tachinidae. This affinity extends from their morphology to their biology, for the larvae of all species are parasitic. Woodlice are used almost exclusively as hosts (Thompson 1934). The best known genera are *Melanophora*, *Rhinophora*, *Phyto*, and *Parafeburia*. *Parafeburia maculata* is a solitary endoparasite of *Porcellio* and *Oniscus*. The planidium-like larva enters the hard-backed host through an articulating membrane, and it obtains oxygen through a respiratory funnel similar to that of Tachinidae.

The Polleniinae are almost unique amongst parasitic insects in attacking earthworms. Other hosts are on record, but confirmation of these is required. *Pollenia rudis*, the cluster-fly, is a very common British species. It is slightly smaller than a blue-bottle, more sluggish in its movements, and its thorax is clothed with dull-golden hairs. Adults frequently congregate inside buildings during the winter months and disperse in the early days of spring. Credit for discovering the life history of *P. rudis* belongs to Keilin (1915). Eggs are laid singly in crevices in soil, and they give rise to an active first instar larva. This larva, if it encounters a host, enters it through one of the body openings, for instance a sperm duct opening, or perhaps (Webb and Hutchinson 1916) through any part of the cuticle. The young *Pollenia* larvae overwinter in a practically immobile state. They are generally located in the coelom of one of the genital segments, and during the winter become encased by layers of the host's blood cells. These are, however, thrown off when feeding activity is resumed in spring. For up to four days the larva moves towards the head of its host, and for the last part of its journey it travels rear-end first. This causes the posterior spiracles of *Pollenia* to arrive first at the worm's prostomium. Chitinous teeth on the anal segment cut through the prostomium to the exterior and this gives the posterior spiracles direct access to the atmosphere. In this situation the larva feeds for up to ten days, when it moults into the second of its three larval instars. The second instar larva grows very rapidly, consuming much of the host's fore-gut, and after another nine days or so it moults into the third instar larva. This larva now moves back along the worm towards its posterior end, consuming most of what remains as it goes. Up to four individuals may develop in a single earthworm but the usual number is one. The most frequent host of *P. rudis* is apparently *Allolobophora chlorotica*, an abundant and widely distributed earthworm.

Turning to the Calliphorinae we have the only other known insect parasite of earthworms. This is *Onesia accepta* which attacks *Microscolex* and other earthworms in Australia (Fuller 1933). The life history is similar to that of *Pollenia* excepting that the female fly is larviparous. *Melinda caerulea* (=*cognata*) is another species with interesting host relations, for the determination of which we are again indebted to Keilin (1919). The fly, which is common in Britain, lays its eggs in the mantle cavity of the land snail, *Helicella virgata*, probably whilst snails are copulating. The young larvae bore their way into the digestive gland. Parasitised snails soon die, and the larva completes its development as a saprovore to eventually pupate inside the mollusc's shell. *M. caerulea* is believed to be host-specific. A second parasite of snails is *Eggisops pecchiolii*, a rare species in Britain which differs from *Melinda* in being larviparous. *Stomorhina lunata* should, strictly, not find a place in a book on parasitic insects, for the larva is a predator in locust egg pods. However it is an interesting species that illustrates further the immense variety in calliphorid life histories. The adults are strong fliers, and the indications are that they accompany migrating swarms of their hosts. As soon as a locust has completed laying a batch of eggs and withdrawn its ovipositor from the ground, the female *Stomorhina* moves to the place and lays its own eggs in the still-soft fluid froth that covers the top of the egg pod. The fly is unable to penetrate a hard pod plug and this fact enables some species of locust to avoid parasitisation (Greathead 1963). Several flies may develop in one egg pod, but so many eggs are destroyed by the larvae moving about that the presence of a single larva in an egg pod is usually sufficient to prevent the emergence of any hoppers. This is very different from the behaviour of the bombyliid *Systoechus*, which destroys only a small proportion of the eggs in a pod. *S. lunata* occasionally reaches southern England, but whether or not it attacks our native grasshoppers is unknown.

Larvae of the fourth subfamily, the Sarcophaginae, are, as their name implies, mostly feeders upon dead flesh. Most species are larviparous, with large uteri in which the eggs are incubated. Some deposit their larvae only on dead animals, others on animal wounds, and a few on healthy animals. This range of larviposition sites will be met again in the discussion of myiasis-causing Calliphoridae in Chapter 14, when we are concerned with vertebrate hosts. Very often individuals of a species will select more than one of the three sites mentioned for larviposition.

Of the British species, *Sarcophaga nigriventris* has been reared from living land snails, grasshoppers, and beetles, *S. melanura* from decaying organic matter and as a parasite of molluscs and vertebrates, *S. haemorrhoidalis*, *S. barbata*, *S. tuberosa*, and *S. carnaria* from a variety of dead or moribund animals, *S. haemorhoa*, *S. setipennis*, *S. filia*, and *S. agnata* from land snails, *S. clathrata* from spiders' egg cocoons, *S. inscisilobata* from dung and grass-hoppers, *S. scoparia* from gypsy moth pupae, and *S. albiceps* and *S. aratrix* from various Lepidoptera and Coleoptera (van Emden 1954). The precise eco-logical requirements of these polyphagous species have yet to be ascertained.

When a living host is attacked, it does not usually survive long. *S. kellyi* larviposits on the wings of locusts in flight; the stricken host falls to the ground and the maggot enters its body through an articulating membrane at the base of the wing. Locusts and grasshoppers are attacked by several other species. *Blaesoxipha laticornis* inserts its incubated eggs into the genital opening of a walking locust (Clausen 1940), but *Helicobia australis* will lay only on dead grasshoppers (Fuller 1938). *Sarcophaga destructor* larvae are able to enter only the bodies of freshly-moulted locusts (Wood 1933).

Nests of bees and wasps are invaded by members of the tribe Milto-grammini. Many of these (*Miltogramma, Metopia*) behave as inquilines (Chapter 13), placing larvae either in the nests where they feed upon the food stores, or sometimes, where wasps are concerned, intercepting a female wasp and laying an egg or larva upon the prey before the wasp carries it into the nest.

Tachinidae (Cyclorrhapha)

The Tachinidae (= Larvaevoridae) (Figure 102) stand out amongst the dipterous families that are exclusively protelean parasites. About 1 500 species are known, and many of these are common insects. As one might anticipate from the size of the family, there is a great deal of diversity in host relationships. Larvae of Lepidoptera and Coleoptera, and also adult beetles, provide the majority of hosts, although a large number of species attack adult exopterygote hosts, especially bugs, grasshoppers, and earwigs. *Loewia* (*Fortisia*) *foeda* is a parasite of centipedes. Only a few species attack other Diptera, but leatherjackets, the larvae of Tipulidae, which generally escape the attentions of parasitic insects, are parasitised by *Siphona* and *Trichopareia*.

Tachinidae are almost without exception solitary endoparasites, and no hyperparasitic species are known. The host does not die until the larval development of the parasite is complete; just occasionally it survives long enough to lay eggs. The host ranges of individual species tend to be less restricted than amongst parasitic Hymenoptera, and extreme examples of poly-phagy are provided by *Compsilura concinnata*, for which well over a hundred

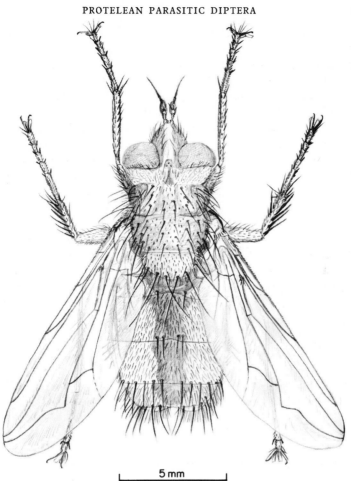

Figure 102. Cyclorrhapha, Tachinidae. Male *Echinomyia fera*.

host species, involving three orders, are recorded in the U.S.A., and *Zenillia* (= *Phryxe*) *nemea*, a British species that attacks Lepidoptera of the families Arctiidae, Geometridae, Lasiocampidae, Lycaenidae, Lymantriidae, Noctuidae, Nymphalidae, Oecophoridae, Pieridae, Pyralidae, Thyatiridae, Tortricidae, Zygaenidae, and doubtless others, as well as the common earwig, *Forficula auricularia*, of the order Dermaptera. In contrast, *Macquartia* species appear to be restricted to leaf beetle (Chrysomelidae) larvae, and *Wagneria* species mainly to noctuid larvae. *Rondanioestrus apivorus* attacks only adult worker honey-bees, laying a larva upon them whilst they are in flight.

Van Emden (1954) divides the Tachinidae into five subfamilies which are listed below with their host groups and examples of genera represented in the British fauna:

PHASIINAE

Hemiptera Heteroptera, occasionally Coleoptera adults, rarely Coleoptera larvae. In this subfamily van Emden and others include as a tribe the Oestrini, which in the present work is treated as a family in Chapter 14.
e.g. *Gymnosoma, Alophora.*

DEXIINAE

Coleoptera larvae and sometimes pupae, occasionally Lepidoptera larvae.
e.g. *Dexia, Trixa.*

MACQUARTIINAE

Lepidoptera larvae, occasionally Coleoptera larvae (Chrysomelidae), Dermaptera, and centipedes.
e.g. *Macquartia, Wagneria, Loewia, Nemoraea, Echinomyia* (Figure 102), *Ernestia, Gymnochaeta.*

TACHININAE

Lepidoptera larvae, occasionally Hymenoptera larvae (Symphyta), Coleoptera larvae (Chrysomelidae, Tenebrionidae) and adults, Diptera larvae (Tipulidae, Syrphidae), Dermaptera, and Orthoptera.
e.g. *Blondelia, Actia, Siphona, Tachina* (= *Exorista*), *Compsilura.*

GONIINAE

Lepidoptera larvae, occasionally Hymenoptera larvae (Symphyta), Coleoptera larvae (Curculionidae, Coccinellidae), and Dermaptera.
e.g. *Gonia, Nemorilla, Carcelia, Zenillia.*

Adult tachinids are usually dull-coloured flies of moderate size, very bristly but otherwise often resembling house-flies in general appearance. They do not have an ovipositor like the parasitic Hymenoptera, and with the exception of a few species which are able to inject their offspring into the body of the host, the Tachinidae have had to devise other means of gaining entry into the host's body. Many species have resorted to larviposition as an alternative to oviposition; the laid larva is always very small. Some species are ovoviviparous, laying fully incubated eggs from which larvae are on the point of hatching. The uterus of species in which the egg is incubated is very richly supplied with tracheae. Incubation of eggs in the uterus, only rarely found in Hymenoptera, is a means of diminishing mortality in the egg stage, and this is very important because many tachinids leave their eggs in exposed situations, unlike parasitic Hymenoptera.

There have been several attempts to subdivide the Tachinidae on the basis of whether a larva, an incubated egg, or an unincubated egg is laid, and the site of deposition of the offspring. Townsend (1908) recognised five groups, listed below with the very approximate number of offspring a female might produce (Clausen 1940). It will be seen that the reproductive potential varies inversely with the likelihood of a larva gaining access to a host.

(i) Eggs laid on or in the host. 100.

(ii) Eggs laid on vegetation or the soil surface. 2000 to 13000.

(iii) Larvae laid on the host. 100.

(iv) Larvae laid in the host. 100.

(v) Larvae laid on vegetation or the soil surface. 250 to 1000.

Pantel (1910) made a further breakdown into ten groups, and Townsend again, in 1934, distinguished no fewer than thirty-five groups, but he included the Oestridae and other Muscoidea. These later groupings are very detailed, but cumbersome to the non-specialist. Van Emden (1954) proposes a probably more readily appreciated scheme and this, with considerable modifications, is presented below.

Group A Larviparous or ovoviviparous, larvae or eggs laid apart from the host on leaves or the surface of soil. The larvae either actively search for a host (Dexiinae), or the progeny are deposited near to a host (most Macquartiinae). The young larvae of *Echinomyia, Nemoraea* and allied genera of Macquartiinae are very small and modified for spending the period of their life before they enter a host in an exposed, drying environment. Their bodies are provided with sclerotised, dark-coloured plates which form a continuous shield when the larva contracts and thereby reduce the transpiration rate. The larva of Dexiinae is more active, and in addition to the hard plates also often possesses two caudal cerci and false legs. It is, in fact, a planidium. Most species in this category produce numerous eggs that are characteristically and neatly arranged in regular rows in the large uterus.

Group B Larviparous or ovoviviparous species which place their progeny directly onto the host's body (some Macquartiinae, some Tachininae).

Group C Oviparous, eggs laid on foliage. The eggs are very small (microtype), usually less than 0.2 millimetres long, and produced at a very rapid rate. The uterus is expanded. The eggs, which are dark-coloured and have a very tough chorion, are viable for six weeks or more. They have to be ingested by the host before their development can proceed. They hatch in the gut of the host, and the larvae then migrate to other parts of the body (most Goniinae). This remarkable life history was first observed by Sasaki in 1887, but it was not generally accepted by entomologists for almost a quarter of a century. The hymenopterous family Trigonalidae has similar habits.

Group D Oviparous species that lay large eggs (macrotype) on the host. The eggs approach one millimetre in length and may be soft-shelled and attached to the hosts' hairs by stalks (*Carcelia*), or thick-shelled and flattened basally so that they may be applied closely to the host cuticle. The uterus is not greatly expanded and the rate of egg production is much lower than in Group C species (some Phasiinae, some Goniinae, some Tachininae).

Group E Oviparous species that insert their thin-shelled eggs into the host by means of a piercing substitute ovipositor or by an ovipositor which operates in conjunction with a separate clasping apparatus which holds the host (most Phasiinae).

Group F   Larviparous or ovoviviparous species that introduce their progeny inside the body of the host through an integumentary lesion made by the piercing apparatus, a separate structure which is not the tip of the ovipositor (some Tachininae, e.g. *Blondelia, Compsilura*, some Goniinae e.g. *Paraphorocera*). The uterus is frequently elongated and coiled, the eggs developing as they pass down it so that distally it may contain young larvae. There is, generally, a low rate of egg production but it is maintained over a long period. However in *Compsilura* and some other forms, the eggs develop simultaneously so that all the progeny are laid in a short time.

The piercing apparatus (Figure 103a), found in females of Group F, is formed from part of the sixth abdominal segment which is produced into a sharp, downwardly curving spine. This spine is grooved on its outer, convex face to enable the ovipositor to slide along it and to guide it into the lesion on

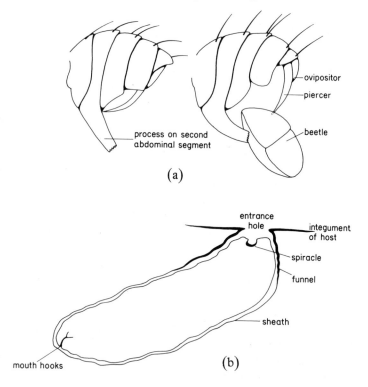

Figure 103. Cyclorrhapha, Tachinidae. (a) Abdomen of female *Chaetophleps setosa* showing (right) probable position of appendages during oviposition in a chrysomelid beetle (from Imms 1931 after Walton). (b) Section through a larva of *Thrixion* attached to the respiratory funnel and enclosed by the sheath inside its host (from Imms 1931 after Pantel).

the host made by the point. By this means females are able to place their eggs or larvae with accuracy, usually in the mid-gut in the space between intestinal epithelium and peritrophic membrane (Clausen 1940). The hosts of most species that possess a piercing apparatus are caterpillars that have a thin, soft integument, but *Chaetophleps setosa* (= *Neocelatoria ferox*) attacks adult Chrysomelidae. The latter tachinid has a ventral, spiny process on the second abdominal segment, and the leaf-beetle is held between this and the point of the very long piercing apparatus (Figure 103a). With this apparatus the fly can puncture the beetle's elytron, although in practice it is seldom required to do so, since hosts are usually attacked whilst in flight or immediately upon alighting whilst their elytra are not covering the relatively thin integument of the abdomen (Walton 1914). A similar type of apparatus is found in some Conopidae (page 197).

There are a few records of species with a piercing apparatus feeding upon the host's fluids in the manner of so many of the parasitic Hymenoptera. The nectar of flowers and honey dew, however, form the major part of the diet of adult Tachinidae.

Species in Group D, laying large, unincubated, thick-shelled eggs directly on the host, prefer hosts that have recently moulted. This reduces the risk of eggs being lost on cast host-skins. Often twenty or thirty eggs are laid on one host, even though the host can probably support only one parasite. There appears to be no avoidance of superparasitism by Tachinidae (Clausen *et al.* 1927). Each species tends to be restricted in its site of oviposition to a localised part of the host's body. *Centeter cinerea* concentrates on the back of the thorax, whilst *C. unicolor* lays most of its eggs on the ventral surface of the abdomen of the chafer hosts. In hatching, depending upon the species, the first instar larva either bores directly through the base of its egg and straight into the host, or it pushes a cap from the top of its egg and attacks the host's cuticle from outside its egg shell. It takes a larva about fifteen minutes to penetrate a host. The point of entry is usually on an intersegmental membrane, and at this time the host may make determined and successful efforts to bite or dislodge its parasite.

Species of Group C lay their eggs, which need to be eaten by the host, only, as one would expect, upon the host's food plant. *Cyzenis* (= *Monochaeta*) *albicans* places its eggs on oak near to fluxes of sap at which the adult fly may feed. The sap fluxes may be started by the feeding activity of the potential caterpillar host (Varley and Gradwell 1958). This species is unusual in apparently being able to develop only in larvae of the winter moth, *Operophtera brumata*, even though its eggs must be consumed by many other species of Lepidoptera whose caterpillars feed upon oak.

First instar larvae of Group A species, deposited apart from the host, must usually rely upon a host passing close to them. The female flies may be stimulated to larviposit or oviposit on leaves on which the host has fed, or sometimes oviposition is stimulated by the odour of the host's frass. This ensures that the distance separating parasite larva from potential host is not too great. *Lydella grisescens* is stimulated to larviposit by the frass of *Ostrinia*, its host, but in the laboratory larviposition is stimulated also by the frass of *Galleria* (Hsiao, Holdaway, and Chiang 1966). *Galleria* is an unsuitable host and it is extremely improbable that it is ever encountered by *Lydella* in natural conditions. *Eupeleteria magnicornis* lays its eggs on foliage on which there are caterpillars. The stimulus to oviposit may be provided by the silken threads left on the surfaces of leaves by caterpillars (Townsend 1908). The young larva hatches immediately after the egg is laid but it remains fastened posteriorly to the basal part of its egg shell. From this anchorage it stretches out in all directions until a caterpillar is contacted. The first instar larvae of many other species share this habit of retaining contact with the remains of the egg, which is itself firmly glued to the substrate. The young larvae are excited by the approach of any object, and they will enter the bodies of both suitable and unsuitable hosts with equal vigour. *Thereseia* is a form with a more active first instar larva. It lays its larvae at entrances to the larval tunnels of its host, the sugar-cane borer, and the tachinid larva burrows through the frass at the tunnel entrance in search of the caterpillar.

The larva of *Dexia rustica* is also very active, and this species is much less precise than *Thereseia* in its oviposition sites. It is a parasite of cockchafers (*Melolontha*) and has been investigated in detail by Walker (1943). The female fly produces large numbers of eggs, laying them in batches in rapid succession on the soil surface. The eggs hatch within a few seconds of being laid and the young larvae burrow into the soil. Walker failed to find any evidence that more eggs are laid on soil above a host than elsewhere. Once a female fly has started to produce eggs, they are laid over a period of several days without any restraint in the absence of hosts. However, the first instar larvae show a greater inclination to burrow into the soil when they are over a host; otherwise many of them tend to stay on the surface of the soil. The larvae are able to burrow to a depth of at least thirty centimetres in search of hosts, but chafer grubs just a little below the soil surface run the greatest risk of parasitisation. There is no avoidance of superparasitism by *Dexia*; indeed the distribution of first instar larvae over the available hosts is aggregated rather than random because the eggs are laid in batches. Up to thirty-seven larvae were found by Walker in a single host and, unusual in the family, occasionally more than one adult was found to be able to develop successfully. The larvae enter the chafer grub at almost any point on its body with the exception of the head capsule, legs and spiracles. However about eight times more enter the dorsal surface than the ventral, since the grub is curved in the shape of a letter C with the dorsal surface outermost and therefore more exposed. Also, the legs can brush away tachinid larvae attempting to enter through the ventral surface.

Species that deposit their offspring upon the host's body, or leave their larvae to find a host, or rely upon their eggs hatching in the host's gut, do not have to face the problem of overcoming the prolonged onslaught of the host's defensive encapsulating reaction. Those species that insert their eggs directly into the host's haemocoel have not only successfully encountered this hazard but have turned it to their own advantage (Salt 1963). There are three larval instars and either all of these, or at least the last two, are enclosed in a structure of host origin known as the respiratory funnel (Figure 103b). The funnel proper is a hard, chitinous, dark-coloured structure whose apex is inserted either through the host's integument to the exterior, or into a large trachea. At its base, the broad end, it is united with a membranous sac inside of which is the tachinid larva. This sac is usually closed distally during the first two larval instars, isolating the parasite within the host's haemocoel, but later it forms an open sleeve. The funnel proper is formed by the host's hypodermal cells, a product of the normal wound-healing process which is modified by the tachinid for its own use. The sheath, however, is formed from the host's haemocytes, and is comparable with the defensive encapsulation of endoparasites by hosts described on page 174. This reaction would normally suffocate the parasite, but the tachinid larva escapes this fate by having access to the atmosphere through its posterior spiracles and funnel. Tachinids thus provide a most interesting exception to the general rule that insect endoparasites in their normal hosts do not elicit a defensive reaction.

By no means all tachinid larvae spend their life in a funnel in the host's haemocoel. There are species that live in nerve ganglia, the intestine, salivary glands, and so forth, at least during the early part of their larval life, and the larvae of several species are known to follow a regular migration route from organ to organ. Respiration of young larvae in organs is effected through the general body surface, but as the larva grows this method becomes inadequate. Older larvae often have a series of sharp hooks arranged about their posterior spiracles. With these hooks they are able to tear the walls of tracheae or air-sacs and reach the air that they contain. *Compsilura concinnata* is an example of such a species.

When a tachinid larva is fully grown, it usually emerges from the remains of its host. It chews its way out but its escape may be facilitated by the production of enzymes which soften the host's cuticle. Pupation, in the majority of species, takes place in the soil or on vegetation. This contrasts with the habits of a large number of parasitic Hymenoptera which pupate inside the host remains, and it is correlated with the absence of biting mouthparts in adult flies. The latter renders escape from an enclosing rigid case, other than a puparium, almost impossible. Exit from the puparium is effected by the inflatable ptilinum on the front of the head pushing away the cap of the puparium. One species, *Diplostichus janitrix*, pupates inside the tough coccon of a sawfly, *Diprion pini*, and overcomes the problem of emergence by cutting a circular groove in the inside wall of the cocoon, near to one end, whilst it is still a larva. The adult fly is easily able to push off the partially severed cap (Robbins 1927).

10. *Biological Control of Insect Pests*

SUCCESSFUL BIOLOGICAL CONTROL is the reduction in numbers of a pest species, through the agency of another organism that has been in some way managed or interfered with by man, to a level at which the pest species ceases to be an economic, medical, or veterinary problem. Relatively few attempts, however, have so far been made at the biological control of medical and veterinary pests. There are other definitions, but the one given above covers the most usually held concept of biological control. The alternative to biological control was, until recently, generally considered to be chemical control in which toxic or otherwise injurious substances (e.g. insecticides) are in some way administered to pests. Other devices for combating pest species include the liberation of sterilised males, alluded to on page 257, the planting of pest resistant strains of crop plants, and various cultural procedures such as the regulation of sowing and harvesting times, crop rotation, the burning of pest-infested refuse, the banding of fruit trees and many others. The use of highly selective attractant chemicals as baits in mechanical traps is another very promising method of control for some pest species.

In this chapter we are only really concerned with the control of pests by using another organism. The controlling organism is very often a parasitic insect, although predators and, more recently, pathogens (e.g. *Bacillus thuringiensis*) are also employed. But before going any further in a discussion of biological control, it is important to make a few points about what might be called the 'natural' control of populations. Biological control is an extension of natural control, and an essentially ecological problem.

In every locality there are rare species and abundant species, and usually each species maintains a fairly constant numerical status over long periods of time. Only in ecosystems containing few species are numerical fluctuations often extreme. Of course, every population fluctuates to some extent, but the fluctuations are about an average population density which, in a stable environment, does not alter much.

The environment imposes a great number of complex and interacting checks upon a population. These prevent the full expression of its capacity for increase, either by eliminating, before they reproduce, all individuals surplus to those required to maintain the population at a steady density, or by depressing the birthrate. If more than the surplus is consistently removed, the population will decline generation after generation until it is extinct. Conversely, if all surplus animals are not destroyed and the birthrate does not decline, the population will increase and, unless there is dispersal, its density will rise. In other words, a population's rates of ingress (birthrate plus rate of immigration) and egress (deathrate plus rate of emigration) must be closely matched. How is a precise mortality or natality achieved such that the population density of a species is more or less stabilised in the absence of a significant amount of individual movement? This is a much-debated question that can be considered only in a summary way here, and since mortality rather than natality is most usually interfered with in biological control procedures, emphasis is laid upon that aspect.

Firstly, some environmental mortality factors act independently of the density of the population. For example, local climatic catastrophes can strike when the population density is high or low. The mortality suffered through such agencies is not obviously related to the density of the population, although in most populations some individuals probably always occupy a small and fairly constant number of especially favourable situations, and these animals are likely to survive when others perish. If this is so, the number of survivors should remain moderately constant whatever the population density before the catastrophe, and the mortality varies directly with the population density. Other mortality factors are more obviously related to population density, increasing in severity as the population rises and decreasing as the population declines. One such density dependent factor could be parasitism. The interaction of a host and parasite population has been expressed in several mathematical models, which show alternating cycles of abundance of the two populations. As the host density increases the parasites, having more available food, are also able to increase, and this they do until, through inflicting a steadily increasing percentage parasitism, they eventually check the rise in the host population and cause it to decline. Subsequently the parasite population falls as well, enabling the host population to recover. Inevitably there is a lag between a change in the host population and a numerical response in the parasite population. Factors which are independent of fluctuating density such as climate could hold a population, at least for a time, between limits below the level at which density dependent factors have an observable effect, or, in a more stable environment, density independent factors can set the general level of abundance of a species about which density is regulated by density dependent factors. However, it seems clear that when a population is held between relatively close levels of density for long periods it is being controlled by density dependent mortality. The changing relative effect of different environmental factors (density dependent and density independent) on the density of a population in different types of environment is represented schematically in Figure 104.

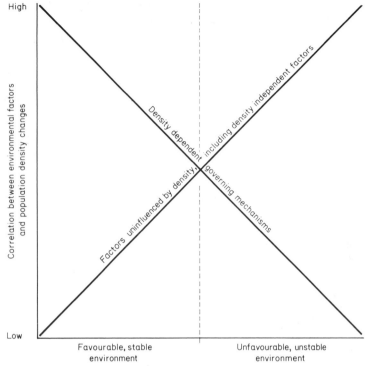

High

Low

Correlation between environmental factors
and population density changes

Density dependent, including density independent factors

Factors uninfluenced by density, governing mechanisms

Favourable, stable
environment

Unfavourable, unstable
environment

Figure 104. The relative effects of different environmental control factors on the densities of populations in two different types of environment (modified after Huffaker and Messenger in DeBach 1964).

Most pests may be said to be man-made, existing only in environments greatly modified by man. The cultivation of a single crop plant (monoculture) may suddenly provide ideal conditions, abundant food and absence of natural enemies, for an endemic insect, which multiplies in the new situation to reach a density that was never previously attained in a more heterogeneous environment with its concomitant variety of population checks. Or, more often, the accidental introduction of an insect species into a new region allows it to multiply unrestrained by the parasites and predators which helped to control it in its native land. When such species become pests, their populations may be reduced to economically acceptable levels by the introduction of parasites or predators. When an insect becomes a pest in its native land the endemic parasites and predators are inadequate, and control may be effected by the introduction of more efficient alien parasites or predators. Similarly, an alien insect pest may be checked by the introduction of species that control it in its native land. This chapter is concerned with showing how, and with what success, parasitic insects have been employed in attempts to reduce the average population density of pests to levels at which they cease to be economically important. The aim of biological control is to stabilise the density of a pest population at a lower level than it held before controlling agents were introduced, preferably at such a level that even at times of peak abundance the former pest no longer inflicts significant economic damage.

The first introductions of insect parasites and predators were made in the last quarter of the nineteenth century, when mainly European and oriental species were imported into the U.S.A. Fortunately, one of the earliest attempts at biological control was an outstanding success and the method gained considerable prestige. This was the almost total annihilation of the cottony cushion scale (*Icerya purchasi*) in the Californian citrus orchards by a predator, the Australian ladybird *Rodolia* (= *Vedalia*) *cardinalis*. A fascinating account of this project has been written by Doutt (1958). Just before the liberation of *Rodolia*, a parasite of *Icerya* in its native Australia, the fly *Cryptochaetum iceryae* was released in California. Although much less spectacular than *Rodolia*, *Cryptochaetum* plays a useful role in keeping down numbers of *Icerya*.

Other early successes for the method of biological control were particularly numerous in the Pacific islands and especially Hawaii. Insecticides have scarcely been needed here at all. For instance, the sugar-cane root chafer, *Anomala orientalis*, became a pest in Hawaii. This beetle was found to be attacked in the Philippines by the wasp *Scolia manilae*. Introduction of the *Scolia* into Hawaii soon brought the chafer under control.

Sometimes parasites of species closely related to the pest species in other parts of the world may be used for biological control. Species of the braconid genus *Opius*, which parasitise the oriental fruit fly, were successfully introduced to control the citrus fruit fly in Hawaii. The moth *Levuana irridescens* was a pest of coconuts in Fiji; another coconut moth, *Artona catoxantha*, was well controlled by a tachinid fly, *Ptychomyia remota*, in Malaya, and in 1923 this fly was introduced into Fiji. It spread very rapidly, and within about a year *Levuana* had ceased to be a pest (Tothill, Taylor, and Paine 1930). The two moths belong to the same family, but are otherwise not very closely related.

It quite often happens that a parasite which is successfully controlling a pest in its native land is a failure when introduced elsewhere to control the same pest species. In Africa three species of *Coccophagus* (Chalcidoidea) are the most numerous parasites of the citrus black scale, *Aleurocanthus woglumi*, but on introduction into California they failed to control the scale although they became established. Another chalcid, however, *Metaphycus helvolus*, quite overshadowed by the *Coccophagus* species in Africa, rapidly brought the black scale under control in most regions of California (Bartlett and van den Bosch in DeBach 1964). The change in status of these parasites on introduction is probably a consequence of climatic differences between the two regions. Even

M. helvolus, which may have originated in Japan, has difficulty in maintaining itself in the dry interior of California. Here the black scale has a single annual generation only, and the life cycles of host and parasite become asynchronous.

Varley and Gradwell (1958) found the tachinid *Cyzenis albicans* to be the most abundant parasite of the winter moth, *Operophtera brumata*, in Wytham Wood near Oxford. Unusual amongst Tachinidae, it appears to be host-specific and its life cycle is synchronised with that of the winter moth. In spite of its abundance, it is ineffective in Wytham in controlling the numbers of its host. It is itself subject to heavy mortality whilst a pupa in the soil, being eaten by predators and attacked by a hyperparasitic Ichneumonid, *Phygadeuon dumetorium*. However when *Cyzenis* was liberated in Canada in an effort to control very heavy winter moth infestations, it reduced the host population density to under one thousandth of the original in a very short time. It will be interesting, in view of the Wytham findings, to see whether *Cyzenis* is able to maintain its control of the winter moth in Canada.

A large proportion of biological control campaigns that are rated as successful have been waged against scale insects using chalcids of the family Aphelinidae. Control of the red scale, *Aonidiella aurantii*, is another example to emerge from the Californian citrus orchards. About the beginning of this century, the scale was partially controlled by *Aphytis chrysomphali*; like *M. helvolus* it is unable to establish itself in inland areas of California. Also, in some climatically suitable coastal districts it is prevented from attacking the scale by ants which harvest the scales' sugary excreta. In 1948 another parasite, *A. lingnanensis*, was introduced and it displaced *A. chrysomphali* from most of its former range. Both of these species of *Aphytis* were brought to America from China. Later still, in 1957, an Indian species *A. melinus* was introduced, and it in turn partially displaced *A. lingnanensis*.

A similar example of species displacement concerns the braconid genus *Opius*, three species of which have been introduced in succession into Hawaii to control the oriental fruit fly, *Dacus dorsalis*. *O. longicaudatus* was replaced by *O. vandenboschi* which was itself replaced by *O. oophilus* (Bess and Haramoto 1958).

These examples highlight the question of whether or not it is advisable to introduce a number of parasite species instead of just one. The evidence to date suggests that it is; with a number of parasite species it is likely that more ecological niches will be filled and the entire range of the pest more nearly occupied. Also in a two-species system of host and monophagous parasite, numerical fluctuations are liable to be great and satisfactory control unlikely. In Hawaii the Mediterranean fruit fly, *Ceratitis capitata*, has been more heavily parasitised since *Opius humilis* was supplemented by the introduction of additional parasite species, even though the parasites destroy each other in cases of multiple parasitism and the contribution made by *O. humilis* to the overall percentage parasitism has decreased (Willard and Mason 1937). There is no certainty, of course, that increasing percentage parasitism of a host denotes decreasing survival in that host.

Five parasites of the citrus blackfly were introduced into Mexico from the oriental region, and the pest has been satisfactorily subjugated in many parts of the country. Different parasite species, however, are dominant under different climatic conditions so that the pattern of parasitism varies geographically (Doutt and DeBach in DeBach 1964).

In England the holly leaf-miner (*Phytomyza ilicis*) is parasitised by the chalcid *Chrysocharis gemma* and, more rarely, by the braconid *Opius ilicis*. When the leaf-miner became a pest after it was accidently imported into Canada, the two British parasite species were released. Because *C. gemma* appeared to be the most effective species in England, it was given priority over the braconid, but in Canada the chalcid lived up to expectations only on Vancouver Island. On the mainland *C. gemma* failed to establish itself. The performance of *O. ilicis* provides a striking contrast, for from an initial release of only ten adults on the mainland of British Columbia, the species has now become well established. A larger introduction onto Vancouver Island did not enable *O. ilicis* to gain ascendancy over *C. gemma* there (Turnbull and Chant 1961).

It would seem from the preceding example that the more the parasite load on a pest can be increased, the greater is the probability of successful control. Some pests, however, have been very satisfactorily controlled by the introduction of a single parasite species. The case of *Scolia manilae* and the sugar-cane root chafer may be cited as an example. Also, not all natural enemies of a pest necessarily contribute towards bringing about a reduction in its numbers. In Fiji the coconut leaf-mining beetle (*Promecotheca reichei*), an indigenous pest, was controlled by native parasites until a predatory mite was accidently introduced into the islands. The mite destroys larvae and pupae of the beetle, but not eggs or adults. This has the effect of converting the beetle population from one of overlapping generations into one that has synchronised stages. Larval stages are removed from the population for long periods, and since it is these stages that the native parasites attack, the beetle population escaped from its previous control. In 1933, however, a chalcid *Pediobius parvulus* was introduced from Java, where it parasitised other species of *Promecotheca*. *P. parvulus* has a life cycle of the same duration as that of *P. reichei* so that it was able to synchronise with its host and achieve successful control (Taylor 1937).

It is very clear that great caution should be exercised before any foreign organisms are introduced into a country. Detailed ecological investigations of both pest and parasite should first be conducted. A very obvious danger is the risk of introducing an obligate hyperparasite mistakenly believed to be a primary parasite. Such an introduction would benefit the pest population, clearly shown in the following diagram from Varley (1959).

Light → PLANT
Light → plant ➡ PEST
Light → PLANT → pest ➡ PARASITE or PREDATOR
Light → plant ➡ PEST → parasite or predator ➡ HYPERPARASITE

These are simple energy chains in which the dominant species are shown in capital letters, the species in lower case being maintained at a density such that they cannot utilise all the food available to them. A species added to the end of the chain spreads its influence to all preceding members. *Quaylea whittieri* is a hyperparasitic chalcid which was introduced into California without adequate preliminary screening, and it has gravely reduced the effectiveness of *Metaphycus lounsburyi* and *Scutellista cyanea*, primary chalcid parasites of the citrus black scale. However many hyperparasites are facultative, attacking indiscriminately both parasitised and unparasitised hosts. As such they are not necessarily detrimental to the control of the pest. Only detailed studies can reveal the habits of such species. The vast complexity of an ecosystem must always be kept in mind; an example, perhaps rather extreme, is the food web associated with the gall wasp *Cynips divisa* (Figure 57).

Another problem to be considered is the existence of sibling species and biological races of parasites. Parasites that appear morphologically identical may have quite different biological characteristics, and some failures in biological control may be the result of the introduction of the wrong race or sibling species of parasite. A parasite species may have forms which are adapted to different environments or even to different host species. The chalcid *Comperiella bifasciata* is known to occur in two different strains, one of which parasitises the yellow scale (*Aonidiella citrina*) and the other the red scale (*A. aurantii*), while one strain of another chalcid, *Prospaltella perniciosa*, attacks the red scale but another strain does not (Doutt and DeBach in DeBach 1964). An introduction of the tachinid earwig parasite *Bigonicheta setipennis* from England to Canada failed to establish itself, but a collection of the same tachinid from France was at once successful (Turnbull and Chant 1961). Several other similar examples could be cited. A parallel situation arises when the pest exists in more than one biological form. Muldrew (1953), working on parasitism of *Pristiphora erichsonii*, the larch sawfly, in Canada, found that in Manitoba and Saskatchewan the ichneumonoid *Mesoleius tenthredinis* was a much less effective parasite than elsewhere. Closer investigation revealed that in these two provinces the host population was encapsulating the parasite eggs, a capacity not developed in other parts of the country. As a means of minimising these obstacles to successful biological control, it is urged (e.g. Allen 1958) that a predator or parasite be imported from different regions throughout its range so as to include a wide range of genetic variation.

Even after thorough investigations of pest and parasite have been made, it is still not possible to predict whether or not the parasite on introduction will succeed as an agent of biological control. As a guide, however, the following qualities are shared by most introduced parasites that have proved successful in biological control.

(i) High host-searching capacity.

(ii) Having a narrowly limited host range (as opposed to being monophagous) so that when the pest population is reduced to low densities, the parasite is able to maintain itself on alternative hosts. This enables a high parasite population to remain available to counteract upsurges in the pest population. It has the effect of damping host-parasite oscillations in density. In addition, alternative hosts may accommodate the parasite population during seasons when the pest is unavailable, or when the pest is attacked with chemical insecticides. However, the available host range should not be so large that the parasite population dissipates itself upon economically harmless species.

(iii) Having a life cycle considerably shorter than that of the pest when the pest population consists of overlapping generations, and having a life cycle synchronised with that of the pest when the pest population is composed of a single developmental stage at any time.

(iv) Able to survive in all habitats occupied by the pest.

(v) Able to be cultured easily in the laboratory.

(vi) Able to quickly subjugate the pest population (within three years of introduction according to Clausen (1951)).

(vii) Egg parasites appear, in general, to be less successful in biological control than parasites which attack later developmental stages.

The very successful application of biological control on oceanic islands merits further consideration. Whilst the method has often been successful in continental regions, California being a case in point, the record for islands probably still remains superior. There are at least three reasons that contribute to this. The first is historical; islands such as Hawaii and Fiji have long been centres of activity in the field of biological control. Hawaiian sugar-planters in particular have done much to promote the method. The second reason concerns the nature of the protected crop. On islands the main crops are very often sugar or copra, and in both cases the commercial product is extracted from the plants. Minor pest damage to these crops is invisible in the product and of negligible importance. In contrast, when the commercial product is a fruit or vegetable, its external appearance is of vital importance in marketing. Even small blemishes made by a low density pest population could render the crop a commercial failure. In other words, island crops are often more economically tolerant of limited pest damage. Finally, the climate of oceanic islands is usually very equable. This often permits continuous development of insects so that populations consist of all stages simultaneously, avoiding the possibility of a parasite being unable to synchronise with its host. It also confers a stability upon the ecosystem that promotes the establishment and survival of a number of controlling agents. It was shown earlier in this chapter that density dependent controls are ascendent in stable environments.

As a broad generalisation, biological control has been more effective in protecting trees than field crops. Varley (1959) summarises the effectiveness of biological control projects in the U.S.A. up to 1950 (Table 9) and showed that

Table 9. The effectiveness of biological control in protecting different crops in the U.S.A. up to 1950 (from Varley 1959).

	Fruit crops	Forest and ornamental trees	Field and garden crops	Total
Number of pest species	33	16	42	91
Number of pests effectively controlled	12	4	2	18

33% of orchard, forest, and ornamental tree pests have been effectively controlled compared to only 5% of pests of field and garden crops. We may attribute this disparity to the relative permanency of orchards and plantations, and also to the frequency with which ecologically disturbing cultural routines are applied to field crops. Pepper and Driggers (1934) have shown that the braconid *Macrocentrus ancylivorus* is most effective in controlling the oriental fruit moth (*Grapholitha molesta*) in those orchards which harbour a varied weed ground flora. Moth larvae of various species, feeding on the weeds, provide an excellent reservoir for the parasite at times when its preferred host is unavailable. This is an example of the greater control of a pest in a heterogeneous than in a monocultural environment; in this instance it is achieved by increasing the natality of the parasite.

The culture of a varied flora, therefore, can in itself be a biological control device, but it is not one that is regularly practised. Rather the reverse in the current tendency towards widespread and sometimes indiscriminate use of weed-killers and the despoilation of roadside verges, hedgerows, and other marginal land.

Thus far we have considered mostly cases of successful application of biological control, although Varley's figures (Table 9) should help to correct this bias. Turnbull and Chant (1961) assess the results of thirty-one biological control projects undertaken in Canada. They recognise the difficulties involved in such assessment and define their criteria as follows: 'We will consider any measurable reduction of damage as a degree of control, but for complete control we will require that the damage be reduced to a level that is unlikely to be intolerable in any foreseeable economic situation'. On this basis, of the thirty-one projects, twelve are rated as successful and three as partially successful. Turnbull and Chant conclude that 'this record cannot be considered good and indicates a need to revise methods.' DeBach (1964) presents a more favourable view of biological control, and listing by countries 225 cases of biological control by entomophagous insects, ranks 66 as completely successful, 88 as substantially successful and 71 as partially successful. No failures are mentioned. DeBach also goes into the question of financial savings and estimates that, at the very least, $110 000 000 were saved in California between 1923 and 1959 as a result of successful biological control projects. Savings continue to accrue each year since the annual cost of insecticide treatment, which is high, and the annual loss from pest damage, are both permanently avoided.

Once biological control has been successfully established in an area it should be permanent, or require the minimum of servicing in the form of occasional reintroductions of the controlling agent, and this is one big advantage that the method has over chemical control. Insecticides have to be regularly applied, often several times during a year. Thus biological control should be much less expensive than chemical control in the long run, even though the initial cost of basic research and location of natural enemies for biological control could be high.

Another reason why biological control is preferable to chemical control is the fact that evolution works against insecticides, and resistant strains of the pest are very liable to evolve during the course of regular insecticidal application. Should this occur, either a new insecticide must be developed or the potency of the original insecticide increased by the incorporation of a synergist. Unfortunately, once resistance to one insecticide has been acquired by an insect, it is often found to have improved resistance to others. This difficulty is only rarely encountered in the application of natural enemies, for even if the pest does develop a degree of immunity to a parasite, the parasite is also likely to evolve to overcome the immunity and restore the equilibrium.

But the most potent argument against the widespread use of insecticides is that the great majority are exceedingly dangerous chemicals, toxic not only to insects but to warm-blooded vertebrates as well. Many are very persistent, stable compounds, their concentration in the environment is being gradually built up, and they are being spread to parts of the world far removed from areas of application. These dangers and some resulting disasters have been well publicised in recent years. A particularly unfortunate incident occurred in 1967 near Bogota when eighty people, including seventy children, died as a result of eating bread somehow contaminated by an insecticide. Bird deaths directly following insecticide treatment of crops have caused a great deal of public concern, and so too has the serious reduction in numbers in recent years of many birds of prey, apparently in part the result of abnormal reproductive behaviour following the accumulation of sublethal doses of insecticides in their tissues. Large, predatory animals at the so-called 'ends' of food chains are especially liable to accumulate harmful quantities. A more direct effect on agriculture follows from the destruction of beneficial insects as well as pests following the use of non-systemic insecticides. Sometimes the pest population is less susceptible to the insecticide than are its natural enemies so that the pest may ultimately benefit from an insecticide application.

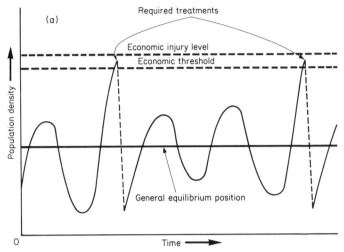

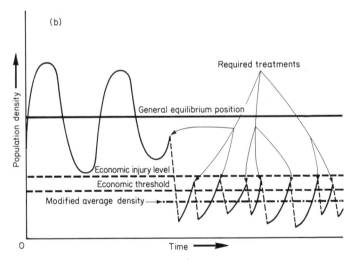

Figure 105. The theory of integrated control. (a) A pest species whose general equilibrium position lies well below the economic threshold; insecticide treatment is reserved for the infrequent occasions when the population exceeds the economic threshold. (b) A severe pest which requires frequent insecticide application to maintain its population density below the economic threshold (from Stern, Smith, van den Bosch, and Hagen 1959).

Nevertheless, the enormous saving in terms of agricultural produce and timber that has been made possible by the timely use of insecticides cannot be over-rated. Insecticides provide an effective and *immediate* remedy to most pest outbreaks. Biological control requires time to take effect, time during which serious pest damage may be incurred.

The ideal solution almost certainly lies in the proper and fully considered use of both biological and chemical control. An integration of both methods is likely to provide the cheapest and most effective way of increasing mans' share of the crops he grows. But before integrated control can be achieved, detailed investigations must be made. 'The prime requisite for integrated pest control is basic ecological knowledge of the entire complex involved, including the extent of biological control of each host insect that occurs in the absence of treatment' (DeBach 1964).

The general equilibrium position of a population is defined as its average density, maintained as long as there are no permanent changes in the environment. Provided that the general equilibrium position is well below the economic threshold (the density of the population at which measures against it must be taken to avoid economic injury), the population will only infrequently attain pest status. The basic idea of integrated control (Stern, Smith, van den Bosch, and Hagen 1959) is to lower the general equilibrium position of the pest by using parasitic and predatory species so that it lies far below the economic threshold, and to reserve chemical treatment for sporadic pest outbreaks when biological control has proved insufficient (Figure 105a). Chemical control should be applied in such a way as to be as little damaging as possible to biological control. A serious pest may have a general equilibrium position well above the economic threshold necessitating the frequent use of insecticides. In such an eventuality steps should be taken to lower the general equilibrium position by permanently altering the pest's environment. This may be achieved by using resistant crop strains, introducing greater crop heterogeneity or making other changes in cultural methods (Figure 105b).

I will close this chapter with two quotations, essentially similar, which probably reflect the cautious optimism of most entomologists towards integrated control. Both emanate from entomologists working in California, the original and still the major centre of biological control.

'It should be emphasized . . . that the development of integrated control is not a panacea that can be applied blindly to all situations, for it will not work if biotic mortality agents are inadequate or if low economic thresholds preclude utilizing biological control. However, it has worked so well in some appropriate situations that there can be no doubt as to its enormous advantages and its promise for the future' (Stern, Smith, van den Bosch, and Hagen 1959).

'Continued striving for complete biological control while utilizing the scientific integration of chemical control with the biological control that already exists in a given faunal complex is now accepted by many progressive entomologists as the major truly scientific approach to pest control. It will continue to constitute the best approach to pest control in the years to come' (DeBach 1964).

11. *Protelean Parasites in the orders Neuroptera, Lepidoptera, and Coleoptera*

OF THE THREE orders Neuroptera, Lepidoptera, and Coleoptera, the first two do not contain any families of typically parasitic species, but both include forms which might be described as parasitic in a broad sense, or at least as predators that have moved in the direction of parasitism in not at once destroying the host and in being relatively sessile and feeding on a very limited number of host individuals. A few Coleoptera behave as typical parasites.

Neuroptera

The Neuroptera are primarily predatory insects, both as larvae and adults. In the family Sisyridae the larvae are aquatic, and species of *Sisyra* and *Climacia* feed upon fresh-water sponges. The eggs are laid in batches on vegetation in the vicinity of water. First instar larvae make their way into the water, where they drift until a sponge is encountered. The sisyrid larva then moves into one of the sponge's ostioles and begins to feed, breaking up the sponge with a pair of attenuated mandibles. Sponges seem generally to be distasteful to other animals, but some further aquatic insects that feed upon them are species of *Sigara* (Hemiptera), *Leptocerus* (Trichoptera), and a few dipterous larvae.

Larvae of species in the family Mantispidae are terrestrial and 'parasitic' inside egg cocoons of spiders. About 170 species are described. They lay stalked eggs in very large numbers on tree trunks and similar situations. The first instar larvae crawl to the ground, where they search for egg cocoons. Host families include Gnaphosidae, Clubionidae, Thomisidae, Lycosidae, and Pisauridae. The larvae feed on the spiders' eggs, and when fully grown they differ very much in form from the active, predatory larvae in other neuropterous families. The fully grown mantispid larva has a small head, enlarged abdomen, and very small legs, features associated with its sessile existence in the midst of food.

Lepidoptera

Lepidoptera, unlike Neuroptera, are essentially phytophagous. A few larvae however, distributed among several families, have become regular predators of scale insects and other Hemiptera Homoptera. Scale insects have been adopted by some caterpillars as suitable prey by virtue of being very slow-moving, abundant on the plants upon which caterpillars live, and in having a soft integument (Balduf 1938). Other lepidopterous larvae, for example those of some Phycitidae, are predators on caterpillars of their own order, and the family Lycaenidae includes several species whose later larval stages are passed in ants' nests feeding on the brood.

Species belonging to the very small Australian family Cyclotornidae have a remarkable life history which involves both ants and homopterous insects. *Cyclotorna monocentra* has been astutely investigated by Dodd (1912). The numerous minute eggs are laid on plants harbouring a species of cicadellid (Hemiptera Homoptera). The active first instar larva, which resembles a pink woodlouse with two long tails, moves about in search of nymphs or adults of the homopteran to which it attaches itself and on which it feeds. After a time it may, like a predator, move off in search of another host. The effect that the caterpillar has on its host is not clear; it certainly does not immediately kill it. Eventually the *Cyclotorna* larva leaves the homopteran and spins a cocoon on a leaf and moults. The second instar larva emerges from the cocoon and adopts a curious posture on the leaf. Dodd's description of this is as follows: 'Resting upon the abdominal legs the front segments were raised a little and the terminal ones turned over the back, so that the tails often projected beyond the head, at the same time the anal parts and claspers were stretched out as far as possible.' In this position the larva may be seized by a worker ant of the genus *Iridomyrmex* and carried to the nest. There the larva becomes very highly coloured in shades of red, yellow, and blue. Like many other myrme-cophilous insects, it provides the ants with a sweet secretion. In addition, the caterpillars sometimes run their mouthparts over the bodies of ants and the ants eagerly solicit this attention. In return for these favours, the caterpillars feed upon the ant larvae. When fully grown, they leave the nest and pupate on the trunks of trees. A tripartite relationship such as this, involving Lepidoptera, Homoptera, and ants, has probably evolved from the trophic connections between ants and Homoptera on the one hand, and between Lepidoptera and Homoptera on the other (Balduf 1938).

Another family whose representatives have developed a strong parasitic tendency is the Epipyropidae. These are very small moths of which about thirty species are known. They occur in all of the major zoogeographical regions but are most numerous in Australia. The larvae are attached to adult Fulgoridae and sometimes other families of Hemiptera Homoptera but their exact host relationships are rather problematical. Westwood (1876) considered that the caterpillars fed either on the waxy secretions or on the honey dew produced by their hosts, and Sternlicht (1966) could find no trace of injury to individuals of a species of issid 'attacked' by larvae of *Heteropsyche schawerdae*. However, the caterpillars have long, thin mandibles which seem to be more suited to piercing the host's integument than to scraping off secretions, and some species at least are known to feed upon body fluid. *Fulgoraecia euribrachydis* lays its eggs on plants infested with its fulgorid host, and the active first instar larva, which can stand erect like a planidium or triungulin, attaches itself to the host's body. It moves to a position beneath the host's wings where it feeds from a single, small, integumentary lesion (Krishnamurti 1933). The host, which is relatively large compared to its parasite, is not visibly affected. This is not true, however, for the hosts of other species. *F. ceroleste*, an African moth, is said (Kirkpatrick 1947) to feed only on secreted wax, yet the fulgorid host dies within a day of a fully grown caterpillar leaving it.

Coleoptera

The order Coleoptera contains more described species than any other, and there is a wealth of variety in their biology. There are, however, relatively few protelean parasitic beetles, and these are restricted to a small number of families.

Many parasitic Diptera and Hymenoptera use their well-developed powers of flight to take them to a position where they can accurately deposit their progeny beside or inside hosts. Controlled, accurate flight is also an asset in enabling a relatively large area to be thoroughly investigated in a search for hosts. Adult beetles in general lack the manoeuvrability in flight that characterises most Diptera and Hymenoptera. Also, beetles do not possess a true ovipositor (although in some there is a substitute ovipositor as in certain Diptera, page 185); many weevils have to use their snouts for inserting their eggs into plant tissue. Because of these deficiencies, a very large proportion of parasitic Coleoptera are hypermetamorphic and have first instar triungulin larvae for gaining access to their hosts.

In their host relationships, several beetles are somewhere between ectoparasites and predators. A gradation of host relationships is exemplified by beetles that as larvae feed upon terrestrial snails. Thus the Lampyridae (glow-worms) are predators, each larva killing a series of snails in rapid succession. A snail is overcome by means of an injected toxin. The Drilidae, on the other hand, feed more slowly as larvae and they do not at once destroy a snail so that, for a while, they behave as ectoparasites. One drilid larva consumes a relatively small number of snails during its extended lifetime.

Likewise the larvae of Carabidae, Rhipiceridae, Cleridae, Colydiidae, and Passandridae are typically predators of other insects, but some species in the genera *Lebia* (Figure 106) and *Pelecium* (Carabidae), *Sandalus* (Rhipiceridae), *Hydnocera* and *Trichodes* (Cleridae), *Deretaphrus*, *Dastarchus*, *Sosylus*, and *Bothrideres* (Colydiidae), and *Catogenus* (Passandridae) are able to complete their development externally upon single host individuals, usually other Coleoptera, sometimes Hymenoptera, and in one case (*Sandalus*) cicadas. The larvae of such species frequently have, in their second and later instars, reduced

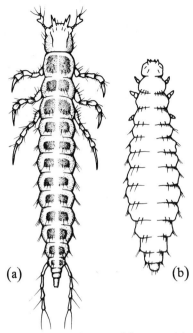

(a)   (b)

Figure 106. Larval stages of a parasitic ground-beetle, *Lebia scapularis*. (a) First instar larva which locates subterranean pupae of a leaf-beetle. (b) Second instar larva, a relatively sedentary feeding stage on the host (not drawn to scale) (from Silvestri 1904).

legs and enlarged abdomens as adaptations to their semi-sessile, 'parasitic' existence. Similar modifications are found in larvae of some Coccinellidae (*Rodolia*) and Anthribidae (*Brachytarsus*) which feed upon the egg masses of scale insects, each larva spending its entire life beneath a single scale.

The very large family Staphylinidae includes a subfamily, the Aleocharinae, some of whose species are ectoparasites of dipterous pupae within puparia. Eggs are laid apart from the host but the first instar larvae (Figure 107) are very active and seek out fly puparia in soil, litter, or, more rarely, on plants. A larva successful in its search will bite a small hole in the puparium, enter it, and commence feeding on the pupa inside. At first very small holes are made in the host integument for the withdrawal of fluid, but these punctures soon turn necrotic. The entrance hole in the puparium wall is sealed with an anal secretion of the aleocharine after entry has been effected. This fulfils three functions; it prevents the easy entry of a second larva, it prevents the entry of harmful micro-organisms, and it reduces the possibility of desiccation (White and Legner 1966). When the larva moults into the second instar, it changes from an active, large-headed insect with a fairly tough cuticle into one in which the cuticle is thin and white, the head relatively small, and the legs

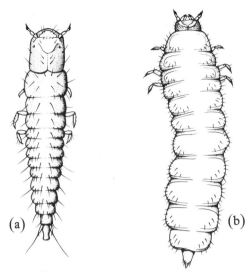

Figure 107. *Aleochara curtula* (Coleoptera, Staphylinidae). (a) First instar larva. (b) Third instar larva. Both in dorsal view (from Clausen 1940 after Kemner).

rudimentary (Figure 107). Eventually the entire contents of the host puparium are consumed and the aleocharine larva, now in its third instar, pupates either inside or outside the puparium depending on the species. Only one parasite can complete its development in a puparium. Fighting occurs if more than one larva enters a puparium, and at most one beetle eventually emerges.

All of the beetles so far mentioned feed as larvae externally upon their prey or hosts. This is true also of the Meloidae (oil beetles) (Figure 108a). Most meloid larvae feed upon grasshopper and locust eggs, but a large number live in the cells of solitary bees. Here they behave as inquilines rather than true parasites, destroying first the egg or young larva of the host and then eating the honey stored in the cell.

The female oil beetle lays several thousand eggs, depositing them in soil in the general vicinity of the hosts' nests (*Meloe*), at the entrances to the nests (*Sitaris, Hornia*), or sometimes in flowers (some *Apalus*). Rau (1930) records that the female *Hornia minutipennis* will lay its eggs before it emerges from the host cell in which it completed its own development. The larvae of Meloidae pass through four stages before becoming prepupae. The first of these is the triungulin larva which locates the host's cell. It is an agile creature, rather flattened, and strongly sclerotised with well-developed tarsal claws and two long posterior hairs. When the eggs are laid at the entrance to a nest, as in the case of *Sitaris*, the triungulins do not enter the old nest but attach themselves instead to emerging bees. The first bees to emerge after the triungulins hatch are mostly males, and these may acquire a large number of triungulins. Some of these transfer to female bees when the hosts copulate and are then carried to new nests. The eggs of *Meloe* are laid further from a host nest and the triungulins climb up vegetation and often congregate in flowers. They can survive for more than a week without feeding, remaining motionless until a visiting insect causes the plant to vibrate. They then rush about and some climb onto the insect. If this happens to be a bee of the correct species, they are carried to its nest and development can proceed. The large number of eggs produced by *Meloe* compensates for the inevitably high mortality suffered at this stage of the life cycle.

Triungulins of species of *Cerocoma* search for host cells themselves and do not rely upon transport by the host bee. Species that as larvae feed upon grasshopper eggs likewise make their own way to their food source.

After the triungulin is carried into the nest on the body of a bee, it still faces the problem of getting into a cell. This it may solve by climbing onto the bee's egg as it is being laid, and dropping with it into the honey in the cell (Fabre 1857). Alternatively, the bee may lay its egg on the wall of the cell and the triungulin can then climb down the cell wall to it before the cell is sealed. The bee's egg is soon eaten by the triungulin and the empty shell floats on the honey, serving as a raft for the triungulin as it feeds. For a colourful account of these remarkable events in the life history of the oil beetle, one can do no better than read the work of J. H. Fabre who added a great deal to our

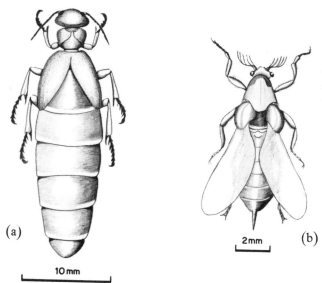

(a)

10 mm

2 mm   (b)

Figure 108. Two beetles whose larvae are parasitic upon larval Hymenoptera. (a) *Meloe majalis* (Meloidae), an oil-beetle parasitic upon bees. (b) *Rhipiphorus subdipterus* (Rhipiphoridae), a parasite of wasps (both after Grassé 1949).

knowledge of these beetles. Soon after eating the host's egg, the triungulin moults into the second instar which is carabid-like in form and floats on the honey store. This larval stage is succeeded by others of more grub-like form with rudimentary legs, small heads, enlarged bodies with indistinct segmentation, and thin, white integuments. Only one meloid larva can develop in a cell.

The Rhipiphoridae (Figure 108b) is another family whose species are parasitic upon the Hymenoptera (although a few species attack cockroaches or beetles), but they differ from meloids and all other beetles in being endoparasitic, at least for a part of their larval life. *Metoecus paradoxus* is a European species which is parasitic upon social wasps (*Vespula*). As in meloids, eggs are laid in soil or on vegetation and the first instar larva is a triungulin. This is transported to the nest on the body of a host worker. In the nest it enters the body of a wasp grub and grows rapidly, but it later emerges from its host, and as a second instar larva feeds as an ectoparasite (Linsley, MacSwain, and Smith 1952). The emergence of the endoparasitic first instar larva occurs at a time when the host, if unparasitised, would have pupated, and it has been suggested (Krombein 1967) that its emergence is triggered by hormonal changes in the host.

Rhipiphorus fasciatus is a larval ectoparasite of prepupae of the bee, *Lasioglossum zephyrum* (Batra 1965). Its triungulins are said to overwinter attached by their mandibles to hibernating adult bees. If this is correct then it is a remarkable feat of endurance for so minute an insect.

Species of *Rhipidius* are parasites of cockroaches, and their life history is simpler than that of *Metoecus*. The female *Rhipidius* is a larviform, wingless beetle which lays its eggs on the ground in areas frequented by the hosts. The triungulins attach themselves to passing cockroaches and enter their bodies. Unlike the larva of *Metoecus*, *Rhipidius* completes its larval feeding as an endoparasite, but it emerges to pupate.

Hypermetamorphosis

Insects that locate their hosts by first instar larvae involve their larval stages in two distinct major functions: finding the host, and feeding upon it. As a result the host-seeking, first instar larva is adapted for an active, exposed existence, whilst subsequent larval instars resemble those of other protelean parasites in being sluggish in movement and modified for living in, or adjacent to, an ample supply of food. An insect that has in its life cycle at least two distinctly different larval types is described as being hypermetamorphic. In Meloidae several different larval types have been described.

The phenomenon of hypermetamorphosis has been mentioned in varying detail in other chapters but its occurrence among insects is here summarised. It occurs in the following orders and families:

Neuroptera: Mantispidae
Coleoptera: Meloidae, Rhipiphoridae, Staphylinidae (Aleocharinae)
Strepsiptera
Diptera: Acroceridae, Bombyliidae, Nemestrinidae, Calliphoridae (some),
    Tachinidae (some)
Hymenoptera: Perilampidae, Eucharitidae, Ichneumonidae (*Euceros*)

The first instar larvae in these families provide an excellent example of
convergent evolution, that is the development independently in several unrelated groups of similar features. This is, of course, because the first instar
larvae have a similar function to perform. They are very small, fusiform
creatures, usually bearing hard sclerotisations and often a pair of long caudal
spines or hairs. Longevity is one of their features, and so is the large numbers
in which they are produced. They are mostly very active animals and several

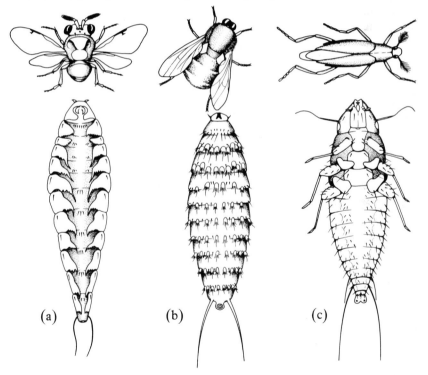

(a)   (b)   (c)

Figure 109. First instar larvae (with the corresponding
adults, not drawn to the same scale) which find hosts for
themselves. (a) Perilampidae (Hymenoptera) and (b) Acroceridae (Diptera) exemplify the planidium type of larva.
(c) A triungulin larva (with well-developed legs) of Rhipiphoridae (Coleoptera). (After various authors.)

of them can jump, stand erect on the tips of their abdomens, or move in a looping, leech-like fashion. Amongst these first instar larvae, however, two types are discernible. That which is found in the Neuroptera, Coleoptera, and Strepsiptera possesses jointed thoracic legs and is known as a triungulin or triungulinid (Figure 109c). The active dipteran or hymenopteran first instar larva is legless and known as a planidium (Figure 109a and b). The planidium uses long setae, thoracic or caudal, in locomotion.

12. Strepsiptera

STREPSIPTERA, OR STYLOPS as they are commonly called, have an historical association with British entomology. In 1811 the order was founded by the Reverend W. Kirby, the 'father of British Entomology' who was made Honorary Life President of the Entomological Society of London on its foundation in 1833. The society adopted, very appropriately, *Stylops kirbii* as the subject of its seal, and in 1932 commenced publication of the journal *Stylops* which was the predecessor of the society's *Proceedings, Series B*.

The Strepsiptera, a small order containing in the region of two hundred species, are as larvae exclusively parasites of other insects. Bees, wasps (both sphecoids and vespoids), and ants amongst the Hymenoptera and Cicadellidae, Cercopidae, Delphacidae, and Pentatomidae in the Hemiptera are the most usual hosts, though the family Stichotrematidae specialises in attacking Orthoptera, and the primitive Mengeidae are parasites of Thysanura. For some time, host-specificity was thought to be the rule in the order, but as data accumulate many species are proving to be oligophagous with a host range extending over a genus or even a family instead of being restricted to one host species. *Elenchus tenuicornis*, which enjoys an almost cosmopolitan distribution, is known to attack fifty-five leaf-hopper species (Baumert 1959). It has even been suggested (Bohart 1941) that some species (Myrmecolacinae) may attack hosts belonging to two orders, larvae entering ants developing into males and those entering Orthoptera developing into females.

Typically the female Strepsiptera have abandoned a free-living existence, and they, like the larvae, spend their entire life inside the host's body. Only the fused head and thorax of the female protrude from the host (Figure 110a). As might be imagined, they have become extensively modified in connection with this unusual habit and are markedly different from the free-living males. Female Strepsiptera are, with the exception of the Mengeidae to which reference will be made later, larviform and devoid of wings, legs, eyes, and antennae. The mouth and anus are minute and non-functional, and the lumen of the gut is obliterated. Food is absorbed through the cuticle of the abdomen, which is embedded in the host. The head and thorax are fused to form a small cephalothorax with a hard cuticle, whilst the abdomen is inflated, white, and soft, enclosed by the persistent last larval exuvia. Its bag-like

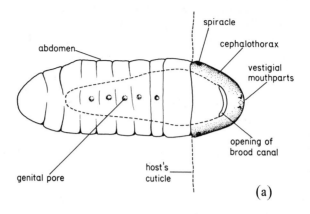

(a)

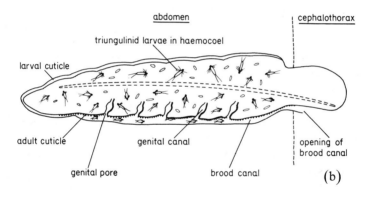

(b)

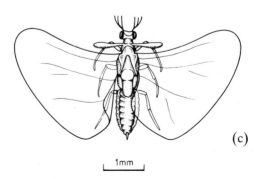

(c)

L__1mm__|

Figure 110. Strepsiptera. (a) Diagrammatic representation of
female in ventral view. (b) The same in longitudinal section.
(c) Male *Eoxenos laboulbenei* (from Parker and Smith 1934).

form contains large numbers of eggs or first instar larvae. In the ventral mid-line of the abdomen there is (except in *Stichotrema*) a series of two to five unpaired genital canal openings. It is through these openings that the active first instar larvae make their way into a brood canal between the female's abdomen and the unshed larval cuticle, and thence to the outside world (Figure 110b). This female reproductive system is unique among insects. The females breathe by tracheae whose spiracles are situated on two lateral projections of the cephalothorax.

Male Strepsiptera (Figure 110c) are equally unusual insects. They are only two or three millimetres in length, brown or black in colour, and equipped with large, membranous, fan-shaped hindwings and a correspondingly large metanotum, but the forewings are reduced to rod-shaped halteres. Like the females they have non-functional mouthparts, and they live for only a few hours. During this brief life male stylops are intensely active and they devote all of their time to agitatedly seeking mates. In this quest they are aided by antennae whose surface, equipped with sense organs, is increased by the addi-tion of side branches, and they have protruding compound eyes. The large hindwings are normally in continual movement, but if the halteres are ampu-tated wing movement ceases. Evidently the halteres exercise a stimulatory effect upon the flight mechanism. The halteres carry batteries of campaniform sensillae, as do the analogous structures in Diptera, and they no doubt serve as stabilising organs, maintaining the insects equilibrium in the air by the detection of any deviation from a straight flight path. Males of most species have well-developed tarsi to enable them to cling to the body of a host in which the female is ensconced. Males are attracted by the host species even, in *Corioxenos antestiae* at least, if this is not stylopised. Copulation is effected by the male inserting his aedeagus into the brood canal of the female where it opens to the exterior as a bow-shaped or crescentic slit ventrally at the base of the cephalothorax. The sperm enter the genital canals and disperse throughout the female's body, fertilising the eggs which are lying free in the body cavity (Schrader 1924). It is thought that some species, for instance of the genus *Halictoxenos*, might be parthenogenetic since no males have ever been found. No males are yet known in the family Stichotrematidae either. However, this is far from being conclusive evidence for parthenogenesis since the flight period of the male is very short. Occasionally observers have reported swarms of male stylops but this is a matter of being at the right place at the right time. Perkins (1918b) records large assemblages of males of *Neostylops aterrima* and *Stylops wilkellae* around thorn bushes in early spring in southern England.

Stylopised bees apparently emerge from their overwintering pupae rather earlier in the year than healthy bees. This is indicated in the data that follow, from Perkins (1918a), concerning the condition of *Andrena wilkella* caught at different times in the spring.

Salt (1927) offers, in explanation of this phenomenon, the opinion that

Dates of capture	% stylopised
April 24–27	100
April 27–May 4	90
May 7	30
May 11	under 10

stylopised bee larvae, in an attempt to compensate for nourishment lost to their parasites, consume more quickly than normal the provisions of honey left for them in their cells by their mother. This forces them to advance the date of their pupation and subsequent emergence. As further evidence, Salt instances the wasp *Polistes* parasitised by *Xenos*. Unlike *Andrena*, the development of stylopised *Polistes* is retarded. *Polistes* does not practice mass provisioning in the manner of *Andrena*; instead, females feed their larvae progressively as they develop and development is protracted by loss of food to the stylops. The development of stylopised Hemiptera, which likewise usually have an unlimited supply of food, is also prolonged.

Perkins (1918b) also discovered that males of *N. aterrima* and *S. wilkellae* usually emerge from their bee hosts early in the morning when the hosts, as adult insects, are first emerging in the spring from their subterranean cells. Sunlight probably stimulates the stylops emergence. Hassan (1939) found that *Elenchus tenuicornis* is another species that emerges early in the morning; its hosts are Delphacidae. Of thirty-nine males whose emergence was observed, thirty appeared before 9.30 a.m. An early start in the day is obviously essential to an insect that will be dead by sundown and which will best fulfil its function in sunshine.

Strepsiptera are larviparous and the first instar larva is a triungulin or triungulinid (Figure 111a). It is very small, about 0.2 millimetres long, and has a dark coloured or yellow cuticle which is well sclerotised, well-developed legs, and one or two pairs of long caudal filaments. Triungulinids are short-lived creatures. Those of *Elenchus tenuicornis* die in about $4\frac{1}{2}$ hours at $14°C$ and 90% R.H., and sooner at higher temperatures and lower humidities (Raatikainen 1967).

Each female stylops can produce an enormous number of such larvae, probably usually about two thousand, though the figure for *Neostylops aterrima* may be as high as seven thousand. Polyembryony has been reported in a few species. Batra (1965) records that four or five larvae may develop from one egg inside the female. This high reproductive potential is an adaptation to offset the relatively poor chance an individual triungulinid has of continuing its development. In the majority of species, most of the larvae emerge from their mother within a single day. Emergence is effected by the movement of the larvae; the female remains passive. The triungulinids make their way to the outside world from the brood canal and through its external opening. This opening is originally covered by a membrane but is pierced by the male during

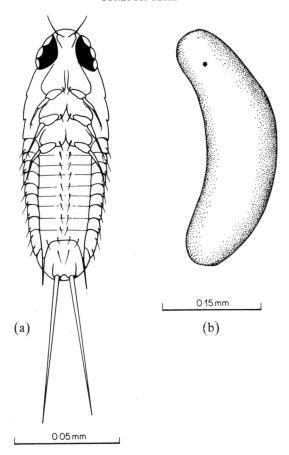

Figure 111. Larvae of *Elenchus templetoni* (Strepsiptera).
(a) First instar triungulinid larva. (b) Second instar larva.
(Both from Williams 1957.)

copulation. The function of the triungulinid is to gain access to a new host. It is assisted in this, at least in many species, by its ability to jump. By holding its abdomen curved with the tail filaments beneath it and directed forwards, the larva is in a position to catapult itself into the air by suddenly straightening its abdomen. The behaviour of the triungulinids of *Corioxenos antestiae* has been observed by Kirkpatrick (1937a). They adopt a motionless but 'alert' posture on vegetation, their fore-parts raised and supported only by the hind pair of legs. In response to movement, the triungulinids leap. Colour also affects this response, for it is greatest when the moving object is black and orange, the main colours of its host, the pentatomid bug *Antestia* (Kirkpatrick 1937b). Which sense organs are involved in colour perception is unknown; the larva does not have compound eyes. Should the triungulinids land on the wrong host, they soon leave it. Having attained the body of a nymph of *Antestia*, the triungulinids penetrate its body at an ecdysis, possibly stimulated to do so by the host's moulting fluid. Triungulinid larvae, unlike all other larval stages except the last, have jaws, and these they use to effect an entry into the host. Hassan (1939) found that the triungulinids of *Elenchus tenui- cornis* required ten minutes to enter their delphacid host.

The first instar nymph of *Antestia* is free from parasitisation because it has had no preceding ecdysis. *Elenchus templetoni*, however, will enter all stages of its host, the delphacid *Dicranotropis muiri*, including first instar nymphs (Williams 1957). *E. templetoni* also differs from *Corioxenos antestiae* in that triungulinids will enter an unsuitable host, the delphacid *Perkinsiella sacchari- cida*. Larvae of *Elenchus* quickly die inside *Perkinsiella*, but there is no evidence that they avoid it. *E. tenuicornis* triungulinids, however, will not enter Cicad- ellidae, (Hassan 1939), or Aphidae (Raatikainen 1967).

The triungulinids of *Corioxenos*, *Elenchus*, *Halictophagus*, and other para- sites of Hemiptera, as well as those which emerge in the nests of social Hymenoptera, reach their hosts directly, but the many species parasitic upon solitary bees and wasps need to invoke the aid of a transporter in order to reach their host larvae, which are hidden away in cells, usually underground. In these species, the triungulinids climb onto flowers and attach themselves to visiting bees and wasps. A female bee containing a mature female *Halictoxenos jonesi* spends much time on flowers, dragging her abdomen amongst the stamens and depositing triungulinids (Batra 1965). On the correct species of Hymenoptera they should be carried to a host larva in its cell, and the triungulinids move from adult to larva when the latter is being fed. This is an example of phoresy. Triungulinids of *Stylops pacifica* may deviate from the normal mode of transport on the surface of a bee, in being drawn into the bee's crop with nectar and regurgitated onto the pollen ball in the bee's nest (Linsley and MacSwain 1957).

The triungulinids invade the abdominal haemocoel of the host, and there they feed. After their first moult, the stylops larvae are no longer triungulinids but instead rather featureless, grub-like creatures that lack mouthparts and legs and scarcely move (Figure 111b). They gain nourishment by diffusion through their thin cuticles. In the few species in which development has been followed, five to seven larval instars have been reported, although six may be the normal number (Greathead 1968). The larvae usually show some sexual dimorphism, especially in their later stages. During the last instar the larva, which has regained mandibles, forces its cephalothorax through an inter- segmental membrane of the host's abdomen. The site of this extrusion appears to be fairly constant for a particular species. *N. aterrima*, for instance, usually emerges between the fourth and fifth abdominal segments. At the time when the stylops larvae make their appearance externally, the host may be a pupa (Hymenoptera), nymph or adult (Hemiptera Homoptera), or an adult (Hemiptera Heteroptera). Female larvae of *Elenchus templetoni* extrude from adult Delphacidae, but male larvae occasionally appear on old nymphs. After extrusion the male larva soon pupates, but the last larval skin is not shed; instead it encloses the pupa and hardens to form a puparium. This puparium is provided with a cap which is pushed off when the adult male emerges. In the development of the female there is no obvious pupal stage, the final instar larva changing into the larviform female shortly after extrusion of the cephalothorax. Again the larval integument is retained, and it becomes partly sclerotised to form a 'pseudopuparium' surrounding the adult female. Females are much longer-lived than their partners, and in most species may be found over a period of several weeks.

The number of stylops which an individual host is able to support depends, as might be expected, very much upon the size of the host. The maxima recorded concern *Xenos* in the wasp *Polistes* (Pierce 1909). Thirty-one *Xenos* larvae were found in one wasp, and fifteen extruded stages in another. There is a tendency for such multibroods to be unisexual. In Hemiptera there are usually no more than five stylops to a host. Williams (1957) records that 68.5% of stylopised *Dicranotropis muiri*, a small leaf-hopper, contained only one extruded *E. templetoni*. The sex of this leaf-hopper and the sex of the parasites it contains are not related. Williams found that 100 male *D. muiri* contained 82 female and 60 male extruded parasites, and that 100 female hosts contained 82 female and 61 male extruded parasites.

Stylops, unlike nearly all other protelean parasitic insects, permit their hosts to continue living for what, in some cases, probably amounts to a normal life span. This is because they feed by absorption and perpetrate scarcely any mechanical injury on the hosts' tissues. Nevertheless, they frequently have a profound affect upon their hosts. The modification of the rate of development of stylopised Hymenoptera has already been mentioned. A much more dramatic result of stylopisation is the appearance in the host of secondary sexual characters of the opposite sex. A stylopised insect may sometimes be described as an intersex.

Pérez (1886) studied forty-seven species of *Andrena* and noted that in stylopised individuals there is a tendency for the abdomen to become more globular, the head smaller than normal, the body pilosity more abundant, longer, finer, and brighter in colour, and the puncturation of the integument closer, more numerous, and superficial. Males of many species of *Andrena* have white or yellow facial markings which are absent or diminished in females. Stylopisation reduces the size of the markings in the male and increases them, or even produces them, in the female (Figure 112). Stylopisation tends also to reduce the pollen-collecting apparatus in female *Andrena* so that their hind legs approach in appearance those of a healthy male; the stylopised male may occasionally show slight indications towards developing the broadened hind basitarsus and tibia of the normal female. The ovipositor of the stylopised female and the genitalia of the stylopised male *Andrena* are reduced. One character, however, which is not modified by stylopisation is the antennal segment number, invariably twelve in females and thirteen in males. Behaviour also seems to be unaffected. Pérez emphasised that the changes do not involve

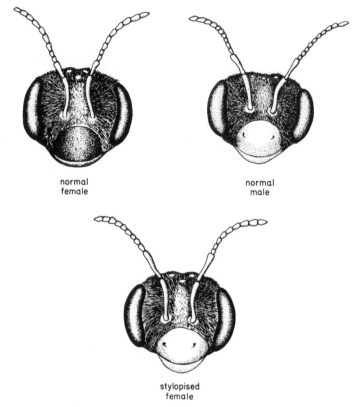

normal
female

normal
male

stylopised
female

Figure 112. Heads of the bee *Andrena chrysosceles* illustrating the development of normal male colouration in a stylopised female (after Smith and Hamm 1914).

merely the loss or degradation of characters towards a common mean condition, but may represent the acquisition of characters of the opposite sex: '*c'est une femelle qui emprunte les attributs du mâle; c'est un mâle qui revêt les caractères de la femelle.*' Pérez' findings have been corroborated by Smith and Hamm (1914), Perkins (1918a), and Salt (1927). Salt (1927, 1931), in addition to confirming Pérez' work, found that wasps of the genera *Odynerus*, *Ancistrocerus*, *Chlorion*, and *Sphex* show a tendency towards sex reversal in their external characters.

Internally, the gonads of both sexes of stylopised *Andrena* are reduced. Stylopised female bees are sterile since their ovaries contain no developing eggs, but live spermatozoa are to be found in the testes of stylopised males. In bugs there is less sexual dimorphism and so the effect of stylopisation is not so pronounced as in bees. Muir (1918) noted that genitalia and gonads of stylopised Delphacidae tend to be abortive. Williams (1957) made similar observations on *Dicranotropis muiri* and concluded that the extent of abnormality probably depends upon the stage at which the host is initially attacked. Female hosts are able to lay eggs only if they contain very young stylops. In this host species, nymphs in which stylops become adult never themselves attain adulthood, and Hassan (1939) found that adult delphacids usually died soon after emergence of *Elenchus tenuicornis*, frequently becoming infested by a fungus. It has been claimed (Lindberg 1960) that there is an unusually high proportion of brachypterous individuals among stylopised delphacids, but this is not substantiated by Raatikainen (1967).

Female hosts are more frequently attacked than males, at least by the species parasitising Hymenoptera, and it is the females of both Hymenoptera and Hemiptera that are most modified by stylopisation. The sex of the stylops and the degree of superparasitism apparently do not always affect the extent of morphological modification of the host. There are indications, however, that the larger male stylops may have a greater effect than the female, and that several stylops tend to have more effect than one. In general, it may be said that those characters of the host that are differentiated latest in development are those which are most likely to be modified by stylopisation.

Although the gonads of a stylopised host are frequently destroyed, this 'parasitic castration' is not itself responsible for the tendency to reversal in the secondary sexual characters. Insects do not have sex hormones produced by the gonads or other endocrine glands circulating throughout their bodies in the blood. This is apparent from the existence of sex mosaics or gynandromorphs, in which part of an individual may show male characters and another part female characters. The sex of each part of the body is determined at the cellular level and not overridden by the influence of circulating sex hormones. Stylopised hosts are weakened by loss of body materials to their parasites, and it is believed that this allows the genes of the alternate sex to gain expression. Salt (1927) cites as evidence for the nutritional deficiency in stylopisation the fact that a tendency towards sex reversal is seen only in solitary bees like *Andrena* and solitary wasps like *Odynerus*, species that have only a limited amount of food left for them by the female parent before the cell is sealed. Social wasps, such as *Polistes*, *Belonogaster*, and *Vespula*, feed their larvae 'on demand' and in these forms, where there is no likelihood of undernourishment, there is no tendency towards sex reversal (Wheeler 1910, Salt 1931).

The effect of stylopisation on the insect host is paralleled in some ways by the effects of *Sacculina* and other parasitic barnacles on their crab hosts (page 271), although sacculinisation often results in a spectacular change in male crabs towards the female condition but little change in the opposite direction.

The family Mengeidae includes some very interesting but atypical strepsi-
pterans. Founded on the fossil genus *Mengea*, males of which were discovered
in baltic amber from northern Europe, it was thought, until recently, that the
Mengeidae were extinct. However, two allied genera, *Eoxenos* and *Mengenilla*,
are now known to be living in the mediterranean region and northern China
(Miyamoto 1960), although it should be mentioned that these two are sometimes placed in a separate family, the Mengenillidae. Perhaps the most striking
feature of these forms is that the females are free-living and, though wingless,
they have functional legs, eyes, and antennae (Figure 113). Also, the head and
thorax are not fused, there being no cephalothorax as in the parasitic females
in other families, and the abdomen has paired spiracles. The males exhibit
typical strepsipterous features, but they are much longer-lived than is usual

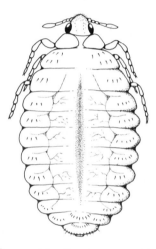

Figure 113. *Eoxenos laboulbenei*, an example of a free-living
female strepsipteran (from Parker and Smith 1933).

in the group. Adult mengeids of both sexes have been found living together
beneath stones, together with their triungulinid larvae, and it is in this situation that the triungulinids encounter their hosts. These are Thysanura, an
apterygote order which includes such insects as silver-fish and fire-brats. It is
interesting that the most unspecialised group of Strepsiptera have as their
hosts some of the most unspecialised of all insects, but this cannot be a case of
co-evolution as in the case of certain vertebrate ectoparasites (Szidat's rule,
page 25) since the Strepsiptera are of much more recent origin than the
Thysanura.

The biology of Mengeidae has been investigated by Ulrich (1943). The larva
in its final instar leaves its host and pupates under a stone. A pupa is formed by
both males and females, quite different from the typical life history in the
Strepsiptera. Mating is also atypical in that the male pierces the integument

of his free-living partner with his aedeagus, and spermatozoa are injected directly into the abdominal haemocoel where the eggs are located. The female has a single vaginal opening, the normal insectan condition, instead of the median series of genital pores, but the vagina is apparently functionless.

Before leaving Strepsiptera, reference must be made to their taxonomic status. Regarded as a distinct order by many authorities, at least until very recently, there is good reason to consider them as merely a highly aberrant group of beetles. Crowson (1955) refers them to the superfamily Stylopoidea in the Coleoptera, basing his opinion on the following points. The sternites of the male abdomen are more extensively sclerotised than the tergites, a feature seen only in beetles among endopterygote insects and associated with the beetles' protective elytra. The triungulinid larva bears a close resemblance to the triungulin larvae of other beetles with endoparasitic larvae (e.g. Meloidae, Rhipiphoridae), although the final stage larva of the primitive Mengeidae is most remarkable in possessing compound eyes. The visual organs of other endopterygote larvae are simple ocelli. Stylops also differ from Coleoptera in having no trochanters in their legs. Crowson homologises the very reduced forewings of the male stylops with the hard beetle elytra or wing-covers, pointing out that their reduction can be correlated with the fact that almost the entire life of the male stylops is spent in flight. Beetle elytra protect the beetle only when it is not flying; in flight they are held away from the body. In having reduced elytra, the stylops avoids carrying excess weight. Beetles such as Staphylinidae and Cantharidae which are often active fliers also have reduced or thin elytra. Crowson considers that muscles which cause the rapid beating of the elytra or halteres of stylops are probably homologous with muscles which move beetle elytra. If the halteres are derived from beetle elytra then they will have lost the flight muscles of an insect forewing, and they could not be expected to reacquire them, since to do so would be a contravention of Dollo's Law; once an organ has been lost in the course of evolution, it can never be regained in its original form.

Male stylops quite strikingly resemble certain heteromerous beetles which have parasitic larvae, especially the Rhipiphoridae (Figure 108b). However this resemblance is most likely a result of convergent evolution rather than indicative of a very close relationship. The heteromerous beetles are characterised by having four tarsal segments to the hind legs and five on each of the other legs; male stylops have from two to five tarsal segments and in the Mengeidae all legs have five tarsal segments. Crowson concludes that 'Stylopoidea must stand as an independent superfamily of Coleoptera-Phytophaga'.

13. *Commensalism, Inquilinism, and Social Parasitism*

THE FAMILIAR OAK-MARBLE GALL is the larval home of a gall wasp, *Andricus kollari*. The gall-wasp larva feeds upon the vegetable tissues surrounding its larval cell in the centre of the gall. The marble gall is also the site of development of two other species of gall wasp, *Synergus reinhardi* and *S. umbraculus*, which lay their eggs in developing galls. The Synergini are themselves incapable of inducing gall formation and are completely dependent upon *A. kollari* for the provision of their larval food. They are smaller than *A. kollari* and one marble gall can support about twenty *S. umbraculus* and about ten *S. reinhardi*.

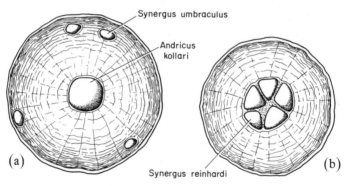

Figure 114. Sections through the centres of oak marble-galls to show the positions occupied by cells of the gall-maker (*Andricus kollari*) and its inquilines (*Synergus umbraculus* and *S. reinhardi*).

When the occupants of individual marble galls are reared, a very striking difference in the host relationships of *S. reinhardi* and *S. umbraculus* becomes apparent. A single marble gall may produce adults of *A. kollari* alone, or of *S. reinhardi* alone, or of *S. umbraculus* plus either *A. kollari* or *S. reinhardi*. *A. kollari* and *S. reinhardi* never emerge from the same gall; *S. umbraculus* and *A. kollari* are able to develop side by side, but *S. reinhardi* destroys *A. kollari*. Cutting open mature marble galls reveals that the cells of *S. umbraculus* lie in the peripheral tissue of the gall, whereas those of *S. reinhardi* occupy a central position where they replace the single chamber of *A. kollari* (Figure

114). *S. reinhardi* larvae grow more rapidly than the *A. kollari* larva, and the latter dies, probably as a result of either starvation or mechanical pressures; it does not appear to be eaten by the *Synergus* larvae.

Because the *Synergus* species do not feed upon the tissue of the gall-maker, they cannot be described as parasites. Instead, they exemplify phenomena that may be thought of as lying somewhere between parasitism and symbiosis. Symbiosis, as here understood, is the intimate association of two organisms for their mutual benefit. Many continental writers, however, employ the term symbiosis to describe all cases, parasitic and otherwise, of two species living together, and the term mutualism is used for the type of association that is here described as symbiosis.

Commensalism is an association in which the advantages are all to one of the partners. It describes the feeding together of two animal species at a food source which has been provided by, or for, one of the species, the host. The commensal may do harm to its host other than depriving it of some food. The species of *Musca* that feed on blood flowing from vertebrate wounds inflicted by horse flies could be described as facultative commensals. Other examples will be discussed later.

Not very distinct from commensalism, and really a form of it, is the association described as inquilinism. An inquiline is a commensal that lives in a very close spatial relationship with its host, either inside its body or in its shelter. It does not feed upon its host but, nevertheless, frequently destroys it. Tapeworms are inquilines rather than true parasites since they do not feed upon their host. The two species of *Synergus* just described are both inquilines, although they differ greatly in their affect upon the host species.

Synergus is of the same family, the Cynipidae, as the host species, and a close taxonomic relationship such as this often exists in inquilinism. All species of *Synergus* are inquilines of oak gall-wasps, and species of another cynipid genus, *Periclistus*, are inquilines of rose gall-wasps. Gall midges (Cecidomyiidae) include among their number the genera *Clinodiplosis* and *Trotteria* whose species are inquilines in galls of species belonging to other genera in the family. *Trotteria sarothamni* is an inquiline of *Asphondylia sarothamni* and it usually destroys its host, but if its eggs are laid in a gall containing the prepupa or pupa of its host, both species survive (Parnell 1964). Species of the phytophagous chalcid genus *Tetramesa*, whose larvae feed in grasses, are parasitised and inquilined (which relationship is the more frequent is not yet clear) by some species of *Eurytoma*, another eurytomid genus. In all of these examples except the last, it seems probable that at some time in the past the species which are now inquilines pursued independent existences like their hosts, but then found it more advantageous to usurp the food source of an allied species with similar habits. In the case of the inquiline cynipids, it is fairly safe to assume that they had a not too distant ancestor which was capable of inducing gall formation. Many of the species are still capable of modifying the structure of their hosts' galls.

Inquilinism would seem to be an obvious road to parasitism. It demands only that the inquiline changes from a phytophage to an entomophage. However there are no clear examples of this change ever having taken place.

An association rather similar to an inquiline relationship in which the host dies is the destruction of *Rhyssella curvipes* by *Pseudorhyssa alpestris* (page 151). The *Pseudorhyssa* larva kills the *Rhyssella* larva so that it can enjoy sole possession of a paralysed wood-wasp larva. It steals the prey of the primary parasite and is sometimes called a cleptoparasite. Similar examples are provided by *Eurytoma pini*, a chalcid that will oviposit only on stem-boring cater-

Figure 115. A commensal beetle, *Lomechusa strumosa*, being fed by a worker ant (dapted from Donisthorpe 1927).

pillars after they have been paralysed by another parasite (Arthur 1961), and by those Gasteruptiidae that oviposit on bee's eggs and feed mainly on the stored pollen and nectar in the cell.

Many insects live as commensals in the nests of social Hymenoptera, pilfering from the food stores of their hosts, or soliciting regurgitated food from workers. The adults of two aleocharine staphylinid beetles, *Lomechusa* and *Atemeles*, mingle with ants in their nests (Donisthorpe 1927). With their antennae they tap the worker ants, and an accosted ant eventually regurgitates a drop of food which is lapped up by the beetle (Figure 115). However, the relationship is not entirely one-sided in its benefits. The beetles have epidermal glands which pour a sweet secretion upon tufts of golden hairs (trichomes). The ants are evidently fond of this secretion, and it may be because of this that the beetles are tolerated in their nests. The relationship thus tends towards symbiosis, such as exists between ants and the aphids which they 'farm'. Sometimes a beetle meets a hostile worker-ant, in which case it defends itself by cocking its abdomen over its back and emitting an obnoxious odour from abdominal stink (or repugnatorial) glands. The beetles bear some resemblance in their form and movements to the ants with which they live. This mimicry may give them some protection in the nest, although it is most likely to benefit them in the presence of predators outside the nest. *Amphotis*, a beetle belonging to the family Nitidulidae, also habitually lives in ants' nests, and solicits food by tapping workers with its antennae. It is a very flattened beetle, and when threatened by an ant it tucks its legs and antennae beneath the shield of its thorax and elytra, applies itself very closely to the ground, and is then quite protected (Donisthorpe 1927). Another interesting commensal is *Metopina*, a phorid fly which is found in nests of ponerine ants in America. The larva of *Metopina* lives encircled about the neck of an ant larva, and when the ant larva is fed by a worker, the fly larva uncoils and shares in the meal. It pupates inside the cocoon spun by the ant larva, and in due course both ant and fly emerge from the cocoon.

A group of myrmecophilous insect species lick or gnaw at the host ants' cuticles, apparently feeding upon integumentary secretions. Such species, akin in some ways in their habits to *Hemimerus* (page 88), are called strigilators (Wheeler 1923) and they include a cockroach (*Attaphila*), a wingless and silent cricket (*Myrmecophila*), and a beetle (*Oxysoma*).

In 1895, Wasmann estimated that there were 1246 myrmecophilous species known, 1177 of which were insects, and the great majority of these belonged to the Coleoptera. He distributed these species under a number of headings which indicate the precise nature of the association between the host ant and its 'guest', but this terminology is really only of interest to the specialist. A useful summary is provided by Caullery (1922). Social insects other than ants are also hosts to commensals. Termite mounds harbour a large number, including springtails (Collembola) which, like *Metopina*, intercept food passing between host individuals. The curious little fly *Braula coeca* (page 75) is another commensal, this time of honey bees.

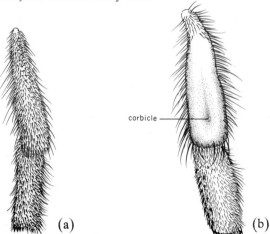

corbicle

(a)　(b)

Figure 116. Hind tibiae and basitarsi of (a) female *Psithyrus barbutellus* and (b) its host, queen *Bombus hortorum*. A pollen basket (corbicle) is present only on the tibia of the *Bombus*.

Amongst social and solitary Hymenoptera the usurpation of the nest of one species by another species is of quite frequent occurrence. Such cuckoo-like behaviour is called social parasitism. *Psithyrus*, a genus of cuckoo bees, is closely allied to *Bombus*, the genus that includes social bumble bees. The female *Psithyrus* emerges from hibernation later in spring than queens of the host species. *Psithyrus* are similar to bumble bees but lack a pollen-collecting apparatus on their hind legs (Figure 116), and are therefore incapable of founding a colony for themselves. They also have a thicker cuticle, sharper mandibles, and a longer and more powerful sting than *Bombus*. The female *Psithyrus* enters a young *Bombus* colony of the appropriate species, probably locating it by scent (Free and Butler 1959). The smaller the colony, the better its chances of successful entry. The *Bombus* workers at first fight the intruder, but their inferior stings and thinner cuticles are not usually sufficiently compensated for by their superior numbers, and so several die and the female *Psithyrus* is generally able to establish herself in the nest. Gradually the *Bombus* workers cease to fight the *Psithyrus* and 'an uneasy truce is established' (Richards 1953). The *Bombus* queen is killed by the cuckoo bee, or occasionally the two insects may continue to live side by side. In either case, the colony produces few, if any, *Bombus* workers, whilst male and female *Psithyrus* emerge in numbers. The eggs laid by a *Bombus* queen which is allowed to live are seemingly often destroyed by the *Psithyrus*. Because the *Bombus* workers are not being replaced, the colony gradually runs down, and the female *Psithyrus*, after mating, enter a relatively early hibernation. The larger the *Bombus* colony successfully invaded by a *Psithyrus*, the greater the number of *Psithyrus* that are eventually produced. However a *Psithyrus* has difficulty in establishing herself in a large colony. There is evidence that when a small colony is entered the *Bombus* queen is not so often immediately killed, and more of her eggs are allowed to develop by the *Psithyrus* so as to raise the worker population to an adequate level.

The *Psithyrus* species often closely resembles in colour the *Bombus* species with which it associates. The significance of this is not clear, although it seems rather unlikely to indicate, as is sometimes suggested, that *Psithyrus* is polyphyletic in origin, each species having evolved independently from its host species of *Bombus* (Richards 1927). The similarity in colour between the two species can scarcely be of value to the *Psithyrus* once it is inside its host's nest, although it probably allows it to approach a nest undetected as an enemy. The common colouration may be an example of Müllerian mimicry; that is, the adoption of similar appearance by two distasteful or dangerous species to their mutal advantage.

Although no species of *Bombus* is known to be an obligate social parasite,*

---

* A recently described European species, *Bombus inexspectatus*, may be an obligate parasite, since no worker caste has been discovered and the females do not seem to gather pollen or produce wax.

an interaction sometimes takes place between closely allied species which may indicate how such a relationship evolves. *Bombus lucorum* and *B. terrestris* nest in similar situations, often in deserted field-mouse nests, and there frequently seems to be a shortage of such sites. Consequently there is both intra- and inter-specific competition for nesting sites. The corpses of queens, stung to death by their more successful competitors, may often be found just inside the entrance to a young colony, and Sladen (1912) reports the successful take-over of newly founded colonies of *B. lucorum* by *B. terrestris*. Richards (1949) suggests that the invasion of the range of a northern, early-nesting species by an allied species with a more southerly distribution would create circumstances favourable to the evolution of a social parasitic relationship. The southern species emerges later from hibernation because higher temperatures are required to rouse it. On emergence it might find all available nesting sites already occupied by the northern species, and be faced with the necessity of fighting for the possession of one. *B. terrestris* has a more southerly range than *B. lucorum*. When it take over a young *B. lucorum* nest, *B. terrestris* is behaving as a temporary social parasite. It is in no way dependent for food upon *B. lucorum*, and the composition of the colony quickly changes from being purely of one species to purely of the other; when the replacement is complete, parasitism of course ceases.

Vespula austriaca is a wasp that behaves as a social parasite of *V. rufa*. The two species are again closely allied but they differ in that *V. austriaca*, like *Psithyrus*, lacks a worker caste. It is an obligate social parasite, not a facultative social parasite like *Bombus terrestris*. *V. austriaca* is almost certainly of southern origin; it usually emerges from hibernation a few weeks after *V. rufa*, and the female enters a nest that is already established by *V. rufa*. There is much fighting, and a large number of the *V. rufa* workers die; the queen, also, is eventually killed. Subsequently the colony produces fewer and fewer host workers and more and more males and females of *V. austriaca*. Eventually there are too few workers to maintain the colony, but many *V. austriaca* emerge. These mate, and the females enter hibernation.

Nest take-over amongst aculeate Hymenoptera is not confined to the social species. There are many solitary bees and wasps which usurp the provisions of their host's nests for the benefit of their own offspring. Such species are inquilines and cleptoparasites. Examples of cleptoparasitic solitary Aculeata included among the British fauna are *Evagetes* and *Ceropales* (Richards and Hamm 1939) (Pompilidae), *Nysson* (Hamm and Richards 1930) (Sphecidae), *Sphecodes* (Halictidae), *Epeolus* (Richards 1937) and *Melecta* (Apidae), and *Coelioxys* (Megachilidae). *Epeolus* attacks *Colletes* (Colletidae), but all of the other genera mostly utilise members of their own families as hosts.

The female *Evagetes* locates the solitary cell or nest of a spider-hunting wasp, generally a species of *Anoplius* (Evans, Lin, and Yoshimoto 1953). The *Evagetes* waits a little distance from a female *Anoplius* which is provisioning its cell with a paralysed spider. After the *Anoplius* has laid its egg on the spider, the *Evagetes* enters the cell, chews up the *Anoplius* egg, and lays its own egg in its place. *Ceropales* does not locate the nest of its host pompilid but oviposits on a paralysed spider being dragged to the nest by the host. The *Ceropales* larva destroys the young host larva in the nest. Another variation in behaviour is found in *Nysson*. The female *Nysson* enters the nest of another sphecid wasp and lays its egg on the prey before the host species oviposits. The egg of the cleptoparasite is concealed in a groove or fold in the integument of the prey so that the host wasp does not detect it. The egg is much smaller than the host egg, and its larva hatches first and destroys the host egg with its powerful mandibles. The bee *Coelioxys* parallels *Nysson* in its behaviour, and so too do the Chrysididae and Sapygidae, which were discussed in Chapter 8 .

In most of these examples it is likely that host and parasite evolved from a common ancestor. Structural resemblances are very close and differences can usually be ascribed to the cleptoparasitic habit. *Evagetes* and *Nysson* both have relatively short, stout antennae, unlike those of their hosts, presumably reflecting their need to find host nests instead of active prey. *Ceropales* has a hooked ovipositor which enables it to insert its egg in the lung-book of a spider.

Sometimes a cleptoparasite will attack more than one host species. *Coelioxys* is closely allied to the leaf-cutting bee *Megachile*, which is its usual host. The more distantly related *Anthophora* is also attacked, however, probably because it sometimes nests in close proximity to *Megachile*.

Bees of the genus *Nomada* are not very closely related to their hosts, which are solitary bees of the genus *Andrena*. This probably indicates that inquilinism in *Nomada* is a very long-established habit. *N. panzeri* attacks a number of *Andrena* species. Individuals originating in nests of different host species differ slightly in their colour patterns, but it is not known whether these differences are genetic or merely a result of differing environmental factors operating upon the immature stages.

An ant society is generally very hostile to alien insects, even to worker ants of the same species but from another nest. However, as with bees and wasps, suitable nesting sites frequently seem to be in short supply, and different species may nest in close proximity. As a rule, the individuals of such species become, with time, tolerant towards one another. This is especially true when the species belong to different subfamilies of the Formicidae, the family which includes all ants.

Formicoxenus nitidulus is a small ant that nests only inside the large mound nests of wood ants of the genus *Formica*. It constructs its galleries in the heart of the wood ants' nest, but they are too narrow to permit the *Formica* workers to enter and destroy the *Formicoxenus* brood. The workers of *Formicoxenus* are not molested by the much larger *Formica*. The food of *Formicoxenus* is not certainly known, although some it probably steals from its large neighbours.

Species of *Leptothorax* may also nest with species of *Formica*, although in this case the association is not obligatory.

More aggressive is the behaviour sometimes shown by *Formica rufa*. The queen *F. rufa* may found a new colony by invading the nest of *F. fusca* and deposing the rightful queen. The *F. fusca* workers rear the first offspring of *F. rufa* but they are gradually replaced by those of the invading species, so that eventually the nest contains only *F. rufa*. Similarly, *Lasius umbratus* queens invade nests of *L. niger*. Here the association is certainly obligatory. Before entering a *L. niger* nest, the queen *L. umbratus* seizes and eats a worker *L. niger*, possibly to ensure that the correct host species has been selected (Sudd 1967). Once inside the nest, however, the queen *L. umbratus* is very submissive and does not retaliate when mauled by the *L. niger* workers. In this way a fatal attack is averted, and the *L. umbratus* queen is slowly accepted by the workers and comes to compete with the *L. niger* queen for food that they bring. Eventually the *L. niger* queen is deserted or killed by her own workers, who then adopt the *L. umbratus* queen. A parallel case concerns *Strumigenys xenos*, an Australian ant parasitic upon *S. perplexa* (Brown 1955). These are examples of temporary social parasitism of one species by another that is closely related, analogous to that of *Bombus terrestris* and *B. lucorum*.

Temporary social parasitism is also practised by *Bothriomyrmex* and *Labauchena*. The queens of these ants invade the nests of their hosts and kill the resident queens, but they are small insects and unable to achieve this by direct combat.

Queens of *Bothriomyrmex decapitans*, a north African species, allow themselves to be carried into nests of *Tapinoma nigerrimum* by workers of that species. Once inside the *Tapinoma* nest, the queen *Bothriomyrmex* takes refuge on the back of the queen *Tapinoma* and starts to chew at the neck of its host, eventually decapitating it. This process takes some time, and before the queen *Tapinoma* has been killed the *Bothriomyrmex* has probably acquired the odour of its host's nest. This allows it to be accepted by the *Tapinoma* workers, which rear its brood. Rather similar behaviour has been observed in the South American ant *Labauchena*. Queens of *Labauchena* are, however, very much smaller than those of the host, *Solenopsis*, and a few of them together invade the host nest. They co-operate in removing the head of the large *Solenopsis* queen, a process that may take up to forty-five days to complete.

Another social ant parasite is *Formica sanguinea*, and this species has the remarkable habit of slave-making or dulosis. The host ant is *F. fusca*. A young queen *F. sanguinea* starts off her colony by invading a nest of *F. fusca*. She seals herself off from the *F. fusca* workers in a corner of the nest to which she has taken a few *F. fusca* pupae. Here she produces her own brood and the *F. fusca* workers tend the parasite's offspring. *F. fusca* is gradually outnumbered by the parasite workers, which are being continuously produced. The sequence of events is so far similar to that found in a case of temporary social parasitism. However, when the number of *F. fusca* workers, which are not being replaced, becomes low, the *F. sanguinea* workers 'press-gang' others into their service by raiding another nest of *F. fusca* and seizing larvae and pupae, which they carry back to their own nest and rear. The resulting *F. fusca* work for the host species, mostly performing functions inside the nest. Repeated slave-making raids are undertaken during the existence of a *F. sanguinea* colony. *F. sanguinea* are perfectly able to run a colony on their own without the enforced assistance of *F. fusca*, and in parts of the species' range they behave as typical temporary social parasites and do not engage in slave-making expeditions. It is only in western Europe, including Britain, that dulosis, which is a form of permanent social parasitism since the colony always includes two species, is practised. The habit may have its origins in normal food-foraging behaviour, a suggestion first advanced by Charles Darwin. Some of the *F. fusca* brood that is captured and carried back to the *F. sanguinea* nest is eaten, and other ants (e.g. *F. rufa*) are known occasionally to raid nests of other species in order to secure provisions which may include some of the brood of the raided species.

Darwin (1859) describes an unsuccessful slave-making foray by *F. sanguinea*: '. . . about a score of the slave-makers . . . approached and were vigorously repulsed by an independent community of the slave-species (*F. fusca*); sometimes as many as three of these ants clinging to the legs of the slave-making *F. sanguinea*. The latter ruthlessly killed their small opponents, and carried their dead bodies as food to their nest, twenty-nine yards distant; but they were prevented from getting any pupae to rear as slaves.'

Whilst *F. sanguinea* is not an obligate slave-maker, at least over part of its range, there are some ants which are entirely dependent upon their slaves. These are species of *Polyergus*, amazon ants, and they are conspicuously modified in having long, pointed, sickle-shaped mandibles that are designed for fighting; they are unable to construct a nest or tend their own brood. A European species is *P. rufescens*. The young queen *P. rufescens*, like its *Formica sanguinea* counterpart, enters a small nest of its host, again usually *F. fusca*, and destroys the rightful queen. When the *P. rufescens* brood emerges, slave-making raids are directed against other *F. fusca* colonies. The slaves do all the work in the nest, building, cleaning, nursing, and even feeding the *Polyergus* workers, but they do not accompany their masters on raids.

A raid by *Polyergus* is a much more direct and determined affair than a *F. sanguinea* raid, reflecting the greater importance of slaves in the economy of the former insect. Instead of small groups of workers, a large army of *Polyergus*, extending sometimes over an area some 3 metres long and 20 centimetres broad, sets off in search of a nest of *F. fusca*. Raids are made in an afternoon fairly late in the season, and the army advances in a more or less straight line at a speed of about 100 metres an hour (Dobrzańska and Dobrzański 1960). Should a nest of *F. fusca* be encountered, the raiders stream into it and carry off the brood. The preparations for such a raid have been the subject of some dispute. For a long time it was held that *Polyergus* scouts first located a *F. fusca* nest and then led the raiding army to it. Observations have shown, however, that the individuals leading the column are constantly changing, and it seems instead that *F. fusca* nests are located from a rather short distance by the army as it advances (Sudd 1967). Once a nest has been found, it may be attacked on two or three consecutive days. A *Polyergus* colony has been known to make forty-four raids in thirty days (Forel 1920), although on average only about one third of these are successful in finding a *F. fusca* nest. In contrast, a nest of *F. sanguinea* may make only three or four raids a year. Those workers of the slave species that resist the ravages of *Polyergus* are killed, but others are not attacked.

Strongylognathus is another genus of slave-making ants. *Tetramorium* is the host. Workers of *Strongylognathus*, like those of *Polyergus*, have the narrow, sickle-shaped mandibles (Figure 117) characteristic of an obligatorily slave-making species. Raids of *S. huberi* differ from those of *Polyergus rufescens* in that the slave-making species is accompanied by its slaves on an expedition.

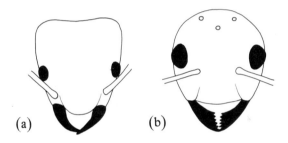

Figure 117. Heads of (a) *Strongylognathus*, a parasitic ant, and (b) *Formica*, a foraging ant, showing the differing shapes of the mandibles.

Indeed, it is the slaves that do any fighting that is necessary, for the mandibles of *S. huberi*, despite their formidable appearance, are supplied with relatively weak muscles and their chief function appears to be one of threat. In *S. testaceus*, a European species, the worker caste is produced in very small numbers and slave-making raids are not attempted. Instead, the queen *S. testaceus* lives in a *Tetramorium caespitum* nest without destroying the *Tetramorium* queen so that the brood of the two species is reared side by side. The *Tetramorium* queen is somehow prevented from producing any caste except workers, whilst the *Strongylognathus* queen produces mostly males and females which are reared by the *Tetramorium* workers.

In *S. testaceus* we have a species which seems to be in the process of losing its worker caste. In the British species, *Anergates atratulus*, this process is complete. No workers at all are produced, and the species may be compared to *Psithyrus* or *Vespula austriaca*. *A. atratulus* is another social parasite of *Tetramorium*. Queens invade the hosts' nests, killing the *Tetramorium* queens and using the *Tetramorium* workers as nurses for their own broods. The *Anergates* queen also is dependent upon *Tetramorium* workers for food since it is unable to eat without assistance. The gaster of the queen *Anergates* becomes enormously distended with eggs, almost like that of a queen termite. All of the eggs give rise to males or queens, and because no more host workers are produced or captured, the colony soon dies out. *Anergates* relies upon a high reproductive rate to compensate for its complete dependence upon short-lived colonies of its host.

Polyergus retains its workers for capturing slaves to prolong the life of its colony, whilst *Strongylognathus testaceus* appears to be moving towards the state of absolute dependence upon its host seen in *Anergates*. *F. sanguinea* still retains a very large measure of independence with its fully functional worker caste, and it would seem to be just embarking upon what may be a very hazardous evolutionary line. Richards (1953) draws attention to the fact that social parasites are generally rare, in numbers of both species and individuals, probably because social parasitism necessitates reliance upon another species whose reproduction is curtailed in the process. In this respect they are comparable with host-specific parasitic Hymenoptera and, generally speaking, these too are less abundant than their polyphagous allies.

14. Flies that Parasitise Vertebrates

DIPTERA ARE THE ONLY INSECTS that regularly infest the bodies of vertebrates; such an infestation is termed myiasis. The maggots feed either upon ingested food or upon healthy or necrotic tissue of the living host. In nearly all cases development proceeds, at least for a while, inside the host. By no means all examples of myiasis are examples of parasitism. Maggots feeding on dead wound tissue or on faecal matter in the rectum are certainly not behaving as parasites. To omit mention of these non-parasitic forms, however, would render more difficult an understanding of the evolution of myiasis.

Dipteran larvae are not infrequently accidently eaten by man, and as the higher Diptera have a cuticle which is very resistant to penetration by toxins, they are quite often carried passively right through the alimentary tract to emerge, still living, from the anus. The cheese-skipper, the larva of *Piophila casei*, which is often a serious pest in cheese and meat, has frequently been reported to pass live through the human gut. Infestations of this type are termed pseudomyiasis since the larvae concerned do not develop inside the 'host', but in practice it is often difficult to distinguish between pseudomyiasis and cases of facultative intestinal myiasis.

A curious case involving *Megaselia scalaris*, a species of Phoridae whose larvae feed in a wide variety of decomposing organic materials (including emulsion paint!), is quoted by Patton and Evans (1929). A man living in Burma passed in his faeces, over a period of about a year, *Megaselia* larvae of various ages, puparia, and even adult flies. It is believed that larvae had been accidentally ingested with food, but the extended duration of the infestation, persisting long after the patient was placed under medical supervision, together with the occurrence of adult flies, has led to the suggestion that the flies bred and passed through a number of generations within the gut. Zumpt (1965) dismisses this suggestion, perhaps rightly, as an impossibility, although it is known that some Phoridae pass several generations on interred corpses.

Myiasis can be either facultative or obligatory. Flies whose larvae are free-living in decomposing organic materials may occasionally gain access to the body of a vertebrate and there develop, at least for a while, on faeces, suppurating wounds, or the like. Several species of Calliphoridae and Muscidae exhibit this facultative myiasis. Obligatory myiasis producing flies live as larvae only upon vertebrates, either externally like *Auchmeromyia* and *Neottiophilum*, or internally like warble flies and bot-flies (Oestridae, Gasterophilidae). Flies such as Hippoboscidae, which feed upon vertebrates only when adult, do not come into the category of myiasis producers.

The subject of myiasis is comprehensively reviewed in a monograph by Dr F. Zumpt (1965) of the South African Institute for Medical Research. In this work only species from the Old World are considered, and 63 are listed as causing facultative myiasis or pseudomyiasis, and 106 as causing obligatory myiasis. These species are distributed among the following families and genera:

Facultative Myiasis
  Anisopidae: *Anisopus*
  Psychodidae: *Psychoda*
  Phoridae: *Megaselia*
  Syrphidae: *Eristalis*
  Piophilidae: *Piophilus*
  Ephydridae: *Teichomyza*
  Muscidae: *Musca, Muscina, Stomoxys, Ophyra, Fannia*
  Calliphoridae & Sarcophagidae: *Lucilia, Calliphora, Protophormia, Chrysomya, Sarcophaga, Wohlfahrtia*
Obligatory Myiasis
  Neottiophilidae: *Neottiophilum*
  Chloropidae: *Batrachomyia*
  Muscidae: *Passeromyia*
  Calliphoridae & Sarcophagidae: *Lucilia, Pachychoeromyia, Auchmeromyia, Cordylobia, Booponus, Elephantoloemus, Protocalliphora, Chrysomya, Wohlfahrtia.* The New World genus *Callitroga* also belongs to this group.
  Gasterophilidae: *Gasterophilus, Gyrostigma, Cobboldia, Platycobboldia, Rodhainomyia, Ruttenia, Neocuterebra.* Plus Cuterebridae: *Cuterebra, Dermatobia* of the New World.
  Oestridae: *Pharyngomyia, Cephenemyia, Pharyngobolus, Tracheomyia, Kirkioestrus, Rhinoestrus, Oestrus, Gedoelstia, Cephalopina, Portschinskia, Oestroderma, Oestromyia, Strobiloestrus, Pavlovskiata, Pallasiomyia, Oedemagena, Przhevalskiana, Hypoderma.*

Zumpt traces the evolution of myiasis, as indicated by the habits of flies living today, and recognises two roots, the saprophagous root and the sanguinivorous root. The saprophagous root originates with a larva of rather generalised feeding habits able to live on a diversity of decaying organic material and sometimes including in its diet the dead tissue associated with septic wounds on vertebrates.

Such an insect is the larva of the greenbottle, *Lucilia sericata*, which most often feeds on carcasses, but also occasionally on purulent wounds. In wounds it consumes only dead tissue, and may even have a beneficial effect in cleansing the affected region. In fact, *L. sericata* maggots had a limited medical use up to about 1930. Larvae reared under sterile conditions were used to speed the healing of septic wounds (Brumpt 1933). In addition to the removal of dead tissue, Stewart (1934) claims that calcium carbonate is secreted by the larva and that this stimulates phagocytosis by creating an alkaline environment. Simmons (1935) found that larval secretions of *L. sericata* were themselves bactericidal, and this was later confirmed in other species. The larva of the bluebottle *Protophormia terraenovae* produces a secretion which inhibits the growth of bacteria, an obvious advantage to a larva that lives in an environment which could be destroyed by bacteria (Pavillard and Wright 1957).

Lucilia cuprina may also feed at suppurating lesions. It is closely allied to *L. sericata*, but unlike that species its diet is not restricted to dead tissue, and it will ingest healthy tissue surrounding a wound, thereby extending the damaged area. The final step along this evolutionary path has been taken by flies whose larvae are incapable of developing in carcasses and other types of decomposing organic matter, and rely entirely upon sores on vertebrates to supply them with their nutritional requirements. *Chrysomya bezziana, Callitroga hominivorax* (=*americana*), and the larviparous *Wohlfahrtia magnifica* are three such obligate parasites. *C. bezziana* and *W. magnifica* both belong to genera whose other species are nearly all feeders, as larvae, in faeces, carcasses, or facultatively in wounds. *C. bezziana* will oviposit in very small sores such as midge or tick bites, and here the first stage larvae feed upon blood and lymph. Later, the second stage larvae invade healthy tissue and the final, third stage larvae are almost completely embedded in living tissue before they are fully fed and drop to the ground to pupate. The maggots or screw-worms of *C. bezziana* attack man and domestic animals in Africa and the Oriental region, not infrequently turning minor injuries into mortal wounds.

Callitroga (=*Cochliomyia*) is a screw-worm of the New World, and it achieved fame as a result of being decimated by a control procedure termed the sterile-male technique. The species was reared in vast numbers in the laboratory and late pupae were irradiated by X-rays. This had the effect of sterilising the males, or at least of inducing lethal mutations in the sex cells, but at the same time it did not alter the males' vitality or longevity. In 1954 and 1955 irradiated males were released weekly on the island of Curaçao in such quantities that they outnumbered normal males. They competed for females with the native males but, of course, were incapable of fathering offspring. The female *C. hominivorax* will copulate once only, whether fertilised or not by the union, and so the large numbers of sterile males had a very detrimental effect upon the population. Males mate several times. By this technique, Curaçao was cleared of screw-worms within five months (Bushland 1959). Knipling (1955) showed on theoretical grounds that a population of

20 million insects, 10 million of which are healthy males, can be virtually wiped out in four generations by releasing 20 million sterile males in each generation. The sterile male technique was used to rid Florida of its screw-worms. Success was achieved in two years at a cost of 8 million dollars which compared favourably with the annual 10 million dollars' worth of damage inflicted by screw-worms (Weidhaas, Schmidt, and Chamberlain 1962). In England the technique has been employed against the green-bottle, *Lucilia sericata*, on the island of Lindisfarne but in this case control was not achieved (Macleod and Donnelly 1961).

A more specialised obligatory parasite is *Lucilia bufonivora*, a species which attacks toads and other amphibians. Eggs are laid on the skin and, on hatching, the young larvae migrate to the nostrils or, occasionally, the orbits. Here they cause extensive lesions and the host is generally killed within a week, the larvae sometimes completing their feeding upon the corpse.

The thin, moist, amphibian skin allows *L. bufonivora* to gain a foothold on previously uninjured hosts, the toad's mucus providing a substrate for the fly larvae to move in without the risk of desiccation. In contrast, blowflies attacking mammals require at least a small pre-existing sore or wound in which an infestation can be initiated. Much research has been done on sheep strike, that is wound myiasis in sheep, and it is known that a succession of species may be involved (Haddow and Thomson 1937, MacLeod 1937). Sheep strikes are generally found, if not in actual injuries, on those regions of the host where the fleece is frequently moistened with urine or faeces. Larvae of primary strike-flies such as *Lucilia sericata* and *Protophormia terraenovae* in Britain, and *Lucilia cuprina* and *Calliphora stygia* in Australia, will live in soiled fleece, where bacterial activity causes inflammation of the host's skin accompanied by lymphal exudations. The larvae feed at the surface of the epidermis which is irritated, causing the stricken sheep to rub or scratch the affected part. This creates larger lesions which can be exploited by more maggots, either of the primary species or of secondary strike-flies. The odour of putrefaction from the primary strike attracts secondary strike-flies like *Lucilia caesar*, *L. illustris*, *Calliphora erythrocephala*, *C. vomitoria*, and several others. Finally, tertiary strike-flies such as species of *Musca* and *Fannia* may afflict severely stricken sheep. Large numbers of sheep are destroyed by these flies and the commercial value of the wool of many more is much reduced. The problem of sheep strike is particularly prevalent in Australia, the dense fleece and wrinkled skin of Merino sheep providing a humid environment and being very liable to soiling.

Zumpt (1965) adds a small branch to his saprophagous root of myiasis. Many musciform flies oviposit or larviposit in vertebrate faeces, occasionally even as faeces are leaving the anus. Such habits could lead to rectal myiasis, in which eggs or larvae are laid around the anus and the larvae crawl into the rectum to complete their feeding inside the body. *Musca domestica* and *Sarco-*

phaga haemorrhoidalis have been recorded as causing rectal myiasis. Similarly, larvae of species of *Psychoda*, *Musca*, *Fannia*, and *Calliphora* may enter the urethra and sometimes even penetrate as far as the bladder. No flies are known to be obligatory causers of rectal or urinogenital myiasis.

Zumpt's sanguinivorous root of myiasis, like the saprophagous root, begins with a relatively unspecialised feeder in decomposing organic matter. Some species, *Muscina stabulans* for example, feed saprophagously during their early larval stages, but later develop a predatory tendency, attacking and eating other maggots sharing their habitat. Zumpt suggests that such a larva, with mouthparts capable of piercing other larvae, could, if it lived in the nest of a bird or mammal, pierce the skin of a vertebrate occupant of the nest and take a blood meal. This facultative ectoparasitism could subsequently develop into an obligatory relationship such as is now the case with species of *Neottiophilum*, *Passeromyia*, and *Protocalliphora* in birds' nests, and species of *Pachychoeromyia* and *Auchmeromyia* in mammalian dwelling places.

Ectoparasitism by larvae of endopterygotes is most exceptional, and in these flies the habit could have developed only in circumstances that allowed the parasite close and prolonged proximity with a host in an almost immobile condition.

The genus *Neottiophilum* contains a single species, the European *N. praestum*, whose larva sucks the blood of nestling passerine birds. It has a single generation a year, and this is synchronised with its hosts, adults hatching in spring to coincide with the birds' breeding season (Keilin 1924). Pupation occurs in the nest. *Passeromyia heterochaeta* lives as a larva in nests of swallows, starlings, weaver birds, and others, mainly in Africa, taking blood meals from the nestlings. For the first two larval instars it lies on the host's body, attached to the plumage by posterior filaments. An allied species, the Australian *P. longicornis*, has developed a more intimate association with its hosts in that after initially feeding externally like *P. heterochaeta*, it later burrows into the nestling's skin and forms a swelling with only the posterior spiracles protruding for respiratory purposes. A similar evolutionary trend is suggested by the habits of the European species of *Protocalliphora*. *P. azurea*, like *Passeromyia heterochaeta*, is an ectoparasitic blood-sucker of nestling birds, particularly tits and warblers. So also are *P. chrysorrhoea* in sand martins' nests, *P. peusi* in crows' nests, and *P. falcozi* which has been found most often in great tits' nests. However *P. lindneri*, which attacks ground-nesting passerines, and *P. braueri* whose principal host appears to be the house sparrow, parallel *P. longicornis* in burrowing subcutaneously and forming boils. The fully grown larvae of all of these species pupate in the nest material. Heavy infestations may kill the host.

Other Calliphoridae have become obligate parasites of mammals. *Auchmeromyia* and *Pachychoeromyia* feed on the blood of African mammals and are ectoparasites behaving like *Passeromyia heterochaeta*. The single species of

Pachychoeromyia, P. praegrandis, and the five described species of *Auchmero-myia,* are all associated with warthogs and aardvarks, sometimes also with the aardwolf. Warthogs and aardwolves frequently make use of aardvark burrows as dwelling places, but the animals are not closely related taxonomically. *A. luteola,* the Congo floor-maggot (Figure 118), in addition regularly includes man's blood in its larval diet. The female fly lays up to 300 eggs singly in dry,

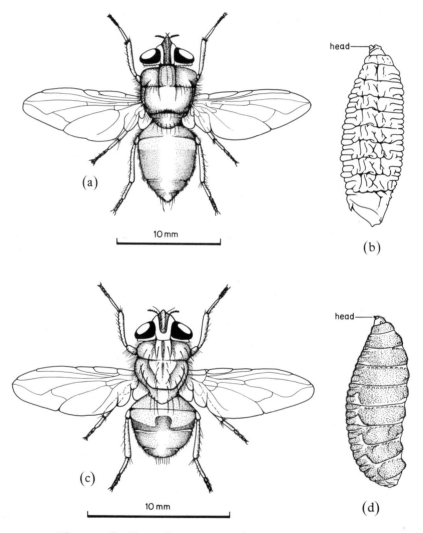

Figure 118. Two dipterous parasites of man. (a) Adult *Auchmeromyia luteola* and (b) its fully grown larva, the Congo floor maggot. (c) Adult *Cordylobia anthropophaga,* the tumbu fly, and (d) its fully grown larva. (All from Smart 1965.)

sandy soil in shaded places, sometimes on the floors of native huts. All larval stages are blood-suckers, and as parasites of man they depend upon people sleeping on the bare ground or on mats. Feeding takes place at night, usually every night, when the feeding maggot scrapes at the host's skin with its mouthparts until the blood capillaries are lacerated.

Parasites that form boils on mammàlian hosts, equivalent to the bird parasite *Passeromyia longicornis*, are found in the genera *Cordylobia*, *Booponus*, and *Elephantoloemus*. *Cordylobia anthropophaga*, the tumbu fly (Figure 118), is undoubtedly the best known species (Blacklock and Thompson 1923). Eggs are laid in sandy soil, usually in the vicinity of urine or faeces, and at first the larvae live just below the soil surface. When a man walks close to where they are lying, they come on to the surface and wave the anterior parts of their bodies about. Should they contact skin, usually of the feet, penetration is rapidly effected, and within a very short time only the posterior spiracles of the maggot are left exposed to the outside world. A boil or abscess forms about the maggot and it rapidly grows. After perhaps as little as eight days the larva may be fully grown. It now leaves the boil and pupates in the ground. As well as man, the tumbu fly will attack rats, dogs, monkeys, and several other mammals, even birds, but survival in non-human hosts is uncertain. *C. rodhaini* has also been recorded as attacking man but the rate of survival is low, and its most usual hosts are rodents. The only other described species of *Cordylobia*, *C. ruandae*, may be a specific parasite of the mouse *Grammomys dolichurus*. All three *Cordylobia* species are known only from Africa. *Booponus* is an oriental genus of four described species which are parasites of Artiodactyla. Like *Cordylobia*, the larvae live in skin boils, but an interesting point of difference between the two genera is that *Booponus* species oviposit on the hairs of their hosts, not on the ground. This is necessary because Artiodactyla do not frequent habitual resting places. *Elephantoloemus indicus*, the final species in this group of mammalian parasites, is a specific parasite of the Indian elephant.

The Chloropidae is a family with diverse larval habits. Most are plant-feeders, often gall-formers on grasses, and included amongst them are agricultural pests such as the frit fly and gout fly. Others feed upon rotting vegetable material, and some are predators of aphids, beetles, and spiders' eggs. However *Batrachomyia*, an Australian genus, is unique in its feeding habits. There are ten described species, all of which are probably obligate parasites of frogs and toads, possibly host-specific, living as final-stage larvae in subcutaneous swellings. Infested hosts often die.

More specialised than the preceding families as larval parasites of vertebrates are the Gasterophilidae and Oestridae, bots and warbles, two allied families all of whose species are obligate parasites.

Gasterophilidae

The adults of this family are large and attractive flies whose larvae, as their family name implies, are 'stomach lovers' or stomach bots. The mouthparts of the adults of both gasterophilids and oestrids are rudimentary.

Gasterophilus is the principal genus. There are nine species, and they chiefly inhabit the Ethiopian and palaearctic regions. They are all parasites of horses, asses, and zebras; occasional parasitism of man does not persist beyond the first larval instar. Although the species tend to share the same hosts, details of their life histories differ. The female *G. pecorum* lays her eggs, which may number over two thousand, in rows upon grasses and other plants. The larvae do not hatch until the eggs are eaten by the host; they may remain viable on vegetation for several months. In contrast, *G. haemorrhoidalis*, *G. intestinalis* (Figure 119a), and *G. nigricornis* lay their eggs upon the hairs of their hosts. *G. haemorrhoidalis* oviposits upon the lips, where moisture stimulates their hatching, and the young larvae burrow into the epidermis of the lip and into the mouth. *G. intestinalis*, on the other hand, lays its eggs upon the legs or back of its host, hatching being stimulated by friction when licked, and the young larvae penetrate the tongue. *G. nigricornis* selects the cheeks for egg-laying, and it differs from the other two species in that its eggs apparently hatch spontaneously without stimulation from the host. The young larvae enter the mouth by burrowing into the epidermis at the corner. The sites occupied by the later larval stages in the alimentary canal of the host also differ specifically. *G. pecorum* occupies the soft palate and adjacent areas in its second larval instar, and the stomach in its third (final) instar. The second instar larva of *G. haemorrhoidalis* moves to the stomach and duodenum, where it moults, and the third instar larva migrates to the rectum of its host. The second stage larvae of both *G. intestinalis* and *G. nigricornis* pass from the mouth (tongue and cheek respectively) to the stomach or duodenum, where they become third stage larvae. In all species, fully grown larvae pass out of the anus and pupate on the ground in the surface layers of soil.

Whereas *Gasterophilus* confines its attentions to the Equidae, *Platycobboldia loxodontis*, *Rodhainomyia roverei*, and *Cobboldia elephantis* live as final stage larvae in the stomachs of elephants, the first two being specific parasites of the African elephant whilst the last-mentioned is known only from the Indian elephant. Only the life cycle of *Platycobboldia* has been elucidated. The eggs are laid about the bases of the tusks, and the first stage larvae enter the mouth. The fully grown larvae move about freely inside the stomach, unlike those of *Gasterophilus* which are attached to the gastric epithelium. Another interesting deviation from *Gasterophilus* is that fully grown *Platycobboldia* larvae do not pass out of the host's anus, but instead migrate up the oesophagus to congregate beneath the tongue prior to being evacuated from the mouth. Pupation again occurs in the soil.

Another distinct group of the Gasterophilidae, comprising the three known species of *Gyrostigma*, is parasitic on rhinoceroses. As far as is known, their eggs are laid upon the host's head and the larvae are attached to the stomach wall, passing out of the anus when fully grown. *Gyrostigma pavesii* and *G. conjungens* parasitise the African rhinoceroses, *Ceratotherium simum* and *Diceros bicornis* respectively, whilst *G. sumatrensis* is known only from larvae collected from a Sumatran rhinoceros (*Didermoceros sumatrensis*). All of these flies are large, and several of them are beautifully coloured. *Platycobboldia loxodontis* has smoky wings, an orange-coloured head, and metallic-blue thorax and abdomen. Some species of *Gasterophilus* are very hairy and resemble bumble bees.

Two other species placed by Zumpt (1965) in the Gasterophilidae are biologically atypical of the family. Both are parasites of the African elephant but their larvae do not dwell in the alimentary tract of their host. *Neocuterebra squamosa* larvae are found in the feet and *Ruttenia loxodontis* larvae develop in boils in the skin of various regions of the body. As Zumpt points out, the latter species occupies in the African elephant the ecological niche filled by the calliphorid *Elephantoloemus indicus* in the Indian elephant.

Cuterebridae

These are New World botflies that cause dermal myiasis of man, rodents, and lagomorphs. There is a fairly clearly defined host-specificity in the family which is probably of ancient origin parasitic upon rodents. *Cuterebra latifrons*, a parasite of the woodrat (*Neotoma fuscipes*), has been investigated in California by Catts (1964). Males are attracted to the summits of hills, where they exhibit territorial behaviour and become spaced out. Females fly to the hilltops to mate and then disperse. Eggs are laid, in small groups of up to ten, near woodrat nests. One fly is recorded (Radovsky and Catts 1960) to lay 270 eggs. Egg hatching is sometimes stimulated by mechanical disturbance. The young larvae adhere to the substrate by a caudal 'holdfast' and they respond to the presence of a mammal by making swaying, questing movements of their anterior regions. Entry into a host is normally effected through a body orifice. In the laboratory, woodrats have been infected *via* their nares after smelling sticks on which were *C. latifrons* larvae. Other rodents were also readily infected in this manner, but development of the fly larvae did not usually proceed to completion. Laboratory mice died at an early stage of infection.

Cuterebra larvae live in subdermal cysts which communicate with the exterior by a pore. The fully grown larva leaves its cyst to pupate in the ground. A frequent site of infestation by *Cuterebra* is the inguinal region; male hosts of *C. emasculator* may have their testes destroyed by the parasite. *C. angustifrons* is chiefly a parasite of the white-footed mouse (*Peromyscus leucopus*) and it has a deleterious effect upon blood composition and movement of its host (Dunaway *et al.* 1967).

Dermatobia hominis (Figure 119d), the human botfly of Central and tropical South America, does not itself search for the whereabouts of a host. It is

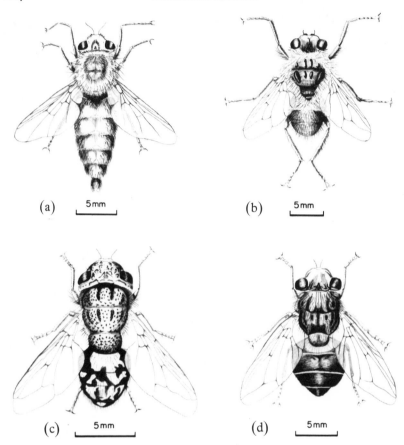

(a)   5mm

(b)   5mm

(c)   5mm

(d)   5mm

Figure 119. Four Diptera Cyclorrhapha whose larvae parasitise vertebrates. (a) *Gasterophilus intestinalis* (Gasterophilidae)
female, the horse botfly. (b) *Hypoderma lineatum* (Oestridae)
female, a warble fly. (c) *Oestrus ovis* (Oestridae) female, the
sheep botfly. (d) *Dermatobia hominis* (Cuterebridae) female,
the human botfly. (Drawings by Arthur D. Cushman reproduced by permission of the United States Department of
Agriculture.)

phoretic and glues its eggs, of which it produces many, to the abdomens of
blood-sucking flies (Figure 120), especially mosquitoes. In this way a batch of
up to one hundred eggs is transported to the host, which is often a cow but
may be a man. How *Dermatobia* evolved the habit of phoresy, so different from
Cuterebra, is unknown. The larvae emerge in response to warmth and each
usually enters the host through a hair follicle. The larva causes a boil or cyst
to form about it in the subcutaneous tissue, and here it attains full size in about
six weeks, remaining all the while in contact with the atmosphere by a small

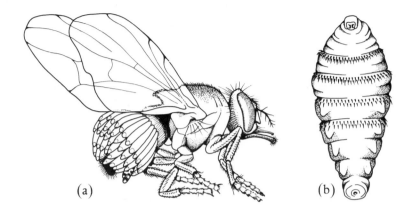

Figure 120.. The human botfly, *Dermatobia* (Cyclorrhapha, Cuterebridae). (a) Eggs carried on the abdomen of a stable fly. (b) The fully grown larva in ventral view. (From Brumpt 1927.)

aperture in the centre of the boil. The fully grown larva eventually emerges through this aperture to pupate in the soil. The larva of *D. hominis* behaves in many ways like a larva of a warble fly (*Hypoderma*) but unlike *Hypoderma* it does not migrate from its original site of entry.

Oestridae

The Oestridae are sometimes classified as a tribe of the Tachinidae (e.g. van Emden 1954). Zumpt (1965) grants them family status and recognises two subfamilies, Oestrinae and Hypoderminae, which have quite distinct larval feeding sites. The larvae of Oestrinae feed in the naso-pharyngeal regions of ungulates, both Perissodactyla and Artiodactyla, with single species attacking marsupials and elephants. The Hypoderminae parasitise Artiodactyla, rodents, and lagomorphs, and as their name suggests, they live subcutaneously, stimulating the formation of small boils.

Among the Oestrinae, the best known species is *Oestrus ovis* (Figure 119c), the sheep bot, a cosmopolitan species now but probably of palaearctic origin. Like others in the subfamily it is larviparous, depositing first-stage larvae in the nostrils of the host. The maggots develop in the nasal and frontal sinuses feeding on mucus and blood. When fully grown, they are sneezed out by the host and pupate on the ground. Sheep and goats are the principal hosts of *O. ovis* but occasional attacks on man have been recorded. Curiously, when attacking man, the larvae are laid, not in the nostrils, but in the eyes; their development does not usually progress beyond the first instar. Other Old World Oestrinae have similar life cycles to *O. ovis*. Their hosts, according to

Zumpt (1965), are as follows:

Deer: *Cephenemyia* (4 species), *Pharyngomyia* (1)

Antelopes, Gazelles: *Pharyngomyia* (1), *Oestrus* (4), *Kirkioestrus* (2), *Rhino-estrus* (2), *Gedoelstia* (2)

Giraffe: *Rhinoestrus* (1)

Camel: *Cephalopina* (1)

Sheep, Goats, Ibex, Tur: *Oestrus* (2), *Rhinoestrus* (1)

Hippopotamus, Bushpig, Warthog: *Rhinoestrus* (3)

Horses, Zebra: *Rhinoestrus* (4)

African Elephant: *Pharyngobolus* (1)

Kangaroo: *Tracheomyia* (1)

The Oestrinae probably originated in the palaearctic region but are now cosmopolitan. *Cephenemyia trompe* is a circumpolar species parasitising reindeer in the palaearctic and caribou in the nearctic.

The best known representatives of the subfamily Hypoderminae are *Hypoderma* species (Figure 119b), large flies often resembling bumble bees, that pass their larval stages as parasites of Bovidae and Cervidae. The common warble fly of cattle is *H. bovis*. The female fly lays its eggs on the hairs of a cow, particularly on the belly and thighs. The act of oviposition is performed very rapidly. On hatching, the young larvae crawl down the hairs to the skin, which they penetrate, and then embark upon a prolonged migration inside the host's body which lasts for about four months. At first the larvae follow the course of nerves until the spinal cord is reached. Here they remain a while before burrowing through fat and muscle towards the skin of the back. A small breathing hole is opened to the outside and the host's tissues swell about each warble larva. In these dorsal swellings the larvae grow and moult twice in three months before becoming fully grown. They leave the swellings to pupate on the ground. There is one generation during a year. The larva of *Hypoderma*, like that of *Protophormia terraenovae*, produces a secretion which has bacteriostatic properties (Beesley 1968).

The life cycle of *H. lineatum* differs from that of *H. bovis* in that the first instar larva migrates through the oesophageal submucosa instead of the spinal nerves. This species is another parasite of cattle, and it lays its eggs mostly on the heels. A few cases of human infestation have been recorded, and although in such cases the larvae are unable to complete their development, they may destroy the eyes.

H. bovis and *H. lineatum* adults both cause cattle to 'gad'. It is as if cattle are aware of the suffering they will endure long after the flies have laid their eggs (page 54).

Other species of *Hypoderma* are parasites of deer, and in most cases they are host-specific. Species of the palaearctic genera *Pavlovskiata*, *Pallasiomyia*, and *Przhevalskiana* attack Asian antelopes or gazelles, sheep, and goats, whilst African antelopes are parasitised by species of *Strobiloestrus*.

Oedemagena tarandi is the reindeer warble fly, a host-specific species that as a larva migrates to the back of its host through the muscles of the hindquarters (Hadwen 1926). The herd of reindeer introduced to the Cairngorms in Scotland were screened for warbles before being released. However, soon after liberation they were found to be infested. The parasite was not *O. tarandi* but *Hypoderma diana*, normally a red-deer parasite. The *H. diana* larvae in reindeer behaved abnormally, making several punctures in the skin, and some reindeer died as a result of infestation (Kettle and Utsi 1955).

Distinct from the group of genera whose hosts are ungulates is another section of the subfamily parasitic upon rodents. These species also cause subcutaneous boils, either on the back or belly, and because they are no smaller than the ungulate parasites, a host can usually support only a single larva at a time. Palaearctic species of *Oestromyia*, *Oestroderma*, and *Portschinskia* attack marmots, pikas, voles, and mice. Their biology is poorly known.

Canthariasis and Scholechiasis

Flies are not the only insects whose larvae may enter the bodies of vertebrates. Lepidopterous caterpillars on vegetables may be eaten accidently by man, but it is rare for them to survive long in the vertebrate gut. Not so casual is the occurrence of larvae of *Tinea vastella* in the horns of African ungulates. This species is probably an obligatory parasite. Such lepidopterous infestation of vertebrates is termed scholechiasis.

Beetles figure more frequently than Lepidoptera among recorded cases of insect infestations of man. Adult dung beetles (Scarabaeidae) sometimes enter the human rectum by way of the anus. These beetles lay their eggs on vertebrate faeces, and it is perhaps not surprising that some follow to the end their search for oviposition sites. Species of *Onthophagus* in India and Ceylon seem especially prone to this habit, and children are the principal victims. Many Dermestidae feed on dried animal skins, and occasionally larvae have been found making feeding galleries in the skin of live, nestling pigeons (Paulian 1943). Coleopterous infestation of vertebrates is termed canthariasis.

15. *Some General Remarks*

THE PRECEDING CHAPTERS are concerned with the numerous and diverse ways in which insects, the largest of all animal groups, have exploited a parasitic mode of life. After their biological variety, perhaps the most impressive feature of parasitic insects is their number. The adaptability of insect morphology and physiology has permitted species of many orders to acquire the specialisations necessary for a parasitic life. Parasites are frequently described as degenerate organisms, but this is a misconception. Rather they are extremely specialised animals that, through natural selection, have lost structures and habits useless to them, had others moulded to greater advantage, and developed new modifications to equip them for their special ways of life. Most parasitic insects belong to the advanced endopterygote orders Hymenoptera and Diptera.

Because parasitism demands a high degree of specialisation, the host ranges of parasitic insects are limited. A few are polyphagous, succeeding in attacking hosts belonging to widely different taxonomic groups, but these parasites are nevertheless limited ecologically, being able to parasitise only those hosts that share some common attribute such as similar habitat, pelage, or nesting site. Many parasites are oligophagous, with a host range embracing only one or two families or a few genera, and in some groups, lice in particular, the majority are monophagous (host-specific).

In several groups there seems to be an evolutionary trend in the direction of host-specificity. When a parasite with an apparently broad host range is studied in detail, it is frequently found to be composed of biologically distinct entities, each associated with a different host or small group of hosts. In other words, many an allegedly polyphagous species is, in reality, an aggregate of monophagous and oligophagous sibling species. An evolutionary trend in the direction of monophagy is perhaps inevitable. In the face of competition from other parasite species, the polyphagous parasite will have a higher survival rate on those hosts where it is a more successful competitor, and it will therefore become more and more adapted to parasitising an increasingly restricted host range. Interspecific competition must prevent the coexistence of a number of polyphagous parasites with similar host ranges. A species could perhaps specialise as an opportunist 'all-rounder', as a polyphage, but to do so it must survive best on different host species at different times. If it consistently survives best on one particular host, the characters associated with parasitism of that host will be selected and emphasised to the almost inevitable detriment of the species' immediate adaptability. Polyphagy could be selected only in a biotically variable environment when the survival rate in different host species varies from generation to generation.

Monophagy is hazardous from an evolutionary point of view. The host may evolve some means of escaping parasitism entirely, or it may become extinct. In either event the monophagous parasite is doomed. Parasites are therefore on a knife edge. Competition frequently compels them to sacrifice genetic variability in the interests of narrow specialisation. But specialisation (and speciation) can easily lead to extinction, selection forces driving parasites into the adoption of a way of life that is successful only from a short term point of view. This is true of all adaptation.

Sometimes, of course, a monophagous parasite may become polyphagous. Liberation from host-specificity may follow the evolution of a particular attribute which allows the parasite to compete successfully outside its original, narrow host-range. This secondary polyphagy allows the cycle polyphagy-monophagy to continue, and it may mark the evolution and subsequent radiation of a new family or other higher taxonomic group. In groups such as the parasitic Hymenoptera, secondary extensions of host ranges must have repeatedly occurred to mask the original host relationships, and in these groups there now exists a very complex and often puzzling array of host associations.

When more than one species evolves from a polyphagous species, the flow of genes through the parental population must have been interrupted so as to isolate the incipient species In parasites such as lice, where there is very little movement between different host species and the populations on each are virtually isolated, widespread host-specificity is the outcome. Species formation in this group must have been dependent to a considerable extent on the speciation of the hosts themselves. Amongst mobile adult protelean parasites, gene flow between some populations attacking different host species may be restricted by sib-mating, as mentioned on page 164, as well as by geographical isolation.

Wheeler (1911) suggested that the animal kingdom would gradually become more and more weighted with parasitic species. Every metazoan is a potential host, and most are hosts to a large number of parasites. Wheeler's hypothesis demands that saturation by parasitic forms has not yet been reached. Certainly, some groups of seemingly possible hosts are not exploited by some groups of parasites. The barrier to streblid parasitisation of non-tropical bats is one that could perhaps be overcome; terrestrial arthropods other than insects could conceivably be more widely parasitised by protelean parasitic insects.

The fossil record

Some insight into present-day relationships between insect parasites and their hosts may be gained by an examination of the fossil record. Exopterygote insect parasites are parasites of birds and mammals. Lice probably evolved soon after the radiation of their host groups in the mesozoic era. No fossil lice have been discovered, but their intimate and universal association with their hosts indicates a relatively long-standing relationship. Similarly, it can be argued that it is likely that lice have been parasites longer than Cimicidae, and Cimicidae longer than *Triatoma*, because in this series there is a decreasingly close association between the parasites and their hosts. The appearance of winged insects, however, predates by a long time that of birds and mammals. The earliest fossil winged insect, *Eopterum devonicum*, is attributed to devonian strata, and in the carboniferous period several orthopteroid groups were living. Lice have affinities with the hemipteroid exopterygote orders (Hemiptera, Psocoptera, Thysanoptera) whose fossil remains date back to the lower permian period. Anoplura and Mallophaga were, therefore, very late amongst Exopterygota in making their appearance.

Turning to endopterygotes, the neuropteroid orders (Neuroptera, Coleoptera) were well represented in the lower permian. The typical larva in these orders is of the active, predatory, campodeiform type. The main radiation of the mecopteroid orders (Trichoptera, Lepidoptera, Diptera, Siphonaptera) occurred in the upper permian (Smart 1963). The mecopteroid larva is typically a caterpillar-like phytophagous or saprophagous insect, and in this form a more accessible host for protelean parasitic insects than the typical neuropteroid larva. Mecopteroid larvae probably appeared in numbers only after considerable coleopteran radiation had already taken place and beetle families were specialising along non-parasitic lines. This could well be a contributory factor to the relative scarcity of protelean parasitic beetles today. Hymenoptera first appeared in the triassic period as Symphyta, a suborder which has retained the phytophagous, caterpillar-like mecopteroid larva, but parasitic Hymenoptera have not been found as fossils in deposits earlier than the jurassic. This is probably a true reflection of a relatively late arrival on the evolutionary scene. It is likely that the parasitic Hymenoptera evolved at a time when potential hosts were very numerous and their subsequent evolution was rapid. Protelean parasitic Diptera probably also appeared first in the mesozoic. No fossil dipteran can be dated earlier than the jurassic period, and the Siphonaptera, an offshoot of Diptera, have been found only in tertiary deposits and later.

Comparisons between parasitic Crustacea and Insecta

Both Crustacea and Insecta are classes within the phylum Arthropoda, but very few insects inhabit the sea whereas crustaceans are predominantly marine organisms.

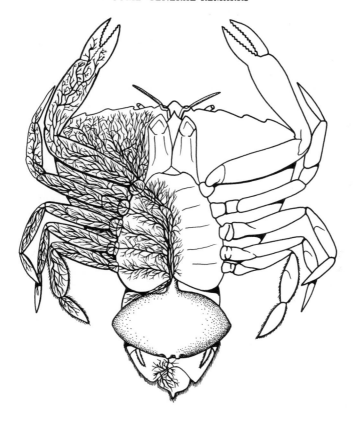

Figure 121. Ventral view of a crab parasitised by *Sacculina* (Crustacea Cirripedia). The ramifications of the absorbtive processes of the parasite are indicated only on the right side of the crab (after Caullery 1922).

Sacculina belongs to the group of Crustacea, the Cirripedia, which also includes free-living barnacles. It starts life, like these free-living barnacles, as a small, planktonic larva but here the similarity ends, for the adult *Sacculina* is a parasite of crabs and has undergone radical specialisation (Figure 121). It appears as a fleshy sac fixed to the lower surface of the crab's abdomen. The sac contains a cavity and in the cavity is a central mass of tissue made up almost entirely of the gonads. Radiating from the sac, and penetrating to all parts of the host, is a system of root-like processes through which the parasite obtains nutriment from the crab. Appendages, gut, and sense organs have been lost by the adult *Sacculina*. These morphological modifications are an extreme example of those that may result in an animal upon taking up a parasitic mode of life.

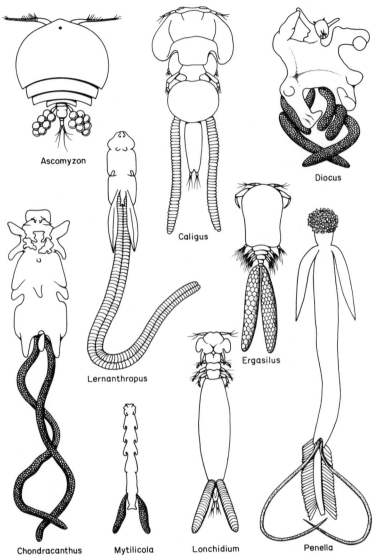

Figure 122. A selection of adult female parasitic Copepoda (Crustacea) (redrawn from various sources).

Other crustaceans with life histories similar to *Sacculina*, that is with a highly specialised, parasitic adult and a free-living larva, are the Branchiura (e.g. *Argulus*) and many genera of Copepoda (e.g. *Lernaea*, *Chondracanthus*, *Caligus* (Figure 122)). In all of these, as in *Sacculina*, the typical planktonic larva is retained no matter how profoundly modified the adult may be. The primary function of the larva is one of dispersal.

This may be considered as the typical parasitic crustacean life-history; a free-living dispersive larva followed by a highly modified, parasitic adult.

In contrast to Crustacea with this form of life cycle are a minority in which the larva is the parasitic stage and the adult is free-living. Protelean parasitism is well-illustrated in the copepod family Monstrillidae (Figure 123). The monstrillid larva lacks a gut, and soon after hatching from the egg it seeks out a host, usually an annelid. The larva enters the body of its host and, once inside, loses its locomotory appendages and develops instead two long, absorbtive processes through which food is obtained. Eventually the larva develops into an adult which also lacks a gut but otherwise resembles a non-parasitic copepod. The adult leaves the host and lives in the plankton. It is therefore the adult and not the larva which is the dispersive phase, a complete reversal of the typical arrangement.

Protelean parasitism is engaged in also by the isopod family Gnathidae (Figure 124), but here the larva is an ectoparasite of fish, and relatively little modified except for the possession of piercing mouthparts. It is probably distributed by its host.

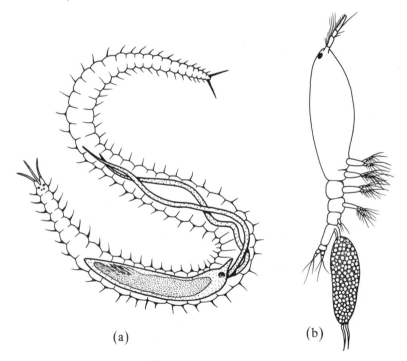

(a)   (b)

Figure 123. Crustacea Copepoda, Monstrillidae. (a) Parasitic larva in a Syllid worm. (b) Free-living adult female. (After Caullery 1922).

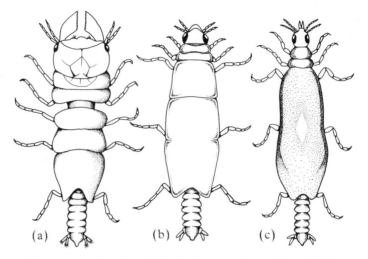

Figure 124. *Gnathia maxillaris* (Crustacea Isopoda). (a) Male. (b) Female. (c) Larva. (From Caullery 1922 after Sars.)

To summarise the situation in Crustacea, it is usual for there to be a free-living planktonic larva through which the species is distributed, and a much-modified, parasitic adult. Protelean parasitism is exceptional. It is engaged in when an alternative method of dispersion to the planktonic larva is available. However, in many groups of Crustacea, there is no reason why the adults should not effectively disperse.

A more important difficulty in crustacean protelean parasitism is the necessity for the larva to change gradually, at successive moults, toward the adult body-form. This same difficulty faces exopterygote insects with their incomplete metamorphosis, and there are no protelean parasitic Exopterygota with free-living adults. Among Endopterygota, the intervention of complete metamorphosis between larva and adult allows divergent morphological specialisation of the immature and mature animals, so that the parasitic mode of life may be fully exploited by the larva. Also, the other possible difficulty in protelean parasitism by Crustacea, the need for the larva to disperse, does not apply to insects since in nearly all insects the winged adult is the dispersive stage. Accordingly, as has been shown, a very large number of species of Endopterygota, particularly Hymenoptera and Diptera, are protelean parasites.

The difference between the typical parasitic life cycle in Crustacea and in endopterygote insects is bridged, among Crustacea, by the Monstrillidae which come closest amongst Crustacea to having a complete metamorphosis. Caullery (1922) states that '. . . *les Monstrillides se comportement comme les Insectes holométaboles. Le parasitisme n'atteint que des organes larvaires, laissant les organes définitifs évoluer comme aux dépens des disques imaginaux.*' However, as in other Crustacea, it is a larval stage (the first instar in monstrillids) that seeks the host, and in many ways comparable to the Monstrillidae are protelean parasitic insects which are hypermetamorphic.

Monstrillids and other parasitic Crustacea seem to be markedly host-specific. This could indicate that the larva, which locates the host, is more rigid in its host-searching behaviour than is the adult of a protelean parasitic insect. Also, a large number of hypermetamorphic insect parasites are monophagous or oligophagous, and none approach the broad host ranges of some of the species in which it is the adult that locates the host. However, in the case of hypermetamorphic insects, the large numbers of first instar larvae that are produced must, between them, contact a variety of possible hosts. The explanation of host-specificity in both these insects and in Crustacea may be, therefore, a lack of ability by the larvae to become established on new hosts rather than an absence of contact with them.

Much of what has been written in this chapter is no more than speculation. Further data are needed. We are still exceedingly ignorant of the lives of many parasites, and this can be rectified only by more research. The parasitic insects are an immense group, and they provide immeasurable scope for the entomologist. Only now is the basic taxonomy of many groups becoming clearly understood, and this is a necessary prelude to biological and ecological studies. In Britain, entomologically perhaps the best studied country in the world, only a very few workers are investigating the enormous numbers of parasitic Hymenoptera, and in this group alone biological information on an overwhelming majority of species is, at best, scanty. I doubt if any group of animals can exceed the scope offered by parasitic insects to the diligent researcher. The field is wide open, the prospect inviting.

Bibliography

ALFORD, D. V. 1968. 'The biology and immature stages of *Syntretus splendidus* (Marshall) (Hymenoptera: Braconidae, Euphorinae), a parasite of adult bumblebees.' *Trans. R. ent. Soc. Lond.*, **120**, 375–393.

ALLEN, H. W. 1959. 'Evidence of adaptive races among oriental fruit moth parasites.' *Proc. Int. Congr. Ent. 10. Montreal* (1956), **4**, 743–749.

ANDREWARTHA, H. G. 1952. 'Diapause in relation to the ecology of insects.' *Biol. Rev.*, **27**, 50–107.

ARNDT, W. 1940. 'Der prozentuale Anteil der Parasiten auf und in Tieren im Rahmen des aus Deutschland bisher bekannten Tierartenbestandes.' *Z. ParasitKde.*, **11**, 684–689.

ARTHUR, A. P. 1961. 'The cleptoparasitic habits and immature stages of *Eurytoma pini* Bugbee (Hymenoptera: Chalcidae), a parasite of the European shoot moth, *Rhyacionia buoliana* (Schiff.) (Lepidoptera: Olethreutidae).' *Can. Ent.*, **93**, 655–660.

ARTHUR, A. P. 1966. 'Associative learning in *Itoplectis conquisitor* (Say) (Hymenoptera: Ichneumonidae).' *Can. Ent.*, **98**, 213–223.

ARTHUR, A. P. 1967. 'Influence of position and size of host shelter on host-searching by *Itoplectis conquisitor* (Hymenoptera: Ichneumonidae).' *Can. Ent.*, **99**, 877–886.

ASH, J. 1952. 'Siphonaptera bred from birds' nests.' *Entomologist's mon. Mag.*, **88**, 217–222.

ASKEW, R. R. 1961a. 'The biology of the British species of the genus *Olynx* Förster (Hymenoptera: Eulophidae), with a note on seasonal colour forms in the Chalcidoidea.' *Proc. R. ent. Soc. Lond. A.*, **36**, 103–112.

ASKEW, R. R. 1961b. 'On the biology of the inhabitants of oak galls of Cynipidae (Hymenoptera) in Britain.' *Trans. Soc. Br. Ent.*, **14**, 237–268.

ASKEW, R. R. 1965. 'The biology of the British species of the genus *Torymus* Dalman (Hymenoptera: Torymidae) associated with galls of Cynipidae (Hymenoptera) on oak, with special reference to alternation of forms.' *Trans. Soc. Br. Ent.*, **16**, 217–232.

ASKEW, R. R. 1968a. 'A survey of leaf-miners and their parasites on laburnum.' *Trans. R. ent. Soc. Lond.*, **120**, 1–37.

ASKEW, R. R. 1968b. 'Considerations on speciation in Chalcidoidea (Hymenoptera).' *Evolution*, **22**, 642–645.

AUSTEN, E. E. 1909. Illustrations of African Blood-sucking Flies other than Mosquitoes and Tsetse-flies. *London: British Museum (Natural History)*, 221 pp.

BACOT, A. W. 1914a. 'The influence of temperature, submersion and burial on the survival of eggs and larvae of *Cimex lectularius*.' *Bull. ent. Res.*, **5**, 111–117.

BACOT, A. W. 1914b. 'A study of the bionomics of the common rat fleas and other species associated with human habitations, with special reference to the influence of temperature and humidity at various periods of the life history of the insect.' *J. Hyg., Camb. (Plague Suppl.)*, **3**, 447–654.

BALDUF, W. V. 1938. 'The rise of entomophagy among Lepidoptera.' *Am. Nat.*, **72**, 358–379.

BANZIGER, H. 1968. 'Preliminary observations on a skin-piercing blood-sucking moth (*Calyptra eustrigata*)) Hmps.) (Lep., Noctuidae)) in Malaya.' *Bull.ent. Res.*, **58**, 159–163.

BARNES, H. F. 1929. 'Gall midges as enemies of aphids.' *Bull. ent. Res.*, **20**, 433–442.

BARNES, H. F. 1930. 'Gall midges (Cecidomyidae) as enemies of the Tingidae, Psyllidae, Aleurodidae and Coccidae.' *Bull. ent. Res.*, **21**, 319–329.

BATES, J. K. 1962. 'Field studies on the behaviour of bird fleas. 1. Behaviour of the adults of three species of bird fleas in the field.' *Parasitology*, **52**, 113–132.

BATRA, S. W. T. 1965. 'Organisms associated with *Lasioglossum zephyrum* (Hymenoptera: Halictidae).' *J. Kans. ent. Soc.*, **38**, 367–389.

BAUMERT, D. 1959. 'Mehrjahrige Zuchten einheimischer Strepsipteren an Homopteren. 2. Imagines, Lebens-zyklus und Artbestimmung von *Elenchus tenuicornis* Kirby.' *Zool. Beitr.* (n.f.), **4**, 343–409.

BEESLEY, W. N. 1968. 'Observations on the biology of the ox warble-fly (*Hypoderma*: Diptera: Oestridae). II. Bacteriostatic properties of larval extracts.' *Ann. trop. Med. Parasit.*, **62**, 8–12.

BENNET-CLARK, H. C. and LUCEY, E. C. A. 1967. 'The jump of the flea: a study of energetics and a model of the mechanism.' *J. exp. Biol.*, **47**, 59–76.

BEQUAERT, J. 1930. 'Notes on Hippoboscidae 2. The subfamily Hippoboscinae.' *Psyche*, **37**, 303–326.

BEQUAERT, J. 1953. 'The Hippoboscidae or louse-flies (Diptera) of mammals and birds. Part 1. Structure, physiology and natural history.' *Entomologica am.* (n.s.), **32/33**, 1–442.

BERG, C. O., FOOTE, B. A., and NEFF, S. E. 1959. 'Evolution of predator-prey relationships in snail-killing Sciomyzid larvae (Diptera).' *Bull. Am. Malacol. Union*, **25ʻ** 10–11.

BESS, H. A. 1939. 'Investigations of the resistance of mealybugs (Homoptera) to parasitization by internal hymenopterous parasites, with special reference to phagocytosis.' *Ann. ent. Soc. Am.*, **32**, 189–226.

BESS, H. A. and HARAMOTO, F. H. 1958. 'Biological control of the oriental fruit fly in Hawaii.' *Proc. Int. Congr. Ent.* 10. Montreal (1956), **4**, 835–840.

BISCHOFF, H. 1927. *Biologie der Hymenopteren*. Berlin: Julius Springer. 598 pp.

BLACKLOCK, D. B. and THOMPSON, M. G. 1923. 'A study on the tumbu fly, *Cordylobia anthropophaga* Grünberg, in Sierra Leone.' *Ann. trop. Med. Parasit.*, **17**, 443–510.

BOHART, G. E. 1941. 'The oviposition of Conopid flies upon smaller Andrenid bees.' *Pan-Pacif. Ent.*, **17**, 95–96.

BOHART, R. M. 1941. 'A revision of the Strepsiptera.' *Univ. Calif. Publs Ent.*, **7**, 91–160.

BORROR, D. J. and DELONG, D. M. 1966. *An Introduction to the Study of Insects*. Revised edition. New York, Chicago, San Francisco, Toronto, London: Holt, Rinehart, and Winston. 819 pp.

BOUCEK, Z. and ASKEW, R. R. 1968. *Index of Entomophagous Insects. Palearctic Eulophidae (excl. Tetrastichinae) (Hym. Chalcidoidea)*. Paris: Le François. 260 pp.

BRACKEN, G. K. 1965. 'Effects of dietary components on fecundity of the parasitoid *Exeristes comstockii* (Cress.) (Hymenoptera: Ichneumonidae).' *Can. Ent.*, 97, 1037–1041.

BRADLEY, W. G. and ARBUTHNOT, K. D. 1938. 'The relation of the host physiology to the development of the Braconid parasite *Chelonus annulipes*.' *Ann. ent. Soc. Am.*, 31, 359–365.

BRINCK, P. 1948. 'Catalogus insectorum Sueciae. IX. Anoplura.' *Opusc. ent.*, 13, 129–156.

BRISTOWE, W. S. 1950. 'Strange mating of *Pipunculus distinctus* Becker (Dipt., Pipunculidae).' *Entomologist's mon. Mag.*, 86, 264.

BROWN, W. L. 1955. 'The first social parasite in the ant tribe Dacetini.' *Insectes soc.*, 2, 181–186.

BRUMPT, E. 1927. *Précis de Parasitologie.* Paris: Masson. 1452 pp.

BRUMPT, E. 1933. 'Utilisation des larves de certaines mouches pour le traitement de l'ostéomyélite et de diverses affections chirurgicales chroniques.' *Annls Parasit. hum. comp.*, 11, 403.

BURR, M. 1911. 'Dermaptera' in Wytsam, P. *Genera Insectorum*, 122, 1–112.

BURR, M. 1954. *The Insect Legion.* 2nd edition. London: Nisbet. 336 pp.

BURR, M. and JORDAN, K. 1912. 'On *Arixenia* Burr, a suborder of Dermaptera.' *Trans. Int. Congr. Ent. 2. Oxford* (1912), 398–421.

BURSELL, E. 1961. 'The behaviour of tsetse flies (*Glossina swynnertoni* Austen) in relation to problems of sampling.' *Proc. R. ent. Soc. Lond. A*, 36, 9–20.

BUSHLAND, R. C. 1959. 'Male sterilization for the control of insects.' *Adv. Pest Control Res.*, 3, 1–26.

BÜTTIKER, W. 1967. 'Biological notes on eye-frequenting moths from N. Thailand.' *Mitt. schweiz. ent. Ges.*, 39 (1966), 151–179.

BUXTON, P. A. 1939. *The Louse.* London:Arnold. 115 pp.

BUXTON, P. A. 1941. 'Some recent work on the louse.' *Proc. R. Soc. Med.*, 34, 193–195.

CAMERON, T. W. M. 1965. *Parasites and Parasitism.* London: Methuen and New York: John Wiley. 322 pp.

CARPENTER, G. D. H. 1915. 'Observations on *Dorylas nigricans* Illig., in Damba and Bugalla Islands.' *Proc. R. ent. Soc. Lond.*, (1914), cvii–cxi.

CATTS, E. P. 1964. 'Biological studies of the rodent botflies, *Cuterebra* spp. (Diptera, Cuterebridae).' *Proc. Int. Congr. Parasit. 1. Rome*, 957.

CAULLERY, M. 1922. *Le Parasitisme et la Symbiose.* 2nd edition. Paris: Librairie Octave Doin. 400 pp.

CHAPMAN, R. F. 1961. 'Some experiments to determine the methods used in host finding by the tsetse fly *Glossina medicorum* Austen.' *Bull. ent. Res.*, 52, 83–97.

CHRISTOPHERS, SIR RICKARD. 1960. *Aedes aegypti (L.): The Yellow Fever Mosquito. Its Life History, Bionomics and Structure.* Cambridge: University Press. 732 pp.

CLARIDGE, M. F. 1961. 'Biological observations on some Eurytomid (Hymenoptera: Chalcidoidea) parasites associated with Compositae, and some taxonomic implications.' *Proc. R. ent. Soc. Lond. A*, 36, 153–158.

CLARIDGE, M. F. and ASKEW, R. R. 1960. 'Sibling species in the *Eurytoma rosae* group (Hym., Eurytomidae).' *Entomophaga*, 5, 141–153.

CLAUSEN, C. P., KING, J. L., and TERANISHI, C. 1927. 'The parasites of *Popillia japonica* in Japan and Chosen (Korea) and their introduction into the United States.' *Bull. U.S. Dep. Agric.*, no. 1429. 55 pp.

CLAUSEN, C. P. 1940. *Entomophagous Insects.* New York and London: McGraw-Hill. 688 pp.

CLAUSEN, C. P. 1951. 'The time factor in biological control.' *J. econ. Ent.*, 44, 1–9.

CLAY, T. 1949a. 'Some problems in the evolution of a group of ectoparasites.' *Evolution*, **3**, 279–299.

CLAY, T. 1949b. 'Piercing mouth-parts in the biting lice (Mallophaga).' *Nature, Lond.*, **164**, 617–619.

CLAY, T. 1964. 'Geographical distribution of the Mallophaga (Insecta).' *Bull. Br. Orn. Club*, **84**, 14–16.

CLAY, T. and MEINERTZHAGEN, R. 1943. 'The relationship between Mallophaga and Hippoboscid flies.' *Parasitology*, **35**, 11–16.

CLOUDSLEY-THOMPSON, J. T. 1957. 'On the habitat and growth stages of *Arixenia esau* Jordan and *A. jacobsoni* Burr (Dermaptera: Arixenioidea), with descriptions of the hitherto unknown adults of the former.' *Proc. R. ent. Soc. Lond. A*, **32**, 1–12.

COE, R. L. 1966. *Diptera Pipunculidae. Handbk Ident. Br. Insects*, **10** (2c). 83 pp.

COLE, L. C. 1945. 'The effect of temperature on the sex ratio of *Xenopsylla cheopis* Rothschild recovered from live rats.' *Publ. Hlth Rep., Wash.*, **60**, 1337–1342.

COLE, L. R. 1959a. 'On the defences of lepidopterous pupae in relation to the oviposition behaviour of certain Ichneumonidae.' *J. Lepid. Soc.*, **13**, 1–10.

COLE, L. R. 1959b. 'On a new species of *Syntretus* Förster (Hym.: Braconidae) parasitic on an adult ichneumonid, with a description of the larva and notes on its life history and that of its host, *Phaeogenes invisor* (Thunberg).' *Entomologist's mon. Mag.*, **95**, 18–21.

COLE, L. R. 1967. 'A study of the life-cycles and hosts of some Ichneumonidae attacking pupae of the green oak-leaf roller moth *Tortrix viridana* (L.) (Lepidoptera: Tortricidae) in England.' *Trans. R. ent. Soc. Lond.*, **119**, 267–281.

CORBET, G. B. 1956a. 'The phoresy of Mallophaga on a population of *Ornithomya fringillina* Curtis (Dipt., Hippoboscidae).' *Entomologist's mon. Mag.*, **92**, 207–211.

CORBET, G. B. 1956b. 'The life-history and host relations of a hippoboscid fly *Ornithomyia fringillina* Curtis.' *J. Anim. Ecol.*, **25**, 403–420.

CORBET, G. B. 1961. 'A comparison of the life-histories of two species of *Ornithomyia* (Dipt., Hippoboscidae).' *Entomologist's Gaz.*, **12**, 24–31.

COWAN, I. M. 1943. 'Notes on life history and morphology of *Cephenemyia jellisoni* Townsend and *Liptotena depressa* Say, two dipterous parasites of the Columbian blacktailed deer (*Odocoileus hemionus columbianus*).' *Can. J. Res. D*, **21**, 171–187.

CROSSKEY, R. W. 1951. 'The morphology, taxonomy and biology of the British Evanioidea (Hymenoptera).' *Trans. R. ent. Soc. Lond.*, **102**, 247–301.

CROSSKEY, R. W. 1962. 'The classification of the Gasteruptiidae (Hymenoptera).' *Trans. R. ent. Soc. Lond.*, **114**, 377–402.

CROWSON, R. A. 1955. *The natural classification of the families of Coleoptera.* London: Nathaniel Lloyd. 187 pp.

CUMBER, R. A. 1949. 'Humble-bee parasites and commensals found within a thirty mile radius of London.' *Proc. R. ent. Soc. Lond. A*, **24**, 119–127.

CUSHMAN, R. A. 1926a. 'Location of individual hosts versus systematic relation of host species as a determining factor in parasitic attack.' *Proc. ent. Soc. Wash.*, **28**, 5–6.

CUSHMAN, R. A. 1926b. 'Some types of parasitism among the Ichneumonidae.' *Proc. ent. Soc. Wash.*, **28**, 29–51.

DANIEL, D. M. 1932. '*Macrocentrus ancylivorus* Rohwer, a polyembryonic

Braconid parasite of the oriental fruit moth.' *Tech. Bull. N.Y. St. agric. Exp. Stn.*, no. 187. 101 pp.

DANILEVSKII, A. S. 1965. *Photoperiodism and Seasonal Development of Insects.* Edinburgh and London: Oliver and Boyd. 283 pp.

DARWIN, C. 1859. *On the Origin of Species by means of Natural Selection.* London: Murray. 502 pp.

DEBACH, P. 1964. *Biological Control of Insect Pests and Weeds.* London: Chapman and Hall. 844 pp.

DICKERSON, G. and LAVOIPIERRE, M. M. J. 1959. 'Studies on the methods of feeding of blood-sucking arthropods. II. The method of feeding adopted by the bed-bug (*Cimex lectularius*) when obtaining a blood-meal from the mammalian host.' *Ann. trop. Med. Parasit.*, **53**, 347–357.

DOBRZANSKA, J. and DOBRZANSKI, J. 1960. 'Quelques nouvelles remarques sur l'éthologie de *Polyergus rufescens* Latr. (Hymenoptère, Formicidae).' *Insectes soc.*, **7**, 1–8.

DODD, F. P. 1912. 'Some remarkable ant-friend Lepidoptera of Queensland.' *Trans. R. ent. Soc. Lond.*, **59** (1911), 577–590.

DOGIEL, V. A. 1964. *General Parasitology* (translated by Z. Kabata). London: Oliver and Boyd. 516 pp.

DONISTHORPE, H. ST. J. K. 1927. *The Guests of British Ants. Their Habits and Life-Histories.* London: Routledge. 244 pp.

DOUTT, R. L. 1958. 'Vice, virtue and the vedalia.' *Bull. ent. Soc. Am.*, **4**, 119–123.

DOUTT, R. L. 1959. 'The biology of parasitic Hymenoptera.' *A. Rev. Ent.*, **4**, 161–182.

DOWNES, J. A. 1958a. 'The feeding habits of biting flies and their significance in classification.' *A. Rev. Ent.*, **3**, 249–266.

DOWNES, J. A. 1958b. 'Assembly and mating in the biting Nematocera.' *Proc. Int. Congr. Ent.* 10. Montreal (1956), **2**, 425–434.

DUNAWAY, P. B., PAYNE, J. A., LEWIS, L. L., and STORY, J. D. 1967. 'Incidence and effects of *Cuterebra* in *Peromyscus*.' *J. Mammal.*, **48**, 38–51.

DUNNET, G. M. 1950. 'Fleas (Siphonaptera) from mammals in Aberdeenshire.' *Scott. Nat.*, **62**, 42–49.

EASON, R. R., PECK, W. B., and WHITCOMB, W. H. 1967. 'Notes on spider parasites, including a reference list.' *J. Kans. ent. Soc.*, **40**, 422–434.

EASTHAM, L. E. S. 1929. 'The post-embryonic development of *Phaenoserphus viator* Hal. (Proctotrypoidea), a parasite of the larva of *Pterostichus niger* (Carabidae), with notes on the anatomy of the larva.' *Parasitology*, **21**, 1–21.

EDWARDS, R. L. 1965. 'Revision of the genus *Aquanirmus* (Mallophaga: Philopteridae), parasitic on grebes (Podicipidae).' *Can. Ent.*, **97**, 920–935.

EIBL-EIBESFELDT, I. and EIBL-EIBESFELDT, E. 1968. 'The workers' bodyguard.' *Anims' Mag.*, **11**, 16–17.

EICHLER, W. 1936. 'Die Biologie der Federlinge.' *J. Orn., Lpz.*, **84**, 471–505.

EICHLER, W. 1948. 'Some rules in ectoparasitism.' *Ann. Mag. nat. Hist.* (ser. 12), **1**, 588–598.

ELTON, C., FORD, E. B., BAKER, J. R., and GARDNER, A. D. 1931. 'The health and parasites of a wild mouse population.' *Proc. zool. Soc. Lond.*, (1931), 657–721.

EVANS, F. C. and FREEMAN, R. B. 1950. 'On the relationships of some mammal fleas to their hosts.' *Ann. ent. Soc. Am.*, **43**, 320–333.

EVANS, G. O. 1950. 'Studies on the bionomics of the sheep ked, *Melophagus ovinus* L. in west Wales.' *Bull. ent. Res.*, **40**, 459–478.

EVANS, H. E., LIN, C. S., and YOSHIMOTO, C. M. 1953. 'A biological study of

Anoplius apiculatus autumnalis (Banks) and its parasite, *Evagetes mohave* (Banks) (Hymenoptera, Pompilidae).' *Jl. N.Y. ent. Soc.*, **61**, 61–78.

EXLEY, D., FORD, B., and ROTHSCHILD, M. 1965. 'The rabbit flea (*Spilopsyllus cuniculi* Dale) as an indicator of hormones in the host.' *Proc. R. ent. Soc. Lond. C*, **30**, 35–36.

FABRE, J. H. 1857. 'Memoirs sur l'hypermetamorphose et les moeurs des Meloides.' *Annls Sci. nat.*, **7**, 299–365.

FERRIÈRE, C. 1926. 'La phorésie chez les insectes.' *Mitt. schweiz. ent. Ges.*, **13**, 489–496.

FERRIS, G. F. 1931. 'The louse of elephants. *Haematomyzus elephantis* Piaget (Mallophaga: Haematomyzidae).' *Parasitology*, **23**, 112–127.

FERRIS, G. F. and USINGER, R. L. 1939. 'Review of the hemipterous family Polyctenidae.' *Microentomology*, **4**, 1–50.

FISHER, R. C. 1961. 'A study in insect multiparasitism. II. The mechanism and control of competition for possession of the host.' *J. exp. Biol.*, **38**, 605–628.

FISHER, R. C. 1965. 'The physiological suppression of insect parasitoids.' *Proc. Int. Congr. Ent.* 12. *London*, 413.

FLANDERS, S. E. 1931. 'The temperature relationships of *Trichogramma minutum* as a basis for racial segregation.' *Hilgardia*, **5**, 395–406.

FLANDERS, S. E. 1946. 'Control of sex and sex-limited polymorphism in the Hymenoptera.' *Q. Rev. Biol.*, **21**, 135–143.

FLANDERS, S. E. 1959. 'Differential host relations of the sexes in parasitic Hymenoptera.' *Entomologia exp. appl.*, **3**, 125–142.

FLANDERS, S. E. 1967. 'Deviate-ontogenies in the Aphelinid male (Hymenoptera) associated with the ovipositional behavior of the parental female.' *Entomophaga*, **12**, 415–427.

FOOTE, B. A., NEFF, S. E., and BERG, C. O. 1960. 'Biology and immature stages of *Atrichomelina pubera* (Diptera: Sciomyzidae).' *Ann. ent. Soc. Am.*, **53**, 129–199.

FOREL, A. 1920. *Les Fourmis de la Suisse.* Zurich: Rotapfel Verlag, 1948. Reprint 2nd edition. 349 pp.

FREE, J. B. and BUTLER, C. G. 1959. *Bumblebees.* London: Collins. 208 pp.

FULLER, M. E. 1933. 'The life history of *Onesia accepta* Malloch.' *Parasitology*, **25**, 342–352.

FULLER, M. E. 1938. 'On the biology and early stages of *Helicobia australis* (Sarcophaginae), a dipterous insect associated with grasshoppers.' *Proc. Linn. Soc. N.S.W.*, **63**, 133–138.

FULTON, P. B. 1933. 'Notes on *Habrocytus cerealellae*.' *Ann. ent. Soc. Am.*, **26**, 536–553.

GABBUTT, P. D. 1961. 'The distribution of some small mammals and their associated fleas from Central Labrador.' *Ecology*, **42**, 518–525.

GLASGOW, J. P. 1963. *The Distribution and Abundance of Tsetse.* London: Pergamon. 256 pp.

GLASGOW, J. P. 1967. 'Recent fundamental work on tsetse flies.' *A. Rev. Ent.*, **12**, 421–438.

GOODIER, R. 1964. 'Blood feeding in the Pangoninae (Dipt., Tabanidae).' *Entomologist's mon. Mag.*, **94** (1963), 133–136.

GRADWELL, G. R. 1958. '*Eulophus nigribasis* Gradwell (Hym., Chalcidoidea) the overwintering form of *E. larvarum* (L.).' *Entomologist's mon. Mag.*, **94**, 234–235.

GRASSÉ, P.-P. 1949. *Traité de Zoologie.* 9. *Insectes (Paléontologie, Géonémie,*

Insectes inférieurs et Coléoptères). Paris: Masson. 1117 pp.

GREATHEAD, D. J. 1958. 'Notes on the life history of *Symmictus flavopilosus* Bigot (Diptera: Nemestrinidae) as a parasite of *Schistocerca gregaria* (Forskål) (Orthoptera: Acrididae).' *Proc. R. ent. Soc. Lond. A*, **33**, 107–119.

GREATHEAD, D. J. 1963. 'A review of the insect enemies of Acridoidea (Orthoptera).' *Trans. R. ent. Soc. Lond.*, **114**, 437–517.

GREATHEAD, D. J. 1968. 'Further descriptions of *Halictophagus pontifex* Fox and *H. regina* Fox (Strepsiptera: Halictophagidae) from Uganda.' *Proc. R. ent. Soc. Lond. B*, **37**, 91–97.

GRIFFITHS, D. C. 1961. 'The development of *Monoctonus paludum* Marshall (Hym., Braconidae) in *Nasonovia ribis-nigri* on lettuce, and immunity reactions in other lettuce aphids.' *Bull. ent. Res.*, **52**, 147–163.

HAARLØV, N. 1964. 'Life cycle and distribution pattern of *Lipoptena cervi* (L.) (Dipt., Hippobosc.) on Danish deer.' *Oikos*, **15**, 93–129.

HAAS, G. E. and DICKIE, R. J. 1959. 'Fleas collected from cottontail rabbits in Wisconsin.' *Trans. Wis. Acad. Sci. Arts Lett.*, **48**, 125–133.

HADDOW, A. J. and CORBET, P. S. 1961. 'Entomological studies from a high tower in Mpanga Forest, Uganda. V. Swarming activity above the forest.' *Trans. R. ent. Soc. Lond.*, **113**, 284–300.

HADDOW, A. J. and THOMSON, R. C. M. 1937. 'Sheep myiasis in south-west Scotland, with special reference to the species involved.' *Parasitology*, **29**, 96–116.

HADWEN, S. 1926. 'Notes on the life-history of *Oedemagena tarandi*, L., and *Cephenomyia trompe*, Modeer.' *J. Parasit.*, **13**, 56.

HAMM, A. H. and RICHARDS, O. W. 1930. 'The biology of the British fossorial wasps of the families Mellinidae, Gorytidae, Philanthidae, Oxybelidae, and Trypoxylidae.' *Trans. R. ent. Soc. Lond.*, **78**, 95–131.

HARDENBERG, J. D. F. 1929. 'Beiträge zur Kenntnis der Pupiparen.' *Zool. Jb.*, **50**, 497–570.

HASKELL, P. T. 1961. *Insect Sounds*. London: Witherby. 189 pp.

HASSAN, A. I. 1939. 'The biology of some British Delphacidae (Homoptera) and their parasites with special reference to the Strepsiptera.' *Trans. R. ent. Soc. Lond.*, **89**, 345–384.

HATCH, M. H. 1958. 'Blind beetles in the fauna of the Pacific Northwest.' *Proc. Int. Congr. Ent.* 10. Montreal (1956), **1**, 207–211.

HAWKING, F. 1964. 'The periodicity of microfilariae.' *Proc. Int. Congr. Parasitology* 1. Rome, 628–629.

HILL, D. S. 1962. 'A study of the distribution and host preferences of three species of *Ornithomyia* (Diptera: Hippoboscidae) in the British Isles.' *Proc. R. ent. Soc. Lond. A*, **37**, 37–48.

HILL, D. S. 1963. 'The life history of the British species of *Ornithomya* (Diptera: Hippoboscidae).' *Trans. R. ent. Soc. Lond.*, **115**, 391–407.

HILL, D. S., HACKMAN, W., and LYNEBORG, L. 1964. 'The genus *Ornithomya* (Diptera: Hippoboscidae) in Fennoscandia, Denmark and Iceland.' *Notul. ent.*, **44**, 33–52.

HINTON, H. E. 1955. 'Protective devices of endopterygote pupae.' *Trans. Soc. Br. Ent.*, **12**, 49–92.

HINTON, H. E. 1964. 'Sperm transfer in insects and the evolution of haemocoelic insemination.' *R. ent. Soc. Lond. Symposium* 2 (*Insect Reproduction*), 95–107.

HOLLAND, G. P. 1955. 'Primary and secondary sexual characteristics of some Ceratophyllinae, with notes on the mechanism of copulation (Siphonaptera).' *Trans. R. ent. Soc. Lond.*, **107**, 233–248.

HOLLAND, G. P. 1958. 'Distribution patterns of northern fleas (Siphonaptera).' *Proc. Int. Congr. Ent.* 10. *Montreal* (1956), 1, 645–658.

HOPKINS, G. H. E. 1949. 'The host-associations of the lice of mammals.' *Proc. zool. Soc. Lond.*, 119, 387–604.

HOPKINS, G. H. E. and CLAY, T. 1952. *A check list of the Genera and Species of Mallophaga.* London: British Museum (Natural History). 362 pp.

HOPKINS, G. H. E. and ROTHSCHILD, M. In press. *An illustrated catalogue of the Rothschild collection of fleas. Vol 5.* London: British Museum (Natural History).

HSIAO, T., HOLDAWAY, F. G., and CHIANG, H. C. 1966. 'Ecological and physiological adaptations in insect parasitism.' *Entomologia exp. appl.*, 9, 113–123.

HUMPHRIES, D. A. 1967a. 'The function of combs in fleas.' *Entomologist's mon. Mag.*, 102, (1966): 232–236.

HUMPHRIES, D. A. 1967b. 'Drinking of water by fleas.' *Entomologist's mon. Mag.*, 102 (1966): 260–262.

HUMPHRIES, D. A. 1967c. 'The mating behaviour of the hen flea, *Ceratophyllus gallinae* (Schrank) (Siphonaptera: Insecta). '*Anim. Behav.*, 15, 82–98.

HUMPHRIES, D. A. 1967d. 'The action of the male genitalia during the copulation of the hen flea, *Ceratophyllus gallinae* (Schrank).' *Proc. R. ent. Soc. Lond. A*, 42, 101–106.

HUMPHRIES, D. A. 1968. 'The host-finding behaviour of the hen flea, *Ceratophyllus gallinae* (Schrank) (Siphonaptera).' *Parasitology*, 58, 403–414.

HYNES, H. B. N. 1947. 'Observations on *Systoechus somali* (Diptera, Bombyliidae) attacking the eggs of the desert locust (*Schistocerca gregaria* (Forskal)) in Somalia.' *Proc. R. ent. Soc. Lond. A*, 22, 79–85.

IMMS, A. D. 1931. *Recent Advances in Entomology.* London: Churchill. 374 pp.

JACKSON, D. J. 1935. 'Giant cells in insects parasitized by hymenopterous larvae.' *Nature, Lond.*, 135, 1040–1041.

JACKSON, D. J. 1958. 'Observations on the biology of *Caraphractus cinctus* Walker (Hymenoptera: Mymaridae), a parasitoid of the eggs of Dytiscidae. I. Methods of rearing and numbers bred on different host eggs.' *Trans. R. ent. Soc. Lond.*, 110, 533–566.

JACKSON, D. J. 1963. 'Diapause in *Caraphractus cinctus* Walker (Hymenoptera: Mymaridae), a parasitoid of the eggs of Dytiscidae (Coleoptera).' *Parasitology*, 53, 225–251.

JACKSON, D. J. 1964. 'Observations on the life-history of *Mestocharis bimacularis* (Dalman) (Hym. Eulophidae), a parasitoid of the eggs of Dytiscidae.' *Opusc. ent.*, 29, 81–97.

JACKSON, D. J. 1966. 'Observations on the biology of *Caraphractus cinctus* Walker (Hymenoptera: Mymaridae), a parasitoid of the eggs of Dytiscidae (Coleoptera). III. The adult life and sex ratio.' *Trans. R. ent. Soc. Lond.*, 118, 23–49.

JAMES, H. C. 1928. 'On the life-history and economic status of certain Cynipid parasites of dipterous larvae, with descriptions of some new larval forms'. *Ann. appl. Biol.*, 15, 287–316.

JANVIER, H. 1933. 'Etude biologique de quelques hyménoptères du Chili.' *Annls Sci. nat., Zool.*, 16, 210–356.

JENKINSON, F. 1903. '*Verrallia aucta* and its host.' *Entomologist's mon. Mag.*, 39, 222–223.

JOBLING, B. 1939. 'On the African Streblidae (Diptera Acalypterae) including the morphology of the genus *Ascodipteron* and a description of a new species'. *Parasitology*, 31, 147–165.

JOBLING, B. 1951. 'A record of the Streblidae from the Philippines and other

Pacific islands, including morphology of the abdomen, host-parasite relationship and geographical distribution, and with descriptions of five new species (Diptera).' *Trans. R. ent. Soc. Lond.*, 103, 211–246.

JOHNSON, B. 1959. 'Effect of parasitization by *Aphidius platensis* Brethes on the developmental physiology of its host, *Aphis craccivora* Koch.' *Entomologia exp. appl.*, 2, 82–99.

JOHNSON, C. G. 1940. 'Development, hatching and mortality of the eggs of *Cimex lectularius* L. (Hemiptera) in relation to climate, with observations on the effects of preconditioning to temperature.' *Parasitology*, 32, 127–173.

JOHNSON, C. G. 1942. 'The ecology of the bed-bug, *Cimex lectularius* L., in Britain.' *J. Hyg., Camb.*, 41, 345–361.

JORDAN, A. M., LEE-JONES, F., and WEITZ, B. 1962. 'The natural hosts of tsetse flies in northern Nigeria.' *Ann. trop. Med. Parasit.*, 56, 430–442.

JORDAN, K., 1910. 'Notes on the anatomy of *Hemimerus talpoides*.' *Novit. zool.*, 16, 327–330.

JORDAN, K. and ROTHSCHILD, N. C. 1906. 'A revision of the Sarcopsyllidae, a family of the Siphonaptera.' *Thompson Yates Johnstone Labs Rep.*, 7, 15–72.

KEILIN, D. 1915. 'Recherches sur les larves des Dipteres cyclorrhaphes.' *Bull. scient. Fr. Belg.*, 49, 15–198.

KEILIN, D. 1919. 'On the life history and larval anatomy of *Melinda cognata* Meigen (Diptera Calliphorinae) parasitic in the snail *Helicella* (*Heliomanes*) *virgata* Da Costa, with an account of the other Diptera living upon Molluscs.' *Parasitology*, 11, 430–455.

KEILIN, D. 1924. 'On the life-history of *Neottiophilum praeustum* (Meigen 1826) (Diptera-Acalypterae) parasitic on birds, with some general considerations on the problem of myiasis in plants, animals and man.' *Parasitology*, 16, 113–126.

KEILIN, D. and NUTTALL, G. H. F. 1919. 'Hermaphroditism and other abnormalities in *Pediculus humanus*.' *Parasitology*, 11, 279–328.

KEILIN, D. and NUTTALL, G. H. F. 1930. 'Iconographic studies of *Pediculus humanus*.' *Parasitology*, 22, 1–10.

KEMPER, H. 1932. 'Beiträge zur biologie der Blattwanze III.' *Z. Morph. Ökol. Tiere*, 24, 491–518.

KERRICH, G. J. 1936. 'Notes on larviposition in *Polyblastus*.' *Proc. R. ent. Soc. Lond. A*, 11, 108–110.

KERRICH, G. J. 1940. 'Notes on some Ichneumonid and Figitid (Hym.) parasites of Neuroptera.' *Entomologist's mon. Mag.*, 76, 15–17.

KESSEL, Q. C. 1924. 'Notes on the Streblinae, a subfamily of the Streblidae (Diptera Pupipara).' *Parasitology*, 16, 405–414.

KETTLE, D. S. and UTSI, M. N. P. 1955. '*Hypoderma diana* (Diptera, Oestridae) and *Lipoptena cervi* (Diptera, Hippoboscidae) as parasites of reindeer (*Rangifer tarandus*) in Scotland with notes on the second-stage larva of *Hypoderma diana*.' *Parasitology*, 45, 116–120.

KING, P. E. 1961. 'A possible method of sex ratio determination in the parasitic hymenopteron *Nasonia vitripennis*.' *Nature, Lond.*, 189, 330–331.

KIRKPATRICK, T. W. 1937a. 'Studies on the ecology of coffee plantations in East Africa. II. The autecology of *Antestia* spp. (Pentatomidae) with a particular account of a strepsipterous parasite.' *Trans. R. ent. Soc. Lond.*, 86, 247–343.

KIRKPATRICK, T. W. 1937b. 'Colour vision in the triungulin larva of a strepsipteron (*Corioxenos antestiae* Blair).' *Proc. R. ent. Soc. Lond. A*, 12, 40–44.

KIRKPATRICK, T. W. 1947. 'Notes on a species of Epipyropidae (Lepidoptera) parasitic on *Metaphaena* species (Hemiptera: Fulgoridae) at Amani, Tanganyika.' *Proc. R. ent. Soc. Lond. A*, **22**, 61–64.

KNIGHT, G. H. 1968. 'Observations on the behaviour of *Bombylius major* L. and *B. discolor* Mik. (Dipt., Bombyliidae) in the Midlands.' *Entomologist's mon. Mag.*, **103** (1967), 177–181.

KNIPLING, E. P. 1955. 'Possibilities of insect control or eradication through the use of sexually sterile males.' *J. econ. Ent.*, **48**, 459–462.

KNUTSON, L. V. 1966. 'Biology and immature stages of malacophagous flies: *Antichaeta analis, A. atriseta, A. brevipennis*, and *A. obliviosa* (Diptera: Sciomyzidae).' *Trans. Am. ent. Soc.*, **92**, 67–101.

KNUTSON, L. V. and BERG, C. O. 1966. 'Parasitoid development in snail-killing Sciomyzid flies.' *Trans. Am. microsc. Soc.*, **85**, 164–165.

KNUTSON, L. V., BERG, C. O., EDWARDS, L. J., BRATT, A. D., and FOOTE, B. A. 1967. 'Calcareous septa formed in snail shells by larvae of snail-killing flies.' *Science, N.Y.*, **156**, 522–523.

KORNHAUSER, S. J. 1919. 'The sexual characteristics of the Membracid *Thelia bimaculata* (Fab.). I. External changes induced by *Aphelopus theliae* (Gahan).' *J. Morph.*, **32**, 531–635.

KRISHNAMURTI, B. 1933. 'On the biology and morphology of *Epipyrops eurybrachidis* Fletcher.' *Nat. Hist. Soc. J. Bombay*, **36**, 944–949.

KROMBEIN, K. V. 1967. *Trap-nesting wasps and bees: life histories, nests and associates.* Washington, D.C.: Smithsonian. 570 pp.

LAARMAN, J. J. 1959. 'Host-seeking behaviour of malaria mosquitoes.' *Proc. Int. Congr. Zool.* **15** (1958), section 8, paper 24.

LAL, K. B. 1934. 'Insect parasites of Psyllidae.' *Parasitology*, **26**, 325–334.

LAVOIPIERRE, M. M. J. 1965. 'Feeding mechanism of blood-sucking arthropods.' *Nature, Lond.*, **208**, 302–303.

LEES, A. D. 1955. 'The Physiology of Diapause in Arthropods.' *Cambridge Monogr. exp. Biol.*, **4**, 151 pp.

LEIUS, K. 1961. 'Influence of food on fecundity and longevity of adults of *Itoplectis conquisitor* (Say) (Hymenoptera: Ichneumonidae).' *Can. Ent.*, **93**, 771–800.

LESTER, H. M. O. and LLOYD, L. 1928. 'Notes on the process of digestion in tsetse-flies.' *Bull. ent. Res.*, **19**, 39–60.

LEWIS, D. J. and DOMONEY, C. R. 1966. 'Sugar meals in Phlebotominae and Simuliidae (Diptera).' *Proc. R. ent. Soc. Lond. A*, **41**, 175–179.

LINDBERG, H. 1960. 'Die Strepsiptere *Elenchus tenuicornis* Kirby und ihre Wirte *Calligypona propinqua* (Fieb.) und *C. anthracina* (Horv.) (Homoptera Araeopidae).' *Commentat. biol.*, **23**, 1–10.

LINSLEY, E. G. and MACSWAIN, J. W. 1957. 'Observations on the habits of *Stylops pacifica* Bohart.' *Univ. Calif. Publs Ent.*, **11**, 395–430.

LINSLEY, E. G., MACSWAIN, J. W., and SMITH, R. F. 1952. 'The life history and development of *Rhipiphorus smithi* with notes on their phylogenetic significance.' *Univ. Calif. Publs Ent.*, **9**, 291–314.

LUBBOCK, J. 1862. 'On two aquatic Hymenoptera, one of which uses its wings in swimming.' *Trans. Linn. Soc. Lond.*, **24**, 135–142.

MAA, T. C. 1959. 'The family Polyctenidae in Malaya (Hemiptera).' *Pacif. Insects*, **1**, 415–422.

MAA, T. C. 1963. 'Genera and species of Hippoboscidae (Diptera): types, synonymy, habitats and natural groupings.' *Pacif. Insects*, **6**, 1–186.

MAA, T. C. 1964. 'A review of the Old World Polyctenidae (Hemiptera: Cimicoidea).' *Pacif. Insects*, **6**, 494–516.

MACLEOD, J. 1937. 'The species of Diptera concerned in cutaneous myiasis of sheep in Britain.' *Proc. R. ent. Soc. Lond. A*, **12**, 127–133.

MACLEOD, J. and DONNELLY, J. 1961. 'Failure to reduce an isolated blowfly population by the sterile males method.' *Entomologia exp. appl.*, **4**, 101–118.

MADDEN, J. L. In press. 'Host finding mechanisms in *Sirex noctilio* F. (Hymenoptera: Siricidae) and two groups of Siricid parasites.' *Proc. Int. Congr. Ent.* 13. *Moscow* (1968).

MARCHAL, P. 1906. 'Recherches sur la biologie et le développement des Hyménoptères parasites. II. Les Platygasters.' *Archs Zool. exp. gén.*, **4**, 485–640.

MARCHAL, P. 1936. 'Recherches sur la biologie et le développement des Hyménoptères: Les Trichogrammes.' *Annls Épiphyt.*, **2**, 447–550.

MASLENNIKOVA, V. A. 1959. 'The relation between the seasonal cycles of geographical populations of *Apanteles glomeratus* L. and its host *Pieris brassicae* L. [In Russian].' *Ent. Obozr.*, **38**, 517–522.

MATTHYSSE, J. G. 1944. 'Biology of the cattle biting louse and notes on cattle sucking lice.' *J. econ. Ent.*, **37**, 436–442.

MEAD-BRIGGS, A. R. and RUDGE, A. J. B. 1960. 'Breeding of the rabbit flea, *Spilopsyllus cuniculi* (Dale); requirement of a 'factor' from a pregnant rabbit for ovarian maturation.' *Nature, Lond.*, **187**, 1136–1137.

MEDWAY, Lord. 1958. 'On the habit of *Arixenia esau* Jordan (Dermaptera).' *Proc. R. ent. Soc. Lond. A*, **33**, 191–195.

MELLANBY, K. 1939. 'Fertilisation and egg production in the bed-bug, *Cimex lectularius* L./The physiology and activity of the bed-bug (*Cimex lectularius* L.) in a natural infestation.' *Parasitology*, **31**, 193–199/200–211.

MIYAMOTO, S. 1960. 'A new strepsipteron from Shansi, North China, *Mengenilla sinensis* Miyamoto.' *Mushi*, **33**, 37–38.

MOLYNEUX, D. 1967. 'Feeding behaviour of the larval rat flea, *Nosopsyllus fasciatus* Bosc.' *Nature, Lond.*, **215**, 779.

MORRIS, H. M. 1922. 'On the larva and pupa of a parasitic Phorid fly, *Hypocera incrassata* Mg.' *Parasitology*, **14**, 70–74.

MORRIS, K. R. S. 1965. 'The ecology of sleeping sickness.' *Discovery, Lond.*, **26**, 22–28.

MUESEBECK, C. F. W. and DOHANIAN, S. M. 1927. 'A study in hyperparasitism, with particular reference to the parasites of *Apanteles melanoscelus* (Ratzeburg).' *Dep. Bull. U.S. Dep. Agric.*, no. 1487. 35 pp.

MUIR, F. 1912. 'Two new species of *Ascodipteron*.' *Bull. Mus. comp. Zool. Harv.*, **54**, 351–366.

MUIR, F. 1918. 'Pipunculidae and Stylopidae in Homoptera.' *Entomologist's mon. Mag.*, **54**, 137.

MUKERJI, D. and SEN-SARMA, P. 1955. 'Anatomy and affinity of the elephant louse *Haematomyzus elephantis* Piaget (Insecta: Rhyncophthiraptera).' *Parasitology*, **45**, 5–30.

MULDREW, J. A. 1953. 'The natural immunity of the larch sawfly (*Pristiphora erichsonii* (Htg.)) to the introduced parasite *Mesoleius tenthredinis* Morley in Manitoba and Saskatchewan.' *Can. J. Zool.*, **31**, 313–332.

MURAKAMI, Y. 1960. 'Seasonal dimorphism in the Encyrtidae (Hymenoptera, Chalcidoidea).' *Acta hymenopt.*, **1**, 199–204.

MURRAY, M. D. 1961. 'The ecology of the louse *Polyplax serrata* (Burm.) on the mouse *Mus musculus* L.' *Aust. J. Zool.*, **9**, 1–13.

MURRAY, M. D. 1963. 'The ecology of lice on sheep. III. Differences between the biology of *Linognathus pedalis* (Osborne) and *L. ovillus* (Neumann)./IV. The establishment and maintenance of populations of *Linognathus ovillus* (Neumann).' *Aust. J. Zool.*, **11**, 153–156/157–172.

MURRAY, M. D. and NICHOLLS, D. G. 1965. 'Studies on the ectoparasites of seals and penguins. I. The ecology of the louse *Lepidophthirus macrorhini* Enderlein on the southern elephant seal, *Mirounga leonina* (L.).' *Aust. J. Zool.*, **13**, 437–454.

MURRAY, M. D., SMITH, M. S. R., and SOUCEK, Z. 1965. 'Studies on the ectoparasites of seals and penguins. II. The ecology of the louse *Antarctophthirus ogmorhini* Enderlein on the Weddell seal, *Leptonychotes weddelli.*' *Aust. J. Zool.*, **13**, 761–771.

MURRAY, M. D. and VESTJENS, W. J. M. 1967. 'Studies on the ectoparasites of seals and penguins. III. The distribution of the tick *Ixodes uriae* White and the flea *Parapsyllus magellanicus heardi* de Meillon on Macquairie Island.' *Aust. J. Zool.*, **15**, 715–725.

NASH, T. A. M. 1948. *Tsetse Flies in British West Africa.* London: H.M. Stationery Office. 78 pp.

NEVILLE, C. and ROTHSCHILD, M. 1967. 'Fleas—insects which fly with their legs.' *Proc. R. ent. Soc. Lond. C*, **32**, 9–10.

NUTTALL, G. H. F. 1917. 'The biology of *Pediculus humanus.*' *Parasitology*, **10**, 80–185.

OGLOBLIN, A. A. 1913. 'Contribution to the biology of Coccinellidae. [In Russian].' *Ent. Obozr.*, **13**, 27–43.

OLDROYD, H. 1964. *The Natural History of Flies.* London: Weidenfeld and Nicolson. 324 pp.

OLDROYD, H. and RIBBANDS, C. R. 1936. 'On the validity of trichiation as a systematic character in *Trichogramma* (Hym., Chalcididae).' *Proc. R. ent. Soc. Lond. B*, **5**, 148–152.

ORMEROD, W. E. 1961. 'The epidemic spread of Rhodesian sleeping sickness.' *Trans. R. Soc. trop. Med. Hyg.*, **55**, 525–538.

PANTEL, J. 1910. 'Recherches sur les Dipteres a larves entombies.' *Cellule*, **26**, 27–216.

PARKER, H. L. and SMITH, H. D. 1933. 'Additional notes on the strepsipteron *Eoxenos laboulbenei* Peyerimhoff. '*Ann. ent. Soc. Am.*, **26**, 217–233.

PARKER, H. L. and SMITH, H. D. 1934. 'Further notes on *Eoxenos laboulbenei* Peyerimhoff with a description of the male.' *Ann. ent. Soc. Am.*, **27**, 468–477.

PARNELL, J. R. 1964. 'Investigations on the biology and larval morphology of the insects associated with the galls of *Asphondylia sarothamni* H. Loew (Diptera: Cecidomyiidae) on broom (*Sarothamnus scoparius* (L.) Wimmer).' *Trans. R. ent. Soc. Lond.*, **116**, 255–273.

PATTON, W. S. and CRAGG, F. W. 1913. *A Textbook of Medical Entomology.* London, Madras, and Calcutta: Christian Literature Society for India. 764 pp.

PATTON, W. S. and EVANS, A. M. 1929. *Insects, Ticks, Mites and Venomous Animals of Medical and Veterinary Importance. Part I. Medical.* Croyden: H. R. Grubb. 786 pp.

PAULIAN, R. 1943. *Les Coléoptères. Formes-Moeurs-Rôle.* Paris: Payot. 396 pp.

PAVILLARD, E. R. and WRIGHT, E. A. 1957. 'An antibiotic from maggots.' *Nature, Lond.*, **180**, 916.

PEMBERTON, C. E. and WILLARD, H. F. 1918. 'A contribution to the biology of fruit-fly parasites in Hawaii.' *J. agric. Res.*, **15**, 419–466.

PEPPER, B. B. and DRIGGERS, B. F. 1934. 'Non-economic insects as intermediate hosts of parasites of the oriental fruit moth.' *Ann. ent. Soc. Am.*, **27**, 593–598.

PEREZ, J. 1886. 'Des effets du parasitisme des *Stylops* sur les apiaires du genre *Andrena*.' *Act. Soc. linn. Bordeaux*, **40**, 21–60.

PERKINS, R. C. L. 1918a. 'Further notes on *Stylops* and stylopized bees.' *Entomologist's mon. Mag.* **54**, 115–129.

PERKINS, R. C. L. 1918b. 'The assembling and pairing of *Stylops*.' *Entomologist's mon. Mag.*, **54**, 129–131.

PICARD, F. 1930. 'Sur le parasitisme d'un Phoride (*Megaselia cuspidata* Schmitz) aux depens d'un myriapode.' *Bull. Soc. zool. Fr.*, **55**, 180–183.

PIEL, P. O. and COVILLARD, P. 1933. 'Contribution à l'étude du *Monema flavescens* et de ses parasites.' *Notes Ent. chin.*, **1** (10), 1–44.

PIERCE, W. D. 1909. 'A monographic revision of the twisted winged insects comprising the order Strepsiptera Kirby.' *Bull. U.S. natn. Mus.*, no. 66, 232 pp.

PILSON, R. D. and PILSON, B. M. 1967. 'Behaviour studies of *Glossina morsitans* Westw. in the field.' *Bull. ent. Res.*, **57**, 227–257.

POPHAM, E. J. 1961a. 'The functional morphology of the mouthparts of the cockroach *Periplaneta americana* L.' *Entomologist*, **94**, 185–192.

POPHAM, E. J. 1961b. 'On the systematic position of *Hemimerus* Walker—a case for ordinal status.' *Proc. R. ent. Soc. Lond. B*, **30**, 19–25.

POPHAM, E. J. 1962. 'The anatomy related to the feeding habits of *Arixenia* and *Hemimerus* (Dermaptera). '*Proc. zool. Soc. Lond.*, **139**, 429–450.

POPHAM, E. J. 1965. 'The functional morphology of the reproductive organs of the common earwig (*Forficula auricularia*) and other Dermaptera with reference to the natural classification of the order.' *J. Zool.*, **146**, 1–43.

POULTON, E. B. 1931. 'Two especially significant examples of insect mimicry.' *Trans. R. ent. Soc. Lond.*, **79**, 395–398.

PSCHORN-WALCHER, H. 1967. 'Biology of the Ichneumonid parasites of *Neodiprion sertifer* (Geoffroy) (Hym.: Diprionidae) in Europe.' *Tech. Bull. Commonw. Inst. biol. Control*, **8**, 7–52.

PUTTLER, B. 1967. 'Interrelationships of *Hypera postica* (Coleoptera: Curculionidae) and *Bathyplectes curculionis* (Hymenoptera: Ichneumonidae) in the eastern United States with particular reference to encapsulation of the parasite eggs by the weevil larvae.' *Ann. ent. Soc. Am.*, **60**, 1031–1038.

RAATIKAINEN, M. 1967. 'Bionomics, enemies and population dynamics of *Javesella pellucida* (F.) (Hom., Delphacidae).' *Annls Agric. Fenniae*, **6** suppl. 2. 149 pp.

RABAUD, E. 1922. 'Note sur le comportement de *Rielia manticida* Kieff., Proctotrupide parasite des ootheques de Mantes.' *Bull. Soc. zool. Fr.*, **47**, 10–15.

RADOVSKY, J. J. and CATTS, E. P. 1960. 'Observations on the biology of *Cuterebra latifrons* Coquillet (Diptera: Cuterebridae).' *J. Kans. ent. Soc.*, **33**, 31–36.

RAU, P. 1930. 'A note on the parasitic beetle, *Hornia minutipennis* Riley.' *Psyche*, **37**, 155–156.

RAW, A. 1968. 'The behaviour of *Leopoldius coronatus* (Rond.) (Dipt. Conopidae) towards its hymenopterous hosts.' *Entomologist's mon. Mag.*, **104**, 54.

REHN, J. A. G. and REHN, J. W. H. 1936. 'A study of the genus *Hemimerus* (Dermaptera, Hemimerina, Hemimeridae).' *Proc. Acad. nat. Sci. Philad.*, **87**, 457–508.

RETTENMEYER, C. W. 1961. 'Observations on the biology and taxonomy of flies found over swarm raids of army ants (Diptera: Tachinidae, Conopidae).' *Kans. Univ. Sci. Bull.*, **42**, 993–1066.

REUTER, O. M. 1913. *Lebensgewohnheiten und Instinkte der Insekten*. Berlin: Friedlander. 488 pp.

RICHARDS, O. W. 1927. 'The specific characters of the British humblebees (Hymenoptera).' *Trans. R. ent. Soc. Lond.*, 75, 233–267.

RICHARDS, O. W. 1937. 'A study of the British species of *Epeolus* Latr. and their races, with a key to the species of *Colletes* (Hymen., Apidae).' *Trans. Soc. Br. Ent.*, 4, 89–130.

RICHARDS, O. W. 1939. 'The British Bethylidae (*s.l.*) (Hymenoptera).' *Trans. R. ent. Soc. Lond.*, 89, 185–344.

RICHARDS, O. W. 1949. 'The evolution of cuckoo bees and wasps.' *Proc. Linn. Soc. Lond.*, 161, 40–41.

RICHARDS, O. W. 1953. *The Social Insects*. London: Macdonald. 219 pp.

RICHARDS, O. W. and HAMM, A. H. 1939. 'The biology of the British Pompilidae (Hymenoptera).' *Trans. Soc. Br. Ent.*, 6, 51–114.

ROBBINS, J. C. 1927. '*Diplostichus janitrix*, Hartig, a Tachinid parasite of the pine sawfly, and its method of emergence.' *Proc. R. ent. Soc. Lond.*, 2, 17–19.

RODHAIN, J. and BEQUAERT, J. 1916. 'Observations sur la biologie de *Cyclopodia greeffi* Karsch (Dipt.), Nyctéribiide parasite d'une chauve-souris Congolaise.' *Bull. Soc. zool. Fr.*, 40, 248–262.

ROHWER, S. A. and CUSHMAN, R. A. 1917. 'Idiogastra, a new suborder of Hymenoptera with notes on the immature stages of *Oryssus*.' *Proc. ent. Soc. Wash.*, 19, 89–99.

ROTHERAM, S. M. 1967. 'Immune surface of eggs of a parasitic insect.' *Nature, Lond.*, 214, 700.

ROTHSCHILD, G. H. L. 1966. 'Notes on two hymenopterous egg parasites of Delphacidae (Hem.).' *Entomologist's mon. Mag.*, 102, 5–9.

ROTHSCHILD, M. 1952. 'A collection of fleas from the bodies of British birds, with notes on their distribution and host preferences.' *Bull. Br. Mus. nat. Hist. (Entomology)*, 2, 187–232.

ROTHSCHILD, M. 1958. 'The bird fleas of Fair Isle.' *Parasitology*, 48, 382–412.

ROTHSCHILD, M. 1965. 'Fleas.' *Scient. Am.*, 213, 44–53.

ROTHSCHILD, M. and CLAY, T. 1952. *Fleas, Flukes and Cuckoos. A Study of Bird Parasites*. London: Collins. 304 pp.

ROTHSCHILD, M. and FORD, B. 1965. 'Observations on gravid rabbit fleas (*Spilopsyllus cuniculi* (Dale)) parasitising the hare (*Lepus europaeus* Pallas), together with further speculations concerning the course of myxomatosis at Ashton, Northants.' *Proc. R. ent. Soc. Lond. A*, 40, 109–117.

ROTHSCHILD, M. and FORD, B. 1966. 'Reproductive hormones of the host controlling the sexual cycle of the rabbit flea (*Spilopsyllus cuniculi* Dale).' *Proc. Int. Congr. Ent.* 12. London (1964), 801–802.

ROTHSCHILD, M. and FORD, B. 1969. 'Does a pheromone-like factor from the nestling rabbit stimulate impregnation and maturation in the rabbit flea?' *Nature, Lond.*, 221, 1169–1170.

ROTHSCHILD, M. and HINTON, H. E. 1968. 'Holding organs on the antennae of male fleas.' *Proc. R. ent. Soc. Lond. A*, 43, 105–107.

SAFAVI, M. 1968. 'Étude biologique et écologique des hyménoptères parasites des oeufs des punaises des céréales.' *Entomophaga*, 13, 381–495.

SALT, G. 1927. 'The effects of stylopization on aculeate Hymenoptera.' *J. exp. Zool.*, 48, 223–331.

SALT, G. 1931. 'A further study of the effects of stylopization on wasps.' *J. exp. Zool.*, 59, 133–166.

SALT, G. 1932. 'Superparasitism by *Collyria calcitrator*, Grav. With an appendix by R. H. Stoy.' *Bull. ent. Res.*, 23, 211–216.

SALT, G. 1934. 'Experimental studies in insect parasitism. II. Superparasitism.' *Proc. R. Soc. B*, **114**, 455–476.

SALT, G. 1937a. 'The sense used by *Trichogramma* to distinguish between parasitized and unparasitized hosts.' *Proc. R. Soc. B*, **122**, 57–75.

SALT, G. 1937b. 'The egg-parasite of *Sialis lutaria*: a study of the influence of the host upon a dimorphic parasite.' *Parasitology*, **29**, 539–553.

SALT, G. 1941. 'The effects of hosts upon their insect parasites.' *Biol. Rev.*, **16**, 239–264.

SALT, G. 1955. 'Experimental studies in insect parasitism. VIII. Host reactions following artificial parasitization.' *Proc. R. Soc. B*, **144**, 380–398.

SALT, G. 1963. 'The defence reactions of insect metazoan parasites.' *Parasitology*, **53**, 527–642.

SALT, G. 1964. 'The Ichneumonid parasite *Nemeritis canescens* (Gravenhorst) in relation to the wax moth *Galleria mellonella* (L.).' *Trans. R. ent. Soc. Lond.*, **116**, 1–14.

SALT, G. 1968. 'The resistance of insect parasitoids to the defence reactions of their hosts.' *Biol. Rev.*, **43**, 200–232.

SAMARINA, G. P., ALEXEYEV, A. N., and SHIRANOVICH, P. I. 1968. 'The study of fertility of the rat fleas (*Xenopsylla cheopis* Rothsch. and *Ceratophyllus* Bosc.) under their feeding on different animals.' *Zool. Zh.*, **47** (2).

SASAKI, C. 1887. 'On the life history of *Ugimyia sericaria* Rondani.' *J. Coll. Sci. imp. Univ. Tokyo*, **1**, 1–46.

SAUNDERS, D. S. 1962. 'The effect of the age of female *Nasonia vitripennis* (Walker) (Hymenoptera, Pteromalidae) upon the incidence of larval diapause.' *J. Insect Physiol.*, **8**, 309–318.

SCHMIEDER, R. G. 1933. 'The polymorphic forms of *Melittobia chalybii* Ashmead and the determining factors involved in their production.' *Biol. Bull. mar. biol. Lab., Woods Hole*, **65**, 338–354.

SCHNEIDER, F. VON. 1950. 'Die Abwehrreaktion des Insektenblutes und ihre Be-einflussung durch die Parasiten.' *Vjschr. naturf. Ges. Zürich*, **95**, 22–24.

SCHNEIDER, F. VON. 1951. 'Einige physiologische Beziehungen zwischen Syrphidenlarven und ihren Parasiten.' *Z. angew. Ent.*, **33**, 150–162.

SCHNEIDERMAN, H. A. and HORWITZ, J. 1958. 'The induction and termination of facultative diapause in the chalcid wasps *Mormoniella vitripennis* (Walker) and *Tritneptis klugii* (Ratzeburg).' *J. exp. Biol.*, **35**, 520–551.

SCHOONHOVEN, L. M. 1962. 'Diapause and the physiology of host-parasite synchronization in *Bupalus piniarius* and *Eucarcelia rutilla*.' *Archs néerl. Zool.*, **15**, 111–174.

SCHRADER, S. H. 1924. 'Reproduction in *Acroschismus wheeleri* Pierce.' *J. Morph.*, **39**, 157–197.

SCUDDER, G. G. E. 1957. 'Reinterpretation of some basal structures in the insect ovipositor.' *Nature, Lond.*, **180**, 340–341.

SEEVERS, C. H. 1955. 'A revision of the tribe Amblyopinini: Staphylinid beetles parasitic on mammals.' *Fieldiana, Zool.*, **37**, 211–264.

SÉGUY, E. 1950. *La Biologie des Diptères*. Paris: Lechevalier. 609 pp.

SELHIME, A. G. and KANAVEL, R. F. 1968. 'Life cycle and parasitism of *Micromus posticus* and *M. subanticus* in Florida.' *Ann. ent. Soc. Am.*, **61**, 1212–1215.

SHANNON, R. C. 1931. 'On the classification of Brazilian Culicidae with special reference to those capable of harboring the yellow fever virus.' *Proc. ent. Soc. Wash.*, **33**, 125–157.

SHARIF, M. 1935. 'On the presence of wing buds in the pupae of Aphaniptera.' *Parasitology*, **27**, 461–464.

SHORT, J. R. T. 1952. 'The morphology of the head of larval Hymenoptera with special reference to the head of Ichneumonoidea, including a classification of the final instar larvae of the Braconidae.' *Trans. R. ent. Soc. Lond.*, 103, 27–84.

SIKES, E. K. and WIGGLESWORTH, V. B. 1931. 'The hatching of insects from the egg, and the appearance of air in the tracheal system.' *Q. Jl microsc. Sci.*, 74, 165–192.

SILVESTRI, F. 1904. 'Contribuzioni alla conoscenza della metamorfosi e dei costumi della *Lebia scapularis* Fourc.' *Redia*, 2, 68–84.

SILVESTRI, F. 1906. 'Contribuzioni alla conoscenza biologica degli imenotteri parassiti. I. Biologia del *Litomastix truncatellus* (Dalm.).' *Boll. Lab. Zool. gen. agr. Portici*, 1, 17–64.

SIMMONDS, F. J. 1948. 'The influence of maternal physiology on the incidence of diapause.' *Phil. Trans. R. Soc. B*, 233, 385–414.

SIMMONS, K. E. L. 1966. 'Anting and the problem of self-stimulation.' *J. Zool., Lond.*, 149, 145–162.

SIMMONS, S. W. 1935. 'The bactericidal properties of excretions of the maggots of *Lucilia sericata*.' *Bull. ent. Res.*, 26, 559–563.

SKAIFE, S. H. 1921. 'On *Braula coeca* Nitzsch, a dipterous parasite of the honey bee.' *Trans. R. Soc., S. Afr.*, 10, 41–48.

SLADEN, F. W. L. 1912. *The Humble-bee, its Life-history and how to Domesticate it.* London: Macmillan. 283 pp.

SMART, J. 1963. 'Explosive evolution and the phylogeny of insects.' *Proc. Linn. Soc. Lond.*, 174, 124–126.

SMART, J. 1965. *A Handbook for the Identification of Insects of Medical Importance.* 4th edition. London: British Museum (Natural History). 303 pp.

SMIT, F. G. A. M. 1957a. 'The recorded distribution and hosts of Siphonaptera in Britain.' *Entomologist's Gaz.*, 8, 45–75.

SMIT, F. G. A. M. 1957b. *Siphonaptera. Handbk Ident. Br. Insects*, 1 (16). 94 pp.

SMITH, G. and HAMM, A. H. 1914. 'Studies in the experimental analysis of sex. Part II—On stylops and stylopization.' *Q. Jl microsc. Sci.*, 60, 435–461.

SMITH, H. S. 1916. 'An attempt to redefine the relationships exhibited by entomophagous insects.' *J. econ. Ent.*, 9, 477–486.

SMITH, K. G. V. 1966. 'The larva of *Thecophora occidensis*, with comments upon the biology of Conopidae (Diptera).' *J. Zool., Lond.*, 149, 263–276.

SMITH, K. G. V. 1967. 'The biology and taxonomy of the genus *Stylogaster* Macquart, 1835 (Diptera: Conopidae, Stylogasterinae) in the Ethiopian and Malagasy regions.' *Trans. R. ent. Soc. Lond.*, 119, 47–69.

SMITH, K. G. V. 1969. 'Further data on oviposition by the genus *Stylogaster* Macquart (Diptera: Conopidae, Stylogasterinae) upon adult Calyptrate Diptera associated with ants and animal dung.' *Proc. R. ent. Soc. Lond. A*, 44, 35–37.

SOUTHWOOD, T. R. E. and LESTON, D. 1959. *Land and Water Bugs of the British Isles.* London and New York: Warne. 436 pp.

SPENCER, G. J. 1958. 'On the Nemestrinidae of British Columbia dry range lands.' *Proc. Int. Congr. Ent.* 10. Montreal (1956), 4, 503–509.

SPRADBERY, J. P. 1968. 'A technique for artificially culturing Ichneumonid parasites of woodwasps (Hymenoptera: Siricidae).' *Entomologia exp. appl.*, 11, 257–260.

SPRADBERY, J. P. In press. 'Factors determining the detection of Siricid woodwasp larvae by Ichneumonid parasites and their application in artificial rearing methods.' *Proc. Int. Congr. Ent.* 13. Moscow (1968).

STARÝ, P. 1966. *Aphid parasites of Czechoslovakia. A review of the Czechoslovak Aphidiidae (Hymenoptera)*. The Hague: Junk. 242 pp.

STEFFAN, A. W. 1965. 'On the epizoic associations of Chironomidae (Diptera) and their phyletic relationships.' *Proc. Int. Congr. Ent.* 12. London (1964), 77–78.

STEFFAN, J.-R. 1961. 'Comportement de *Lasiochalcidia igiliensis* Ms., Chalcidide parasite de Fourmilions.' *C.r. Séanc. Acad. Sci.*, 253, 2401–2403.

STERN, V. M., SMITH, R. F., VAN DEN BOSCH, R., and HAGEN, K. S. 1959. 'The integration of chemical and biological control of the spotted alfalfa aphid. The integrated control concept.' *Hilgardia*, 29, 81–101.

STERNLICHT, M. 1966. '*Heteropsyche schawerdae* n.comb. (Zerny) (Epipyropidae, Lepidoptera) in Israel.' *Israel J. Zool.*, 15, 51–56.

STEWART, M. A. 1934. 'The rôle of *Lucilia sericata* Meig. larvae in osteomyelitis wounds.' *Ann. trop. Med. Parasit.*, 28, 445–454.

STRICKLAND, E. H. 1923. 'Biological notes on parasites of prairie cutworms.' *Bull. Dep. Afric. Dom. Can.*, no. 26. 40 pp.

STUCKENBERG, B. R. 1963. 'A study of the biology of the genus *Stylogaster*, with the description of a new species from Madagascar (Diptera: Conopidae).' *Revue Zool. Bot. afr.*, 68, 251–275.

SUDD, J. H. 1967. *An Introduction to the Behaviour of Ants*. London: Edward Arnold. 200 pp.

SUTER, P. 1965. 'Life cycle of *Echidnophaga gallinacea*.' *Proc. Int. Congr. Ent.* 12. London (1964), 830.

SWYNNERTON, C. F. M. 1936. 'The tsetse flies of East Africa. A first study of their ecology, with a view to their control.' *Trans. R. ent. Soc. Lond.*, 84, 1–579.

SYMMONS, S. 1952. 'Comparative anatomy of the mallophagan head.' *Trans. zool. Soc. Lond.*, 27, 349–436.

TAYLOR, J. S. 1932. 'Report on cotton insect and disease investigations. Pt. II. Notes on the American bollworm (*Heliothis obsoleta* Fabr.) on cotton and on its parasite (*Microbracon brevicornis* Wesm.).' *Sci. Bull. Dep. Agric. For. Un. S. Afr.*, no. 113, 18 pp.

TAYLOR, T. H. C. 1937. *The Biological Control of an Insect in Fiji. An account of the Coconut Leaf-mining Beetle and its Parasite Complex*. London: Imperial Institute of Entomology. 239 pp.

TETLEY, H. 1918. 'The structure of the mouth-parts of *Pangonia longirostris* in relation to the probable feeding-habits of the species.' *Bull. ent. Res.*, 8, 253–267.

TETLEY, J. H. 1958. 'The sheep ked, *Melophagus ovinus* L. I. Dissemination potential.' *Parasitology*, 48, 353–363.

THEODOR, O. 1964. 'On the relationships between the families of the Pupipara.' *Proc. Int. Congr. Parasit.* 1. Rome. 999–1000.

THEODOR, O. 1967. *An illustrated catalogue of the Rothschild collection of Nycteribiidae*. London: British Museum (Natural History). 506 pp.

THIENEMANN, A. 1954. '*Chironomus*. Leben, Verbreitung und wirtschaftliche Bedeutung der Chironomiden.' *Die Binnengewässer*, Stuttgart, 20. 834 pp.

THOMPSON, G. B. 1955. 'Contribution towards a study of the ectoparasites of British birds and mammals. No. 4.' *Ann. Mag. nat. Hist.*, 12, 25–35.

THOMPSON, G. B. 1957. 'The parasites of British birds and mammals. XXX. Mallophaga on birds' eggs.' *Entomologist's mon. Mag.*, 93, 189–190.

THOMPSON, G. H. and SKINNER, E. R. 1961. *The Alder Wood Wasp and its Insect Enemies*. Commonwealth Forestry Institute film.

THOMPSON, W. R. 1934. 'The Tachinid parasites of woodlice.' *Parasitology*, 26, 378–448.

THORPE, W. H. 1931. 'The biology, post-embryonic development, and economic importance of *Cryptochaetum iceryae* (Diptera, Agromyzidae), parasitic on *Icerya purchasi.*' *Proc. zool. Soc. Lond.*, 1930, 929–971.

THORPE, W. H. 1941. 'The biology of *Cryptochaetum* (Diptera) and *Eupelmus* (Hymenoptera), parasites of *Aspidoproctus* (Coccidae) in East Africa.' *Parasitology*, 33, 149–168.

THORPE, W. H. and JONES, F. G. W. 1937. 'Olfactory conditioning in a parasitic insect and its relation to the problem of host selection.' *Proc. R. Soc. B*, 124, 56–81.

TIMBERLAKE, P. H. 1916. 'Note on an interesting case of two generations of a parasite reared from the same individual host.' *Can. Ent.*, 48, 89–91.

TINLEY, K. L. 1964. 'Some observations on certain Tabanid flies in North-Eastern Zululand (Diptera: Tabanidae).' *Proc. R. ent. Soc. Lond. A*, 39, 73–75.

TOTHILL, J. D. 1922. 'The natural control of the fall webworm (*Hyphantria cunea* Drury) in Canada.' *Tech. Bull. Dep. Agric. Can.*, no. 3, 107 pp.

TOTHILL, J. D., TAYLOR, T. H. C., and PAINE, R. W. 1930. *The Coconut Moth in Fiji. A History of its Control by means of Parasites.* London: Imperial Bureau of Entomology. 269 pp.

TOWNES, H. 1969. *The Genera of Ichneumonidae, Part I.* Mem. Am. Ent. Inst., no. 11. 300 pp.

TOWNSEND, C. H. T. 1908. 'A record of results from rearings and dissections of Tachinidae.' *Tech. Ser. Bur. Ent. U.S.*, 12, 95–118.

TOWNSEND, C. H. T. 1934. *Manual of Myiology. Part VIII. Oestroid Generic Diagnoses and Data.* São Paulo, Brazil: Itaquaquecetuba.

TRAUB, R. 1950. 'Siphonaptera from Central America and Mexico: a morphological study of the aedeagus with descriptions of new genera and species.' *Fieldiana, Zool. Mem.*, 1, 1–127.

TRAUB, R. 1966. 'Some examples of convergent evolution in Siphonaptera.' *Proc. R. ent. Soc. Lond. C*, 31, 37–38.

TURNBULL, A. L. and CHANT, D. A. 1961. 'The practice and theory of biological control of insects in Canada.' *Can. J. Zool.*, 39, 697–753.

ULRICH, W. 1943. 'Die Mengeiden und die Phylogenie der Strepsipteren.' *Z. ParasitKde.*, 13, 62–101.

USINGER, R. L. 1966. *Monograph of Cimicidae (Hemiptera-Heteroptera).* Maryland: Entomological Society of America. 585 pp.

USINGER, R. L. and MYERS, J. G. 1929. 'Facultative bloodsucking in phytophagous Hemiptera.' *Parasitology*, 21, 472–480.

VAN EMDEN, F. I. 1954. *Diptera Cyclorrhapha, Calyptrata. Handbk Ident. Br. Insects*, 10 (4a). 133 pp.

VAN VUUREN, L. 1935. 'Waarnemingen omtrent *Phanurus beneficiens* (Zehnt.) (Hym. Scelionidae) op *Schoenobius bipunctifer* Walker.' *Ent. Meded. Ned.-Indië*, 1, 29–33.

VARLEY, G. C. 1941. 'On the search for hosts and the egg distribution of some chalcid parasites of the knapweed gall-fly.' *Parasitology*, 33, 47–66.

VARLEY, G. C. 1959. 'The biological control of agricultural pests.' *Jl R. Soc. Arts*, 107, 475–490.

VARLEY, G. C. and BUTLER, C. G. 1933. 'The acceleration of development in insects by parasitism.' *Parasitology*, 25, 263–268.

VARLEY, G. C. and GRADWELL, G. R. 1958. 'Oak defoliators in England.' *Proc. Int. Congr. Ent.* 10. Montreal (1956), 4, 133–136.

VERHOEFF, K. W. 1902. 'Uber die verwandschaftliche Stellung von *Hemimerus*.' *S.b. ges. naturf. Fr. Berl.*, 1902, 87–89.

VERRALL, G. H. 1901. *British Flies VIII. Platypezidae, Pipunculidae and Syrphidae of Great Britain.* London: Gurney and Jackson. 691 pp.

VIGGIANI, G. 1964. 'La specializzazione entomoparassitica in alcuni Eulofidi (Hym., Chalcidoidea).' *Entomophaga*, 9, 111–118.

VINSON, S. B. 1968. 'Source of a substance in *Heliothis virescens* (Lepidoptera: Noctuidae) that elicits a searching response in its habitual parasite *Cardiochiles nigriceps* (Hymenoptera Braconidae).' *Ann. ent. Soc. Am.*, 61, 8–10.

WALKER, E. P. 1964. *Mammals of the World.* Baltimore: John Hopkins. 1500 pp.

WALKER, F. 1871. *Hemimerus talpoides* in *Catalogue Dermaptera Saltatoria*, part 5, supplement 2. London: British Museum (Natural History).

WALKER, I. 1959. 'Die Abwehrreaktion des Wirtes *Drosophila melanogaster* gegen die zoophage Cynipide *Pseudeucoila bochei* Weld.' *Revue suisse Zool.*, 66, 569–632.

WALKER, M. G. 1943. 'Notes on the biology of *Dexia rustica* F., a dipterous parasite of *Melolontha melolontha* L.' *Proc. zool. Soc. Lond. A*, 113, 126–176.

WALOFF, N. 1967. 'Biology of three species of *Leiophron* (Hymenoptera: Braconidae, Euphorinae) parasitic on Miridae on broom.' *Trans. R. ent. Soc. Lond.*, 119, 187–213.

WALTON, W. R. 1914. 'A new Tachinid parasite of *Diabrotica vittata*.' *Proc. ent. Soc. Wash.*, 16, 11–14.

WASMANN, E. 1895. 'Die myrmecophilen und termitophilen.' *Proc. Int. Congr. Zool. 3. Leyden.*

WATERSTON, J. 1914. 'An account of the bird-lice of the genus *Docophorus* (Mallophaga) found on British auks.' *Proc. R. phys. Soc. Edinb.*, 19, 149–158.

WATERSTON, J. 1922. 'Observations on the life-history of a Liotheid (Mallophaga) parasite of the curlew (*Numenius arquata* Linn.).' *Entomologist's mon. Mag.*, 58, 243–247.

WATERSTON, J. 1926. 'On the crop contents of certain Mallophaga.' *Proc. zool. Soc. Lond.*, 1926, 1017–1020.

WEBB, J. L. and HUTCHINSON, R. H. 1916. 'A preliminary note on the bionomics of *Pollenia rudis* Fabr. in America.' *Proc. ent. Soc. Wash.*, 18, 197–199.

WEIDHAAS, D. E., SCHMIDT, C. H., and CHAMBERLAIN, W. F. 1962. *Radioisotopes and Radiation in Entomology.* Vienna: International Atomic Energy Agency.

WEIS-FOGH, T. 1960. 'A rubber-like protein in insect cuticle.' *J. exp. Biol.*, 37, 889–907.

WEITZ, B. 1963. 'The feeding habits of *Glossina*.' *Bull. Wld Hlth Org.*, 28, 711–729.

WELLMAN, F. C. 1908. 'Some Angolan insects of economic or pathologic importance.' *Ent. News*, 19, 224–230.

WESTWOOD, J. O. 1876. 'Notes on the habits of a lepidopterous insect parasitic on *Fulgoria candelaria*.' *Trans. ent. Soc. Lond.*, 1876, 519–524.

WHEELER, W. M. 1910. 'Effects of parasitic and other kinds of castration upon insects.' *J. exp. Zool.*, 8, 377–438.

WHEELER, W. M. 1911. 'Insect parasitism and its peculiarities.' *Pop. Sci. Mon.*, 79, 431–449.

WHEELER, W. M. 1923. *Social Life among the Insects.* London: Constable. 375 pp.

WHITE, E. B. and LEGNER, E. F. 1966. 'Notes on the life history of *Aleochara taeniata*, a Staphylinid parasite of the house fly, *Musca domestica*.' *Ann. ent. Soc. Am.*, 59, 573–577.

WHITING, A. R. 1967. 'The biology of the parasitic wasp *Mormoniella vitripennis* (*Nasonia brevicornis*) (Walker).' *Q. Rev. Biol.*, 42, 333–470.

WIGGLESWORTH, V. B. 1941. 'The sensory physiology of the human louse *Pediculus humanus corporis* DeGeer (Anoplura).' *Parasitology*, 33, 67–109.

WIGGLESWORTH, V. B. 1964. *The Life of Insects*. London: Weidenfeld and Nicolson. 360 pp.

WILLARD, H. F. and MASON, A. C. 1937. 'Parasitization of the Mediterranean fruit fly in Hawaii, 1914–1933.' *Circ. Dep. Agric. U.S.*, no. 439, 17 pp.

WILLIAMS, J. R. 1957. 'The sugar-cane Delphacidae and their natural enemies in Mauritius.' *Trans. R. ent. Soc. Lond.*, 109, 65–110.

WILSON, B. H. 1967. 'Feeding, mating and oviposition studies of the horse flies *Tabanus lineola* and *T. fuscicostatus* (Diptera: Tabanidae).' *Ann. ent. Soc. Am.*, 60, 1102–1106.

WOOD, O. H. 1933. 'Notes on some dipterous parasites of *Schistocerca* and *Locusta* in the Sudan.' *Bull. ent. Res.*, 24, 521–530.

WOODWARD, T. E. 1951. 'A case of persistent attacks on a human by *Lyctocoris campestris* (F.) (Hem., Anthocoridae).' *Entomologist's mon. Mag.*, 87, 44.

WRIGHT, R. H. and KELLOGG, F. E. 1962. 'Response of *Aedes aegypti* to moist convection currents.' *Nature, Lond.*, 194, 402–403.

WYNNE-EDWARDS, V. C. 1962. *Animal Dispersion in relation to Social Behaviour*. Edinburgh and London: Oliver and Boyd. 653 pp.

ZINNA, G. 1961. 'Specializzazione entomoparassiti negli Aphelinidae: Studio morfologico, etologico e fisiologico del *Coccophagus bivittatus* Compere, nuovo parassita del *Coccus hesperidum* L. per l'Italia.' *Boll. Lab. Ent. agr. Filippo Silvestri*, 19, 301–358.

ZINNA, G. 1962. 'Specializzazione entomoparassiti negli Aphelinidae: Interdipendenze biocenotiche tra due specie associate. Studie morfologico, etologico e fisiologico del *Coccophagoides similis* (Masi) et *Azotus matritensis* Mercet.' *Boll. Lab. Ent. agr. Filippo Silvestri*, 20, 73–184.

ZUMPT, F. 1965. *Myiasis in Man and Animals in the Old World. A Textbook for Physicians, Veterinarians and Zoologists*. London: Butterworths. 267 pp.

Author Index

Subject Index

Page references in italics refer to illustrations

54116